Lessons in Digital Estimation Theory

PRENTICE-HALL SIGNAL PROCESSING SERIES

Alan V. Oppenheim, Editor

ANDREWS and HUNT *Digital Image Restoration*
BRIGHAM *The Fast Fourier Transform*
BURDIC *Underwater Acoustic System Analysis*
CASTLEMAN *Digital Image Processing*
COWAN and GRANT *Adaptive Filters*
CROCHIERE and RABINER *Multirate Digital Signal Processing*
DUDGEON and MERSEREAU *Multidimensional Digital Signal Processing*
HAMMING *Digital Filters, 2nd ed.*
HAYKIN, ED. *Array Signal Processing*
JAYANT and NOLL *Digital Coding of Waveforms*
KINO *Acoustic Waves: Devices, Imaging, and Analog Signal Processing*
LEA, ED. *Trends in Speech Recognition*
LIM, ED. *Speech Enhancement*
MARPLE *Digital Spectral Analysis with Applications*
MCCELLAN and RADER *Number Theory in Digital Signal Processing*
MENDEL *Lessons in Digital Estimation Theory*
OPPENHEIM, ED. *Applications of Digital Signal Processing*
OPPENHEIM, WILLSKY, with YOUNG *Signals and Systems*
OPPENHEIM and SCHAFER *Digital Signal Processing*
RABINER and GOLD *Theory and Applications of Digital Signal Processing*
RABINER and SCHAFER *Digital Processing of Speech Signals*
ROBINSON and TRIETEL *Geophysical Signal Analysis*
STEARNS and DAVID *Signal Processing Algorithms*
TRIBOLET *Seismic Applications of Homomorphic Signal Processing*
WIDROW and STEARNS *Adaptive Signal Processing*

Lessons in Digital Estimation Theory

Jerry M. Mendel

Department of Electrical Engineering
University of Southern California
Los Angeles, California

Prentice-Hall, Inc., Englewood Cliffs, New Jersey 07632

Library of Congress Cataloging-in-Publication Data

MENDEL, JERRY M., (date)
Lessons in digital estimation theory.

Bibliography: p.
Includes index.
1. Estimation theory. I. Title.
QA276.8.M46 1986 511′.4 86-9365
ISBN 0-13-530809-7

To my parents, Eleanor and Alfred Mendel
and
my wife, Letty Mendel

Editorial/production supervision: Gretchen K. Chenenko
Cover design: Lundgren Graphics
Manufacturing buyer: Gordon Osbourne

Printed in the United States of America

10 9 8 7 6 5 4 3 2 1

ISBN 0-13-530809-7 025

PRENTICE-HALL INTERNATIONAL (UK) LIMITED, *London*
PRENTICE-HALL OF AUSTRALIA PTY. LIMITED, *Sydney*
PRENTICE-HALL CANADA INC., *Toronto*
PRENTICE-HALL HISPANOAMERICANA, S.A., *Mexico*
PRENTICE-HALL OF INDIA PRIVATE LIMITED, *New Delhi*
PRENTICE-HALL OF JAPAN, INC., *Tokyo*
PRENTICE-HALL OF SOUTHEAST ASIA PTE. LTD., *Singapore*
EDITORA PRENTICE-HALL DO BRASIL, LTDA., *Rio de Janeiro*

Contents

Preface

Estimation theory is widely used in many branches of science and engineering. No doubt, one could trace its origin back to ancient times, but Karl Friederich Gauss is generally acknowledged to be the progenitor of what we now refer to as estimation theory. R. A. Fisher, Norbert Wiener, Rudolph E. Kalman, and scores of others have expanded upon Gauss' legacy, and have given us a rich collection of estimation methods and algorithms from which to choose. This book describes many of the important estimation methods and shows how they are interrelated.

Estimation theory is a product of need and technology. Gauss, for example, needed to predict the motions of planets and comets from telescopic measurements. This "need" led to the method of least squares. Digital computer technology has revolutionized our lives. It created the "need" for recursive estimation algorithms, one of the most important ones being the Kalman filter. Because of the importance of digital technology, this book presents estimation theory from a digital viewpoint. In fact, it is this author's viewpont that estimation theory is a natural adjunct to classical digital signal processing. It produces time-varying digital filter designs that operate on random data in an optimal manner.

This book has been written as a collection of lessons. It is meant to be an *introduction* to the general field of estimation theory, and, as such, is not encyclopedic in content or in references. It can be used for self-study or in a one-semester course. At the University of Southern California, we have covered all of its contents in such a course, at the rate of two lessons a week. We have been doing this since 1978.

Approximately one half of the book is devoted to parameter estimation and the other half to state estimation. For many years there has been a tendency to treat state estimation as a stand-alone subject and even to treat parameter estimation as a special case of state estimation. Historically, this is incorrect. In the musical "Fiddler on the Roof," Tevye argues on behalf of *tradition* . . . "Tradition! . . ." Estimation theory also has its tradition, and it begins with Gauss and parameter estimation. In Lesson 2 we show that state estimation is a special case of parameter estimation, i.e., it is the problem of estimating random parameters, when these parameters change from one time to the next. Consequently, the subject of state estimation flows quite naturally from the subject of parameter estimation.

Most of the book's important results are summarized in theorems and corollaries. In order to guide the reader to these results, they have been summarized for easy reference in Appendix A.

Problems are included for most lessons, because this book is meant to be used as a textbook. The problems fall into two groups. The first group contains problems that ask the reader to fill in details, which have been "left to the reader as an exercise." The second group contains problems that are related to the material in the lesson. They range from theoretical to computational problems.

This book is an outgrowth of a one-semester course on estimation theory, taught at the University of Southern California. Since 1978 it has been taught by four different people, who have encouraged me to convert the lecture notes into a book. I wish to thank Mostafa Shiva, Alan Laub, George Papavassilopoulos, and Rama Chellappa for their encouragement. Special thanks goes to Rama Chellappa, who provided supplementary Lesson A on the subject of sufficient statistics and statistical estimation of parameters. This lesson fits in very nicely just after Lesson 14.

While writing this text, the author had the benefit of comments and suggestions from many of his colleagues and students. I especially wish to acknowledge the help of Guan-Zhong Dai, Chong-Yung Chi, Phil Burns, Youngby Kim, Chung-Chin Lu, and Tom Hebert. Special thanks goes to Georgios Giannakis. The book would not be in its present form without their contributions.

Additionally, the author wishes to thank Marcel Dekker, Inc. for permitting him to include material from Mendel, J. M., 1973, *Discrete Techniques of Parameter Estimation : the Equation-Error Formulation,* in Lessons 1–9, 11, 18, and 24; and Academic Press, Inc. for permitting him to include material from Mendel, J. M., *Optimal Seismic Deconvolution: an Estimation-Based Approach,* copyright © 1983 by Academic Press, Inc., in Lessons 11–17, 19–22, and 26.

JERRY M. MENDEL
Los Angeles, California

Lesson 1

Introduction, Coverage, and Philosophy

INTRODUCTION

This book is all about estimation theory. It is useful, therefore, for us to understand the role of estimation in relation to the more global problem of modeling. Figure 1-1 decomposes modeling into four problems: representation, measurement, estimation, and validation. As Mendel (1973, pp. 2–4) states, "The *representation problem* deals with how something should be modeled. We shall be interested only in mathematical models. Within this class of models we need to know whether the model should be static or dynamic, linear or nonlinear, deterministic or random, continuous or discretized, fixed or varying, lumped or distributed . . . , in the time-domain or in the frequency-domain . . . , etc.

"In order to verify a model, physical quantities must be measured. We distinguish between two types of physical quantities, *signals*, and *parameters*. Parameters express a relation between signals. . . .

"Not all signals and parameters are measurable. The *measurement problem* deals with which physical quantities should be measured and how they should be measured.

"The *estimation problem* deals with the determination of those physical quantities that cannot be measured from those that can be measured. We shall distinguish between the estimation of signals (i.e., states) and the estimation of parameters. Because a subjective decision must sometimes be made to classify a physical quantity as a signal or a parameter, there is some overlap between signal estimation and parameter estimation. . . .

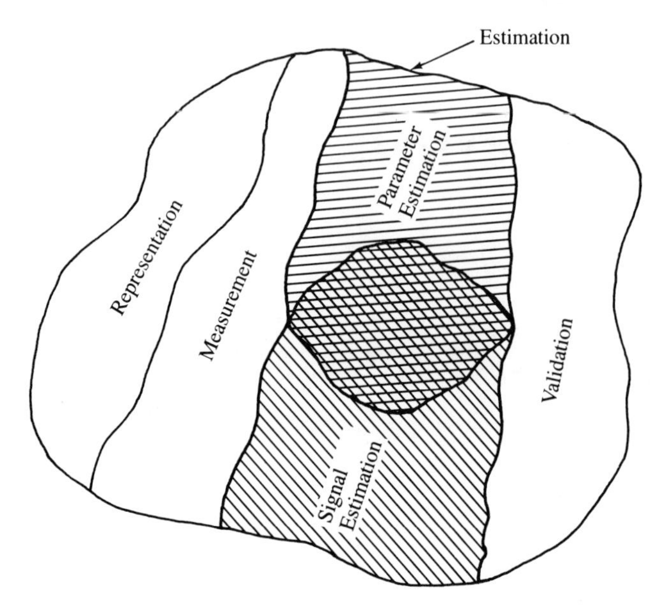

Figure 1-1 Modeling Problem (reprinted from Mendel, 1973, p. 4, by courtesy of Marcel Dekker, Inc).

"After a model has been completely specified, through choice of an appropriate mathematical representation, measurement of measurable signals, estimation of nonmeasurable signals, and estimation of its parameters, the model must be checked out. The *validation problem* deals with demonstrating confidence in the model. Often, statistical tests involving confidence limits are used to validate a model."

In this book we shall be interested in parameter estimation, state estimation, and combined state and parameter estimation. In Lesson 2 we provide six examples each of which can be categorized either as a parameter or state estimation problem. Here we just mention that: the problem of identifying the sampled values of a linear and time-invariant system's impulse response from input/output data is one of parameter estimation; the problem of reconstructing a state vector associated with a dynamical system, from noisy measurements, is one of state estimation (state estimates might be needed to implement a linear-quadratic-Gaussian optimal control law, or to perform postflight data analysis, or signal processing such as deconvolution).

COVERAGE

This book focuses on a wide range of estimation techniques that can be applied either to linear or nonlinear models. Both parameter and state estimation techniques are treated. Some parameter estimation techniques are for deterministic parameters, whereas others are for random parameters, however, state estimation techniques are for random states. Table 1-1 summarizes the book's coverage.

Four lessons (Lessons 3, 4, 5, and 8) are devoted to least-squares estimation because it is a very basic and important technique, and, under certain often-occurring conditions, other techniques reduce to it. Consequently, once we understand the nuances of least squares and have established that a different technique has reduced to least squares, we do not have to restudy the nuances of that technique.

In order to fully study least-squares estimators we must establish their small and large sample properties. What we mean by such properties is the subject of Lessons 6 and 7.

Having spent four lessons on least-squares estimation, we cover best linear unbiased estimation (BLUE) in one lesson, Lesson 9. We are able to do this because BLUE is a special case of least squares.

In order to set the stage for maximum-likelihood estimation, which is covered in Lesson 11, we describe the concept of likelihood and its relationship to probability in Lesson 10.

Lesson 12 provides a transition from our study of estimation of deterministic parameters to our study of estimation of random parameters. It provides much useful information about elements of Gaussian random vari-

TABLE 1-1 Estimation Techniques

I. LINEAR MODELS
   A. Parameter Estimation
      1. Deterministic parameters
         a. Least-squares (batch and recursive processing)
         b. Best linear unbiased estimation (BLUE)
         c. Maximum-likelihood
      2. Random Parameters
         a. Mean-squared
         b. Maximum a posteriori
         c. BLUE
         d. Weighted least squares
   B. State Estimation
      1. Mean-squared prediction
      2. Mean-squared filtering (Kalman filter/Kalman-Bucy filter)
      3. Mean-squared smoothing

II. NONLINEAR MODELS
   A. Parameter Estimation
      Iterated least squares for deterministic parameters
   B. State Estimation
      Extended Kalman filter
   C. Combined State and Parameter Estimation
      1. Extended Kalman filter
      2. Maximum-likelihood

ables. To some readers, this lesson may be a review of material already known to them.

General results for both mean-squared and maximum a posteriori estimation of random parameters are covered in Lesson 13. These results are specialized to the important case of the linear and Gaussian model in Lesson 14. Best linear unbiased and weighted least-squares estimation are also revisited in Lesson 14. Lesson 14 is quite important, because it gives conditions under which mean-squared, maximum a posteriori, best-linear unbiased, and weighted least-squares estimates of random parameters are identical. Lesson A, which is a supplemental one, is on the subject of sufficient statistics and statistical estimation of parameters. It fits in very nicely after Lesson 14.

Lesson 15 provides a transition from our study of parameter estimation to our study of state estimation. It provides much useful information about elements of discrete-time Gauss-Markov random processes, and also establishes the *basic state-variable model,* and its statistical properties, for which we derive a wide variety of state estimators. To some readers, this lesson may be a review of material already known to them.

Lessons 16 through 22 cover state estimation for the Lesson 15 basic state-variable model. Prediction is treated in Lesson 16. The important innovations process is also covered in that lesson. Filtering is the subject of Lessons 17, 18, and 19. The mean-squared state filter, commonly known as the Kalman filter, is developed in Lesson 17. Five examples which illustrate some interesting numerical and theoretical aspects of Kalman filtering are presented in Lesson 18. Lesson 19 establishes a bridge between mean-squared estimation and mean-squared digital signal processing. It shows how the steady-state Kalman filter is related to a digital Wiener filter. The latter is widely used in digital signal processing. Smoothing is the subject of Lessons 20, 21 and 22. Fixed-interval, fixed-point, and fixed-lag smoothers are developed in Lessons 20 and 21. Lesson 22 presents some applications which illustrate interesting numerical and theoretical aspects of fixed-interval smoothing. These applications are taken from the field of digital signal processing and include minimum-variance deconvolution, maximum-likelihood deconvolution, and recursive waveshaping.

Lesson 23 shows how to modify results given in Lessons 16, 17, 19, 20 and 21 from the basic state-variable model to a state-variable model that includes the following effects:

1. nonzero mean noise processes and/or known bias function in the measurement equation,
2. correlated noise processes,
3. colored noise processes, and
4. perfect measurements.

Lesson 24 provides a transition from our study of estimation for linear models to estimation for nonlinear models. Because many real-world systems are continuous-time in nature and nonlinear, this lesson explains how to linearize and discretize a nonlinear differential equation model.

Lesson 25 is devoted primarily to the extended Kalman filter (EKF), which is a form of the Kalman filter that has been "extended" to nonlinear dynamical systems of the type described in Lesson 24. The EKF is related to the method of iterated least squares (ILS), the major difference between the two being that the EKF is for dynamical systems whereas ILS is not. This lesson also shows how to apply the EKF to parameter estimation, in which case states and parameters can be estimated simultaneously, and in real time.

The problem of obtaining maximum-likelihood estimates of a collection of parameters that appears in the basic state-variable model is treated in Lesson 26. The solution involves state and parameter estimation, but calculations can only be performed off-line, after data from an experiment has been collected.

The Kalman-Bucy filter, which is the continuous-time counterpart to the Kalman filter, is derived from two different viewpoints in Lesson 27. We include this lesson because the Kalman-Bucy filter is widely used in linear stochastic optimal control theory.

PHILOSOPHY

The digital viewpoint is emphasized throughout this book. Our estimation algorithms are digital in nature; many are recursive. The reasons for the digital viewpoint are:

1. much real data is collected in a digitized manner, so it is in a form ready to be processed by digital estimation algorithms, and
2. the mathematics associated with digital estimation theory are simpler than those associated with continuous estimation theory.

Regarding (2), we mention that very little knowledge about random processes is needed to derive digital estimation algorithms, because digital (i.e., discrete-time) random processes can be treated as vectors of random variables. Much more knowledge about random processes is needed to design continuous-time estimation algorithms.

Suppose our underlying model is continuous-time in nature. We are faced with two choices: develop a continuous-time estimation theory and then implement the resulting estimators on a digital computer (i.e., discretize the continuous-time estimation algorithm), or discretize the model and develop a discrete-time (i.e., digital) estimation theory that leads to estimation algorithms readily implemented on a digital computer. If both approaches lead to algorithms that are implemented digitally, then we advocate the *principle of simplicity* for their development, and this leads us to adopt the second-choice. *For estimation, our modeling philosophy is, therefore, discretize the model at the front end of the problem.*

Estimation theory has a long and glorious history (e.g., see Sorenson, 1970); however, it has been greatly influenced by technology, especially the computer. Although much of estimation theory was developed in the mathematics, statistical, and control theory literatures, we shall adopt the following viewpoint towards that theory: *estimation theory is the extension of classical signal processing to the design of digital filters that processs uncertain data in an optimal manner.* In fact, estimation algorithms are just filters that transform input streams of numbers into output streams of numbers.

Most of classical digital filter design (e.g., Oppenheim and Schafer, 1975; Hamming, 1983; Peled and Liu, 1976) is concerned with designs associated with deterministic signals, e.g., low-pass and bandpass filters, and, over the years specific techniques have been developed for such designs. The resulting filters are usually "fixed" in the sense that their coefficients do not change as a function of time. Estimation theory, on the other hand, leads to filter structures that are time-varying. These filters are designed (i.e., derived) using time-domain performance specifications (e.g., smallest error variance), and, as mentioned above, process random data in an optimal manner. *Our philosophy about estimation theory is that it can be viewed as a natural adjunct to digital signal processing theory.*

Example 1-1

At one time or another we have all used the sample mean to compute an "average." Suppose we are given a collection of k measured values of quantity X, namely $x(1)$, $x(2), \ldots, x(k)$. The sample mean of these measurements, $\bar{x}(k)$, is

$$\bar{x}(k) = \frac{1}{k}\sum_{j=1}^{k} x(j) \tag{1-1}$$

A recursive formula for the sample mean is obtained from (1-1), as follows:

$$\begin{aligned} \bar{x}(k+1) &= \frac{1}{k+1}\sum_{j=1}^{k+1} x(j) = \frac{1}{k+1}\left[\sum_{j=1}^{k} x(j) + x(k+1)\right] \\ \bar{x}(k+1) &= \frac{k}{k+1}\bar{x}(k) + \frac{1}{k+1}x(k+1) \end{aligned} \tag{1-2}$$

This recursive version of the sample mean is used for $k = 0, 1, \ldots$ by setting $\bar{x}(0) = 0$.

Observe that *the sample mean, as expressed in (1-2), is a time-varying recursive digital filter whose input is measurement* $x(k)$. In later lessons we show that the sample mean is also an optimal estimation algorithm; thus, although the reader may not have been aware of it, the sample mean, which he or she has been using since early schooldays, is an estimation algorithm. □

Lesson 2

The Linear Model

INTRODUCTION

In order to estimate unknown quantities (i.e., parameters or signals) from measurements and other a priori information, we must begin with model representations and express them in such a way that attention is focused on the explicit relationship between the unknown quantities and the measurements. Many familiar models are linear in the unknown quantities (denoted $\boldsymbol{\theta}$), and can be expressed as

$$\mathcal{Z}(k) = \mathcal{H}(k)\boldsymbol{\theta} + \mathcal{V}(k) \tag{2-1}$$

In this model, $\mathcal{Z}(k)$, which is $N \times 1$, is called the *measurement vector*; $\boldsymbol{\theta}$, which is $n \times 1$, is called the *parameter vector*; $\mathcal{H}(k)$, which is $N \times n$ is called the *observation matrix*; and, $\mathcal{V}(k)$, which is $N \times 1$, is called the *measurement noise* vector. Usually, $\mathcal{V}(k)$ is random. By convention, the argument "k" of $\mathcal{Z}(k)$, $\mathcal{H}(k)$, and $\mathcal{V}(k)$ denotes the fact that the last measurement used to construct (2-1) is the kth. All other measurements occur "before" the kth.

Strictly speaking, (2-1) represents an "affine" transformation of parameter vector $\boldsymbol{\theta}$ rather than a linear transformation. We shall, however, adhere to traditional estimation-theory literature, by calling (2-1) a "linear model."

EXAMPLES

Some examples that illustrate the formation of (2-1) are given in this section. What distinguishes these examples from one another are the nature of and interrelationships between $\boldsymbol{\theta}$, $\mathcal{H}(k)$ and $\mathcal{V}(k)$. The following situations can occur.

A. $\boldsymbol{\theta}$ is deterministic
  1. $\mathcal{H}(k)$ is deterministic.
  2. $\mathcal{H}(k)$ is random.
    a. $\mathcal{H}(k)$ and $\mathcal{V}(k)$ are statistically independent.
    b. $\mathcal{H}(k)$ and $\mathcal{V}(k)$ are statistically dependent.

B. $\boldsymbol{\theta}$ is random
  1. $\mathcal{H}(k)$ is deterministic.
  2. $\mathcal{H}(k)$ is random.
    a. $\mathcal{H}(k)$ and $\mathcal{V}(k)$ are statistically independent.
    b. $\mathcal{H}(k)$ and $\mathcal{V}(k)$ are statistically dependent.

Example 2-1 Impulse Response Identification

It is well known that the output of a single-input single-output, linear, time-invariant, discrete-time system is given by the following convolution-sum relationship

$$y(k) = \sum_{i=-\infty}^{\infty} h(i)u(k-i) \tag{2-2}$$

where $k = 1, 2, \ldots, N$, $h(i)$ is the system's impulse response (IR), $u(k)$ is its input and $y(k)$ its output. If $u(k) = 0$ for $k < 0$, and the system is causal, so that $h(i) = 0$ for $i \leq 0$, and, $h(i) \simeq 0$ for $i > n$, then

$$y(k) = \sum_{i=1}^{n} h(i)u(k-i) \tag{2-3}$$

Signal $y(k)$ is measured by a sensor which is corrupted by additive measurement noise, $v(k)$, i.e., we only have access to measurement $z(k)$, where

$$z(k) = y(k) + v(k) \tag{2-4}$$

and $k = 1, 2, \ldots, N$. We now collect these N measurements as follows:

$$\underbrace{\begin{pmatrix} z(N) \\ z(N-1) \\ \vdots \\ z(n) \\ \vdots \\ z(2) \\ z(1) \end{pmatrix}}_{\mathcal{Z}(N)} = \underbrace{\begin{pmatrix} u(N-1) & u(N-2) & u(N-3) & \cdots & u(N-n) \\ u(N-2) & u(N-3) & u(N-4) & \cdots & u(N-n-1) \\ \vdots & \vdots & \vdots & \ddots & \vdots \\ u(n-1) & u(n-2) & u(n-3) & \cdots & u(0) \\ \vdots & \vdots & \vdots & \ddots & \vdots \\ u(1) & u(0) & 0 & \cdots & 0 \\ u(0) & 0 & 0 & \cdots & 0 \end{pmatrix}}_{\mathcal{H}(N-1)}$$

$$\times \underbrace{\begin{pmatrix} h(1) \\ h(2) \\ \vdots \\ h(n) \end{pmatrix}}_{\boldsymbol{\theta}} + \underbrace{\begin{pmatrix} v(N) \\ v(N-1) \\ \vdots \\ v(n) \\ \vdots \\ v(2) \\ v(1) \end{pmatrix}}_{\mathcal{V}(N)} \tag{2-5}$$

Clearly, (2-5) is in the form of (2-1).

In this application the n sampled values of the IR, $h(i)$, play the role of unknown parameters, i.e., $\theta_1 = h(1)$, $\theta_2 = h(2), \ldots, \theta_n = h(n)$, and these parameters are deterministic. If input $u(k)$ is deterministic and is known ahead of time (or can be measured) without error, then $\mathcal{H}(N-1)$ is deterministic so that we are in case A.1. Often, however, $u(k)$ is random so that $\mathcal{H}(N-1)$ is random; but $u(k)$ is in no way related to measurement noise $v(k)$, so we are in case A.2.a. □

Example 2-2 Identification of the Coefficients of a Finite-Difference Equation

Suppose a linear, time-invariant, discrete-time system is described by the following nth-order finite-difference equation

$$y(k) + \alpha_1 y(k-1) + \cdots + \alpha_n y(k-n) = u(k-1) \tag{2-6}$$

This model is often referred to as an all-pole or autoregressive (AR) model. It occurs in many branches of engineering and science, including speech modeling, geophysical modeling, etc. Suppose, also, that N perfect measurements of signal $y(k)$ are available. Parameters $\alpha_1, \alpha_2, \ldots, \alpha_n$ are unknown and are to be estimated from the data. To do this, we can rewrite (2-6) as

$$y(k) = -\alpha_1 y(k-1) - \cdots - \alpha_n y(k-n) + u(k-1) \tag{2-7}$$

and collect $y(1), y(2), \ldots, y(N)$ as we did in Example 2-1. Doing this, we obtain

$$\underbrace{\begin{pmatrix} y(N) \\ y(N-1) \\ \vdots \\ y(n) \\ \vdots \\ y(2) \\ y(1) \end{pmatrix}}_{\mathcal{Z}(N)} = \underbrace{\begin{pmatrix} y(N-1) & y(N-2) & y(N-3) & \cdots & y(N-n) \\ y(N-2) & y(N-3) & y(N-4) & \cdots & y(N-n-1) \\ \vdots & \vdots & \vdots & & \vdots \\ y(n-1) & y(n-2) & y(n-1) & \cdots & y(0) \\ \vdots & \vdots & \vdots & & \vdots \\ y(1) & y(0) & 0 & \cdots & 0 \\ y(0) & 0 & 0 & \cdots & 0 \end{pmatrix}}_{\mathcal{H}(N-1)} \times \underbrace{\begin{pmatrix} -\alpha_1 \\ -\alpha_2 \\ \vdots \\ -\alpha_n \end{pmatrix}}_{\boldsymbol{\theta}} + \underbrace{\begin{pmatrix} u(N-1) \\ u(N-2) \\ \vdots \\ u(n) \\ \vdots \\ u(1) \\ u(0) \end{pmatrix}}_{\mathcal{V}(N-1)} \tag{2-8}$$

which, again, is in the form of (2-1).

In this example $\theta = \text{col}(-\alpha_1, -\alpha_2, \ldots, -\alpha_n)$, and these parameters are deterministic. If input $u(k-1)$ is deterministic, then the system's output $y(k)$ will also be deterministic, so that both $\mathcal{H}(N-1)$ and $\mathcal{V}(N-1)$ are deterministic. This is a very special case of case A.1, because usually $\mathcal{V}$ is random. If, however, $u(k-1)$ is random then $y(k)$ will also be random; but, the elements of $\mathcal{H}(N-1)$ will now depend on those in $\mathcal{V}(N-1)$, because $y(k)$ depends upon $u(0), u(1), \ldots, u(k-1)$. In this situation we are in case A.2.b. □

Example 2-3 Identification of the Initial Condition Vector in an Unforced State Equation Model

Consider the problem of identifying the $n \times 1$ initial condition vector $\mathbf{x}(0)$ of the linear, time-invariant, discrete-time system

$$\mathbf{x}(k+1) = \mathbf{\Phi}\mathbf{x}(k) \tag{2-9}$$

from the N measurements $z(1), z(2), \ldots, z(N)$, where

$$z(k) = \mathbf{h}'\mathbf{x}(k) + v(k) \tag{2-10}$$

The solution to (2-9) is

$$\mathbf{x}(k) = \mathbf{\Phi}^k\mathbf{x}(0) \tag{2-11}$$

so that

$$z(k) = \mathbf{h}'\mathbf{\Phi}^k\mathbf{x}(0) + v(k) \tag{2-12}$$

Collecting the N measurements, as before, we obtain

$$\underbrace{\begin{pmatrix} z(N) \\ z(N-1) \\ \vdots \\ z(1) \end{pmatrix}}_{\mathscr{Z}(N)} = \underbrace{\begin{pmatrix} \mathbf{h}'\mathbf{\Phi}^N \\ \mathbf{h}'\mathbf{\Phi}^{N-1} \\ \vdots \\ \mathbf{h}'\mathbf{\Phi} \end{pmatrix}}_{\mathscr{H}(N)} \underbrace{\mathbf{x}(0)}_{\theta} + \underbrace{\begin{pmatrix} v(N) \\ v(N-1) \\ \vdots \\ v(1) \end{pmatrix}}_{\mathscr{V}(N)} \tag{2-13}$$

Once again, we have been led to (2-1), and we are in case A.1. □

Example 2-4 State Estimation

State-variable models are widely used in control and communication theory, and in signal processing. Often, we need the entire state vector of a dynamical system in order to implement an optimal control law for it, or, to implement a digital signal processor. Usually, we cannot measure the entire state vector, and our measurements are corrupted by noise. In state estimation, our objective is to estimate the entire state vector from a limited collection of noisy measurements.

Here we consider the problem of estimating $n \times 1$ state vector $\mathbf{x}(k)$, at $k = 1, 2, \ldots, N$ from a scalar measurement $z(k)$, where $k = 1, 2, \ldots, N$. The model for this example is

$$\mathbf{x}(k+1) = \mathbf{\Phi}\mathbf{x}(k) + \boldsymbol{\gamma}u(k) \tag{2-14}$$

$$z(k) = \mathbf{h}'\mathbf{x}(k) + v(k) \tag{2-15}$$

We are keeping this example simple by assuming that the system is time-invariant and has only one input and one output; however, the results obtained in this example are easily generalized to time-varying and multichannel systems or multichannel systems alone.

If we try to collect our N measurements as before, we obtain

$$\left.\begin{aligned} z(N) &= \mathbf{h}'\mathbf{x}(N) + v(N) \\ z(N-1) &= \mathbf{h}'\mathbf{x}(N-1) + v(N-1) \\ &\cdots \\ z(1) &= \mathbf{h}'\mathbf{x}(1) + v(1) \end{aligned}\right\} \tag{2-16}$$

Observe that a different (unknown) state vector appears in each of the N measurement equations; thus, there does not appear to be a common "$\boldsymbol{\theta}$" for the collected measurements. Appearances can sometimes be deceiving.

So far, we have not made use of the state equation. Its solution can be expressed as

$$\mathbf{x}(k) = \boldsymbol{\Phi}^{k-j}\mathbf{x}(j) + \sum_{i=j+1}^{k} \boldsymbol{\Phi}^{k-i}\boldsymbol{\gamma}u(i-1) \tag{2-17}$$

where $k \geq j+1$. *We now focus our attention on the value of* $\mathbf{x}(\mathrm{k})$ *at* $\mathrm{k} = \mathrm{k}_1$, *where* $1 \leq \mathrm{k}_1 \leq \mathrm{N}$. Using (2-17), we can express $\mathbf{x}(N), \mathbf{x}(N-1), \ldots, \mathbf{x}(k_1+1)$ as an explicit function of $\mathbf{x}(k_1)$, i.e.,

$$\mathbf{x}(k) = \boldsymbol{\Phi}^{k-k_1}\mathbf{x}(k_1) + \sum_{i=k_1+1}^{k} \boldsymbol{\Phi}^{k-i}\boldsymbol{\gamma}u(i-1) \tag{2-18}$$

where $k = k_1+1, k_1+2, \ldots, N$. In order to do the same for $\mathbf{x}(1), \mathbf{x}(2), \ldots, \mathbf{x}(k_1-1)$, we solve (2-17) for $\mathbf{x}(j)$ and set $k = k_1$,

$$\mathbf{x}(j) = \boldsymbol{\Phi}^{j-k_1}\mathbf{x}(k_1) - \sum_{i=j+1}^{k_1} \boldsymbol{\Phi}^{j-i}\boldsymbol{\gamma}u(i-1) \tag{2-19}$$

where $j = k_1-1, k_1-2, \ldots, 2, 1$. Using (2-18) and (2-19), we can reexpress (2-16) as

$$\left.\begin{array}{c}
z(k) = \mathbf{h}'\boldsymbol{\Phi}^{k-k_1}\mathbf{x}(k_1) + \mathbf{h}'\displaystyle\sum_{i=k_1+1}^{k} \boldsymbol{\Phi}^{k-i}\boldsymbol{\gamma}u(i-1) + v(k) \\
k = N, N-1, \ldots, k_1+1 \\
z(k_1) = \mathbf{h}'\mathbf{x}(k_1) + v(k_1) \\
z(l) = \mathbf{h}'\boldsymbol{\Phi}^{l-k_1}\mathbf{x}(k_1) - \mathbf{h}'\displaystyle\sum_{i=l+1}^{k_1} \boldsymbol{\Phi}^{l-i}\boldsymbol{\gamma}u(i-1) + v(l) \\
l = k_1-1, k_1-2, \ldots, 1
\end{array}\right\} \tag{2-20}$$

These $\mathbf{N}$ equations can now be collected together, to give

$$\underbrace{\begin{pmatrix} z(N) \\ z(N-1) \\ \vdots \\ z(1) \end{pmatrix}}_{\mathscr{Z}(N)} = \underbrace{\begin{pmatrix} \mathbf{h}'\boldsymbol{\Phi}^{N-k_1} \\ \mathbf{h}'\boldsymbol{\Phi}^{N-1-k_1} \\ \vdots \\ \mathbf{h}'\boldsymbol{\Phi}^{1-k_1} \end{pmatrix}}_{\mathscr{H}(N,k_1)} \underbrace{\mathbf{x}(k_1)}_{\boldsymbol{\theta}}$$

$$\underbrace{+ \mathbf{M}(N,k_1)\begin{pmatrix} u(N-1) \\ u(N-2) \\ \vdots \\ u(0) \end{pmatrix} + \begin{pmatrix} v(N) \\ v(N-1) \\ \vdots \\ v(1) \end{pmatrix}}_{\mathscr{V}(N,k_1)} \tag{2-21}$$

where the exact structure of matrix $\mathbf{M}(N, k_1)$ is not important to us at this point. Observe that the state at $k = k_1$ plays the role of parameter vector $\boldsymbol{\theta}$ and that both $\mathscr{H}$ and $\mathscr{V}$ are different for different values of k_1.

If $\mathbf{x}(0)$ and the system input $u(k)$ are deterministic, then $\mathbf{x}(k)$ is deterministic for all k. In this case $\boldsymbol{\theta}$ is deterministic, but $\mathcal{V}(N, k_1)$ is a superposition of deterministic and random components. On the other hand, if either $\mathbf{x}(0)$ or $u(k)$ are random then $\boldsymbol{\theta}$ is a vector of *random parameters*. This latter situation is the more usual one in state estimation. It corresponds to case B.1. □

Example 2-5 A Nonlinear Model

Many of the estimation techniques that are described in this book in the context of linear model (2-1) can also be applied to the estimation of unknown signals or parameters in nonlinear models, when such models are suitably linearized. Suppose, for example, that

$$z(k) = f(\theta, k) + v(k) \tag{2-22}$$

where, $k = 1, 2, \ldots, N$, and the structure of nonlinear function $f(\theta, k)$ is known explicitly. To see the forest from the trees in this example we assume θ is a scalar parameter.

Let θ^* denote a *nominal value* of θ, $\delta\theta = \theta - \theta^*$, and $\delta z = z - z^*$, where

$$z^*(k) = f(\theta^*, k) \tag{2-23}$$

Observe that the *nominal measurements*, $z^*(k)$, can be computed once θ^* is specified, because $f(\cdot, k)$ is assumed to be known.

Using a first-order Taylor series expansion of $f(\theta, k)$ about $\theta = \theta^*$, it is easy to show that

$$\delta z(k) = \left.\frac{\partial f(\theta, k)}{\partial \theta}\right|_{\theta = \theta^*} \delta\theta + v(k) \tag{2-24}$$

where $k = 1, 2, \ldots, N$. It is easy to see how to collect these N equations, to give

$$\underbrace{\begin{pmatrix} \delta z(N) \\ \delta z(N-1) \\ \vdots \\ \delta z(1) \end{pmatrix}}_{\mathcal{Z}(N)} = \underbrace{\begin{pmatrix} \partial f(\theta^*, N)/\partial\theta^* \\ \partial f(\theta^*, N-1)/\partial\theta^* \\ \vdots \\ \partial f(\theta^*, 1)/\partial\theta^* \end{pmatrix}}_{\mathcal{H}(N, \theta^*)} \delta\theta + \underbrace{\begin{pmatrix} v(N) \\ v(N-1) \\ \vdots \\ v(1) \end{pmatrix}}_{\mathcal{V}(N)} \tag{2-25}$$

in which $\partial f(\theta^*, k)/\partial\theta^*$ is short for "$\partial f(\theta, k)/\partial\theta$ evaluated at $\theta = \theta^*$."

Observe that $\mathcal{H}$ depends on θ^*. We will discuss different ways for specifying θ^* in Lesson 25. □

Example 2-6 Deconvolution (Mendel, 1983b)

In Example 2-1 we showed how a convolutional model could be expressed as the linear model $\mathcal{Z} = \mathcal{H}\theta + \mathcal{V}$. In that example we assumed that both input and output measurements were available, and, we wanted to estimate the sampled values of the system's impulse response. Here we begin with the same convolutional model, written as

$$z(k) = \sum_{i=1}^{k} \mu(i)h(k-i) + v(k) \tag{2-26}$$

where $k = 1, 2, \ldots, N$. Noisy measurements $z(1), z(2), \ldots, z(N)$ are available to us, and we assume that we know the system's impulse response $h(j)$, $\forall j$. What is not known is the input to the system $\mu(1), \mu(2), \ldots, \mu(N)$. *Deconvolution is the signal processing procedure for removing the effects of* h(j) *and* v(j) *from the measurements so that one is left with an estimate of* μ(j).

In deconvolution we often assume that input $\mu(j)$ is white noise, but is not necessarily Gaussian. This type of deconvolution problem occurs in reflection seismology. We assume further that

$$\mu(k) = r(k)q(k) \tag{2-27}$$

where $r(k)$ is white Gaussian noise with variance σ_r^2, and $q(k)$ is a random event location sequence of zeros and ones (a Bernoulli sequence). Sequences $r(k)$ and $q(k)$ are assumed to be statistically independent.

We now collect the N measurements, but in such a way that $\mu(1), \mu(2), \ldots, \mu(N)$ are treated as the unknown parameters. Doing this we obtain the following linear deconvolution model:

$$\underbrace{\begin{pmatrix} z(N) \\ z(N-1) \\ \vdots \\ z(2) \\ z(1) \end{pmatrix}}_{\mathfrak{Z}(N)} = \underbrace{\begin{pmatrix} h(N-1) & h(N-2) & \cdots & h(1) & h(0) \\ h(N-2) & h(N-3) & \cdots & h(0) & 0 \\ \vdots & \vdots & \ddots & \vdots & \vdots \\ h(1) & h(0) & \cdots & 0 & 0 \\ h(0) & 0 & \cdots & 0 & 0 \end{pmatrix}}_{\mathcal{H}(N-1)}$$

$$\times \underbrace{\begin{pmatrix} \mu(1) \\ \mu(2) \\ \vdots \\ \mu(N-1) \\ \mu(N) \end{pmatrix}}_{\boldsymbol{\theta}} + \underbrace{\begin{pmatrix} v(N) \\ v(N-1) \\ \vdots \\ v(2) \\ v(1) \end{pmatrix}}_{\mathcal{V}(N)} \tag{2-28}$$

We shall often refer to $\boldsymbol{\theta}$ as $\boldsymbol{\mu}$.

Using (2-27), we can also express $\boldsymbol{\theta} = \boldsymbol{\mu}$ as

$$\boldsymbol{\mu} = \mathbf{Q_q r} \tag{2-29}$$

where

$$\mathbf{r} = \text{col}\,(r(1), r(2), \ldots, r(N)) \tag{2-30}$$

and

$$\mathbf{Q_q} = \text{diag}\,(q(1), q(2), \ldots, q(N)) \tag{2-31}$$

In this case (2-28) can be expressed as

$$\mathfrak{Z}(N) = \mathcal{H}(N-1)\mathbf{Q_q r} + \mathcal{V}(N) \tag{2-32}$$

When event locations $q(1), q(2), \ldots, q(N)$ are known, then we can view (2-32) as a linear model for determining $\mathbf{r}$.

Regardless of which linear deconvolution model we use as our starting point for determining $\boldsymbol{\mu}$, we see that deconvolution corresponds to case B.1. Put another way, we have shown that *the design of a deconvolution signal processing filter is isomorphic to the problem of estimating random parameters in a linear model.* Note, however, that the dimension of $\boldsymbol{\theta}$, which is $N \times 1$, increases as the number of measurements increase. In all other examples $\boldsymbol{\theta}$ was $n \times 1$ where n is a fixed integer. We return to this point in Lesson 14 where we discuss convergence of estimates of $\boldsymbol{\mu}$ to their true values.

In Lesson 14, we shall develop minimum-variance and maximum-likelihood deconvolution filters. Equation (2-28) is the starting point for derivation of the former filter, whereas Equation (2-32) is the starting point for derivation of the latter filter. □

NOTATIONAL PRELIMINARIES

Equation (2-1) can be interpreted as a *data generating model*; it is a mathematical representation that is associated with the data. Parameter vector $\boldsymbol{\theta}$ is assumed to be unknown and is to be estimated using $\mathcal{Z}(k)$, $\mathcal{H}(k)$ and possibly other a priori information. We use $\hat{\boldsymbol{\theta}}(k)$ to denote the estimate of constant parameter vector $\boldsymbol{\theta}$. Argument k in $\hat{\boldsymbol{\theta}}(k)$ denotes the fact that the estimate is based on measurements up to and including the kth. In our preceding examples, we would use the following notation for $\hat{\boldsymbol{\theta}}(k)$:

Example 2-1 [see (2-5)]: $\hat{\boldsymbol{\theta}}(N)$ with components $\hat{h}(i \mid N)$
Example 2-2 [see (2-8)]: $\hat{\boldsymbol{\theta}}(N)$
Example 2-3 [see (2-13)]: $\hat{\mathbf{x}}(0 \mid N)$
Example 2-4 [see (2-21)]: $\hat{\mathbf{x}}(k_1 \mid N)$
Example 2-5 [see (2-25)]: $\widehat{\delta\boldsymbol{\theta}}(N)$
Example 2-6 [see (2-28)]: $\hat{\boldsymbol{\theta}}(N)$ with components $\hat{\mu}(i \mid N)$

The notation used in Examples 1, 3, 4, and 6 is a bit more complicated than that used in the other examples, because we must indicate the time point at which we are estimating the quantity of interest (e.g., k_1 or i) as well as the last data point used to obtain this estimate (e.g., N). We often read $\hat{\mathbf{x}}(k_1 \mid N)$ as "the estimate of $\mathbf{x}(k_1)$ conditioned on N" or as "$\mathbf{x}$ hat at k_1 conditioned on N."

In state estimation (or deconvolution) three situations are possible depending upon the relative relationship of N to k_1. For example, when $N < k_1$ we are estimating a future value of $\mathbf{x}(k_1)$, and we refer to this as a *predicted estimate.* When $N = k_1$ we are using all past measurements and the most recent measurement to estimate $\mathbf{x}(k_1)$. The result is referred to as a *filtered estimate.* Finally, when $N > k_1$ we are estimating an earlier value of $\mathbf{x}(k_1)$ using past, present and future measurements. Such an estimate is referred to as a *smoothed* or *interpolated* estimate. Prediction and filtering can be done in real time whereas smoothing can never be done in real time. We will see that the impulse responses of predictors and filters are causal, whereas the impulse response of a smoother is noncausal.

We use $\tilde{\boldsymbol{\theta}}(k)$ to denote *estimation error*, i.e.,

$$\tilde{\boldsymbol{\theta}}(k) = \boldsymbol{\theta} - \hat{\boldsymbol{\theta}}(k) \tag{2-33}$$

In state estimation, $\tilde{\mathbf{x}}(k_1|N)$ denotes state estimation error, and, $\tilde{\mathbf{x}}(k_1|N) = \mathbf{x}(k_1) - \hat{\mathbf{x}}(k_1|N)$. In deconvolution $\tilde{\mu}(i|N)$ is defined in a similar manner.

Very often we use the following *estimation model* for $\mathcal{Z}(k)$,

$$\hat{\mathcal{Z}}(k) = \mathcal{H}(k)\hat{\boldsymbol{\theta}}(k) \tag{2-34}$$

To obtain (2-34) from (2-1), we assume that $\mathcal{V}(k)$ is zero-mean random noise that cannot be measured. In some applications (e.g., Example 2-2) $\hat{\mathcal{Z}}(k)$ represents a predicted value of $\mathcal{Z}(k)$. Associated with $\hat{\mathcal{Z}}(k)$ is the error $\tilde{\mathcal{Z}}(k)$, where

$$\tilde{\mathcal{Z}}(k) = \mathcal{Z}(k) - \hat{\mathcal{Z}}(k) \tag{2-35}$$

satisfies the equation

$$\tilde{\mathcal{Z}}(k) = \mathcal{H}(k)\tilde{\boldsymbol{\theta}}(k) + \mathcal{V}(k) \tag{2-36}$$

In those applications where $\hat{\mathcal{Z}}(k)$ is a predicted value of $\mathcal{Z}(k)$, $\tilde{\mathcal{Z}}(k)$ is known as a *prediction error*. Other names for $\tilde{\mathcal{Z}}(k)$ are *equation error* and *measurement residual*.

In the rest of this book we develop specific structures for $\hat{\boldsymbol{\theta}}(k)$. These structures are referred to as *estimators*. *Estimates* are obtained whenever data is processed by an estimator. Estimator structures are associated with specific estimation techniques, and these techniques can be classified according to the natures of $\boldsymbol{\theta}$ and $\mathcal{H}(k)$, and what a priori information is assumed known about noise vector $\mathcal{V}(k)$. See Lesson 1 for an overview of all the different estimation techniques that are covered in this book.

PROBLEMS

2-1. Suppose $z(k) = \theta_1 + \theta_2 k + v(k)$, where $z(1) = 0.2$, $z(2) = 1.4$, $z(3) = 3.6$, $z(4) = 7.5$, and $z(5) = 10.2$. What are the explicit structures of $\mathcal{Z}(5)$ and $\mathcal{H}(5)$?

2-2. According to thermodynamic principles, pressure P and volume V of a given mass of gas are related by $PV^\gamma = C$, where γ and C are constants. Assume that N measurements of P and V are available. Explain how to obtain a linear model for estimation of parameters γ and $\ln C$.

2-3. (Mendel, 1973, Exercise 1–16(a), pg. 46). Suppose we know that a relationship exists between y and $x_1, x_2, \ldots, x_n$ of the form

$$y = \exp(a_1 x_1 + a_2 x_2 + \cdots + a_n x_n)$$

We desire to estimate $a_1, a_2, \ldots, a_n$ from measurements of y and $x = \text{col}\,(x_1, x_2, \ldots, x_n)$. Explain how to do this.

2-4. (Mendel, 1973, Exercise 1–17, pp. 46–47). The efficiency of a jet engine may be viewed as a linear combination of functions of inlet pressure $p(t)$ and operating temperature $T(t)$; that is to say,

$$E(t) = C_1 + C_2 f_1[p(t)] + C_3 f_2[T(t)] + C_4 f_3[p(t), T(t)] + \nu(t)$$

where the structures of f_1, f_2, and f_3 are known a priori and $\nu(t)$ represents modeling error of known mean and variance. From tests on the engine a table of values of $E(t)$, $p(t)$, and $T(t)$ are given at discrete values of t. Explain how C_1, C_2, C_3, and C_4 are estimated from these data.

Lesson 3

Least-Squares Estimation: Batch Processing

INTRODUCTION

The method of least squares dates back to Karl Gauss around 1795, and is the cornerstone for most estimation theory, both classical and modern. It was invented by Gauss at a time when he was interested in predicting the motion of planets and comets using telescopic measurements. The motions of these bodies can be completely characterized by six parameters. The estimation problem that Gauss considered was one of inferring the values of these parameters from the measurement data.

We shall study least-squares estimation from two points of view: the classical batch-processing approach, in which all the measurements are processed together at one time, and the more modern recursive processing approach, in which measurements are processed only a few (or even one) at a time. The recursive approach has been motivated by today's high-speed digital computers; however, as we shall see, the recursive algorithms are outgrowths of the batch algorithms. In fact, as we enter the era of very large scale integration (VLSI) technology, it may well be that VLSI implementations of the batch algorithms are faster than digital computer implementations of the recursive algorithms.

The starting point for the method of least squares is the linear model

$$\mathcal{Z}(k) = \mathcal{H}(k)\boldsymbol{\theta} + \mathcal{V}(k) \tag{3-1}$$

where $\mathcal{Z}(k) = \text{col}\,(z(k), z(k-1), \ldots, z(k-N+1))$, $z(k) = \mathbf{h}'(k)\boldsymbol{\theta} + v(k)$, and the estimation model for $\mathcal{Z}(k)$ is

$$\hat{\mathcal{Z}}(k) = \mathcal{H}(k)\hat{\boldsymbol{\theta}}(k) \tag{3-2}$$

We denote the (weighted) least-squares estimator of $\boldsymbol{\theta}$ as $[\hat{\boldsymbol{\theta}}_{WLS}(k)]\hat{\boldsymbol{\theta}}_{LS}(k)$. In this lesson and the next two we shall determine explicit structures for this estimator.

NUMBER OF MEASUREMENTS

Suppose that $\boldsymbol{\theta}$ contains n parameters and $\mathcal{Z}(k)$ contains N measurements. If $N < n$ we have fewer measurements than unknowns and (3-1) is an underdetermined system of equations that does not lead to unique or very meaningful values for $\theta_1, \theta_2, \ldots, \theta_n$. If $N = n$, we have exactly as many measurements as unknowns, and as long as the n measurements are linearly independent, so that $\mathcal{H}^{-1}(k)$ exists, we can solve (3-1) for $\boldsymbol{\theta}$, as

$$\boldsymbol{\theta} = \mathcal{H}^{-1}(k)\mathcal{Z}(k) - \mathcal{H}^{-1}(k)\mathcal{V}(k) \tag{3-3}$$

Because we cannot measure $\mathcal{V}(k)$, it is usually neglected in the calculation of (3-3). For small amounts of noise this may not be a bad thing to do but for even moderate amounts of noise this will be quite bad. Finally, if $N > n$ we have more measurements than unknowns, so that (3-1) is an overdetermined system of equations. The extra measurements can be used to offset the effects of the noise; i.e., they let us "filter" the data. Only this last case is of real interest to us.

OBJECTIVE FUNCTION AND PROBLEM STATEMENT

Our method for obtaining $\hat{\boldsymbol{\theta}}(k)$ is based on minimizing the objective function

$$J[\hat{\boldsymbol{\theta}}(k)] = \tilde{\mathcal{Z}}'(k)\mathbf{W}(k)\tilde{\mathcal{Z}}(k) \tag{3-4}$$

where

$$\tilde{\mathcal{Z}}(k) = \text{col}[\tilde{z}(k), \tilde{z}(k-1), \ldots, \tilde{z}(k-N+1)] \tag{3-5}$$

and *weighting matrix* $\mathbf{W}(k)$ must be symmetric and positive definite, for reasons explained below.

No general rules exist for how to choose $\mathbf{W}(k)$. The most common choice is a diagonal matrix such as

$$\mathbf{W}(k) = \text{diag}[\mu^{k-N+1}, \mu^{k-N+2}, \ldots, \mu^{k}]$$

When $|\mu| < 1$ recent measurements are weighted more heavily than past ones. Such a choice for $\mathbf{W}(k)$ provides the weighted least-squares estimator with an "aging" or "forgetting" factor. When $|\mu| > 1$ recent measurements are weighted less heavily than past ones. Finally, if $\mu = 1$, so that $\mathbf{W}(k) = \mathbf{I}$, then all measurements are weighted by the same amount. When $\mathbf{W}(k) = \mathbf{I}$, $\hat{\boldsymbol{\theta}}(k) = \hat{\boldsymbol{\theta}}_{LS}(k)$, whereas for all other $\mathbf{W}(k)$, $\hat{\boldsymbol{\theta}}(k) = \hat{\boldsymbol{\theta}}_{WLS}(k)$.

Our objective is to determine the $\hat{\boldsymbol{\theta}}_{WLS}(k)$ that minimizes $J[\hat{\boldsymbol{\theta}}(k)]$.

DERIVATION OF ESTIMATOR

To begin, we express (3-4) as an explicit function of $\hat{\boldsymbol{\theta}}(k)$, using (3-2):

$$\begin{aligned} J[\hat{\boldsymbol{\theta}}(k)] &= \tilde{\mathfrak{Z}}'(k)\mathbf{W}(k)\tilde{\mathfrak{Z}}(k) = [\mathfrak{Z}(k) - \hat{\mathfrak{Z}}(k)]'\mathbf{W}(k)[\mathfrak{Z}(k) - \hat{\mathfrak{Z}}(k)] \\ &= [\mathfrak{Z}(k) - \mathcal{H}(k)\hat{\boldsymbol{\theta}}(k)]'\mathbf{W}(k)[\mathfrak{Z}(k) - \mathcal{H}(k)\hat{\boldsymbol{\theta}}(k)] \\ &= \mathfrak{Z}'(k)\mathbf{W}(k)\mathfrak{Z}(k) - 2\mathfrak{Z}'(k)\mathbf{W}(k)\mathcal{H}(k)\hat{\boldsymbol{\theta}}(k) \\ &\quad + \hat{\boldsymbol{\theta}}'(k)\mathcal{H}'(k)\mathbf{W}(k)\mathcal{H}(k)\hat{\boldsymbol{\theta}}(k) \end{aligned} \tag{3-6}$$

Next, we take the vector derivative of $J[\hat{\boldsymbol{\theta}}(k)]$ with respect to $\hat{\boldsymbol{\theta}}(k)$, but before doing this recall from vector calculus, that:

If $\mathbf{m}$ and $\mathbf{b}$ are two $n \times 1$ nonzero vectors, and $\mathbf{A}$ is an $n \times n$ symmetric matrix, then

$$\frac{d}{d\mathbf{m}}(\mathbf{b}'\mathbf{m}) = \mathbf{b} \tag{3-7}$$

and

$$\frac{d}{d\mathbf{m}}(\mathbf{m}'\mathbf{A}\mathbf{m}) = 2\mathbf{A}\mathbf{m} \tag{3-8}$$

Using these formulas, we find that

$$\frac{dJ[\hat{\boldsymbol{\theta}}(k)]}{d\hat{\boldsymbol{\theta}}(k)} = -2[\mathfrak{Z}'(k)\mathbf{W}(k)\mathcal{H}(k)]' + 2\mathcal{H}'(k)\mathbf{W}(k)\mathcal{H}(k)\hat{\boldsymbol{\theta}}(k) \tag{3-9}$$

Setting $dJ[\hat{\boldsymbol{\theta}}(k)]/d\hat{\boldsymbol{\theta}}(k) = \mathbf{0}$ we obtain the following formula for $\hat{\boldsymbol{\theta}}_{\text{WLS}}(k)$,

$$\boxed{\hat{\boldsymbol{\theta}}_{\text{WLS}}(k) = [\mathcal{H}'(k)\mathbf{W}(k)\mathcal{H}(k)]^{-1}\mathcal{H}'(k)\mathbf{W}(k)\mathfrak{Z}(k)} \tag{3-10}$$

Note, also, that

$$\boxed{\hat{\boldsymbol{\theta}}_{\text{LS}}(k) = [\mathcal{H}'(k)\mathcal{H}(k)]^{-1}\mathcal{H}'(k)\mathfrak{Z}(k)} \tag{3-11}$$

Comments

1. For (3-10) to be valid, $\mathcal{H}'(k)\mathbf{W}(k)\mathcal{H}(k)$ must be invertible. This is always true when $\mathbf{W}(k)$ is positive definite, as assumed, and $\mathcal{H}(k)$ is of maximum rank.
2. How do we know that $\hat{\boldsymbol{\theta}}_{\text{WLS}}(k)$ minimizes $J[\hat{\boldsymbol{\theta}}(k)]$? We compute $d^2J[\hat{\boldsymbol{\theta}}(k)]/d\hat{\boldsymbol{\theta}}^2(k)$ and see if it is positive definite [which is the vector calculus analog of the scalar calculus requirement that $\hat{\theta}$ minimizes $J(\hat{\theta})$ if $dJ(\hat{\theta})/d\hat{\theta} = 0$ and $d^2J(\hat{\theta})/d\hat{\theta}^2$ is positive]. Doing this, we see that

$$\frac{d^2J[\hat{\boldsymbol{\theta}}(k)]}{d\hat{\boldsymbol{\theta}}^2(k)} = 2\mathcal{H}'(k)\mathbf{W}(k)\mathcal{H}(k) > 0 \tag{3-12}$$

because $\mathcal{H}'(k)\mathbf{W}(k)\mathcal{H}(k)$ is invertible.

3. Estimator $\hat{\boldsymbol{\theta}}_{\text{WLS}}(k)$ processes the measurements $\mathfrak{Z}(k)$ linearly; thus, it is referred to as a *linear estimator*. It processes the data contained in $\mathcal{H}(k)$ in a very complicated and nonlinear manner.
4. When (3-9) is set equal to zero we obtain the following system of *normal equations*

$$[\mathcal{H}'(k)\mathbf{W}(k)\mathcal{H}(k)]\hat{\boldsymbol{\theta}}_{\text{WLS}}(k) = \mathcal{H}'(k)\mathbf{W}(k)\mathfrak{Z}(k) \tag{13-13}$$

This is a system of n linear equations in the n components of $\hat{\boldsymbol{\theta}}_{\text{WLS}}(k)$.

In practice, one does not compute $\hat{\boldsymbol{\theta}}_{\text{WLS}}(k)$ using (3-10), because computing the inverse of $\mathcal{H}'(k)\mathbf{W}(k)\mathcal{H}(k)$ is fraught with numerical difficulties. Instead, the normal equations are solved using stable algorithms from numerical linear algebra that involve orthogonal transformations (see, e.g., Stewart, 1973; Bierman, 1977; and Dongarra, et al., 1979). Because it is not the purpose of this book to go into details of numerical linear algebra, we leave it to the reader to pursue this important subject.

Based on this discussion, we must view (3-10) as a useful "*theoretical*" formula and not as a useful computational formula. Remember that this is a book on estimation *theory*, so for our purposes, theoretical formulas are just fine.

5. Equation (3-13) can also be reexpressed as

$$\tilde{\mathfrak{Z}}'(k)\mathbf{W}(k)\mathcal{H}(k) = \mathbf{0}' \tag{3-14}$$

which can be viewed as an *orthogonality condition* between $\tilde{\mathfrak{Z}}(k)$ and $\mathbf{W}(k)\mathcal{H}(k)$. Orthogonality conditions play an important role in estimation theory. We shall see many more examples of such conditions throughout this book.

6. Estimates obtained from (3-10) will be random! This is because $\mathfrak{Z}(k)$ is random, and, in some applications even $\mathcal{H}(k)$ is random. It is therefore instructive to view (3-10) as a complicated transformation of vectors or matrices of random variables into the vector of random variables $\hat{\boldsymbol{\theta}}_{\text{WLS}}(k)$. In later lessons, when we examine the properties of $\hat{\boldsymbol{\theta}}_{\text{WLS}}(k)$, these will be statistical properties, because of the random nature of $\hat{\boldsymbol{\theta}}_{\text{WLS}}(k)$.

Example 3-1 (Mendel, 1973, pp. 86–87)

Suppose we wish to calibrate an instrument by making a series of uncorrelated measurements on a constant quantity. Denoting the constant quantity as θ, our measurement equation becomes

$$z(k) = \theta + v(k) \tag{3-15}$$

where $k = 1, 2, \ldots, N$. Collecting these N measurements, we have

$$\begin{pmatrix} z(N) \\ z(N-1) \\ \vdots \\ z(1) \end{pmatrix} = \begin{pmatrix} 1 \\ 1 \\ \vdots \\ 1 \end{pmatrix}\theta + \begin{pmatrix} v(N) \\ v(N-1) \\ \vdots \\ v(1) \end{pmatrix} \tag{3-16}$$

Clearly, $\mathcal{H} = \text{col}\,(1, 1, \ldots, 1)$; hence;

$$\hat{\theta}_{LS}(N) = \frac{1}{N}\sum_{i=1}^{N} z(i) \tag{3-17}$$

which is the sample mean of the N measurements. We see, therefore, that *the sample mean is a least-squares estimator.* □

Example 3-2 (Mendel, 1973)

Figure 3-1 depicts simplified third-order pitch-plane dynamics for a typical, high-performance, aerodynamically controlled aerospace vehicle. Cross-coupling and body-bending effects are neglected. Normal acceleration control is considered with feedback on normal acceleration and angle-of-attack rate. Stefani (1967) shows that if the system gains are chosen as

$$K_{Ni} = \frac{C_2}{100\, M_\delta Z_\alpha} \tag{3-18}$$

$$K_{\dot{\alpha}} = \frac{C_1 - 100\left(\dfrac{Z_\alpha 1845}{\mu}\right) + M_\alpha}{100\; M_\delta} \tag{3-19}$$

and

$$K_{Na} = \frac{C_2 + 100\, M_\alpha}{100\, M_\delta Z_\alpha} \tag{3-20}$$

then

$$\frac{N_a}{N_i}(s) = \frac{C_2}{s^3 + 100s^2 + C_1 s + C_2} \tag{3-21}$$

Stefani assumes $Z_\alpha\, 1845/\mu$ is relatively small, and chooses $C_1 = 1400$ and $C_2 = 14{,}000$. The closed-loop response resembles that of a second-order system with a bandwidth of 2 Hz and a damping ratio of 0.6 that responds to a step command of input acceleration with zero steady-state error.

In general, M_α, M_δ, and Z_α are dynamic parameters and all vary through a large range of values. Also, M_α may be positive (unstable vehicle) or negative (stable vehicle). System response must remain the same for all values of M_α, M_δ, and Z_α; thus, it is necessary to estimate these parameters so that K_{Ni}, $K_{\dot{\alpha}}$, and K_{Na} can be adapted to keep C_1 and C_2 invariant at their designed values. For present purposes we shall assume that M_α, M_δ, and Z_α are frozen at specific values.

From Fig. 3-1,

$$\ddot{\theta}(t) = M_\alpha \alpha(t) + M_\delta \delta(t) \tag{3-22}$$

and

$$N_a(t) = Z_\alpha \alpha(t) \tag{3-23}$$

Our attention is directed at the estimation of M_α and M_δ in (3-22). We leave it as an exercise for the reader to explore the estimation of Z_α in (3-23).

Our approach will be to estimate M_α and M_δ from the equation

$$\ddot{\theta}_m(k) = M_\alpha \alpha(k) + M_\delta \delta(k) + v_{\ddot{\theta}}(k) \tag{3-24}$$

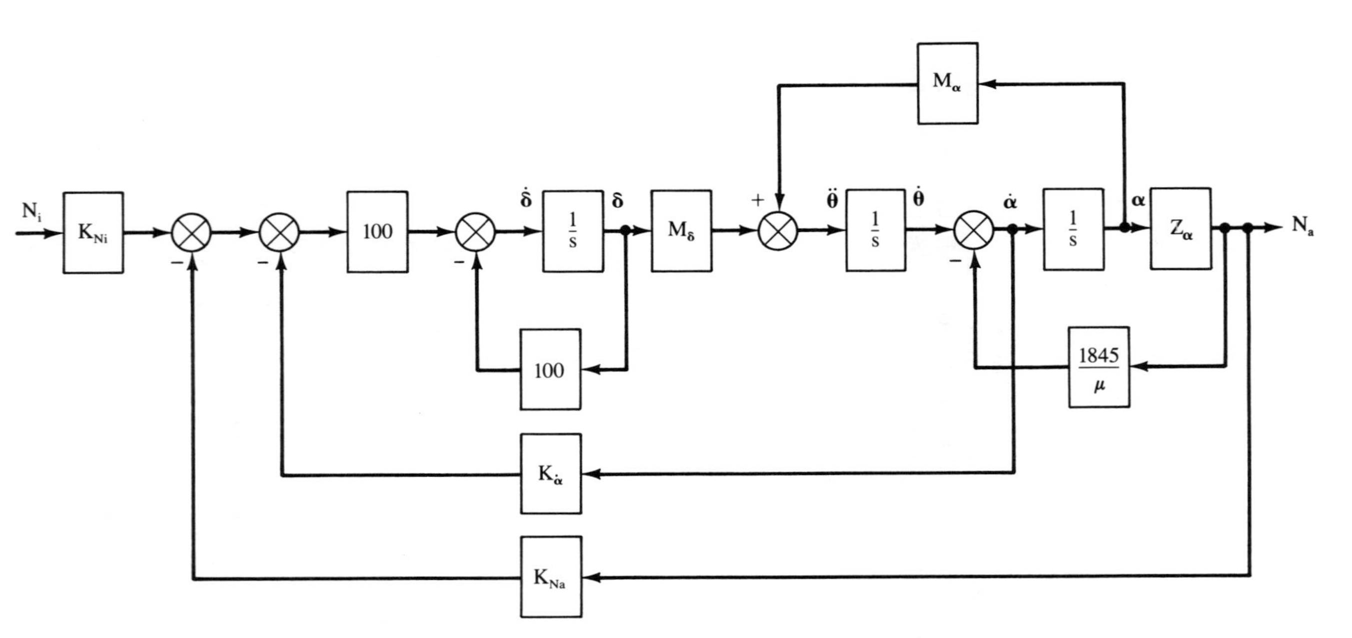

Figure 3-1 Pitch-plane dynamics and nomenclature: N_i, input normal acceleration along the negative Z axis; K_{Ni}, gain on N_i; δ, control-surface deflection; M_δ, control-surface effectiveness; $\ddot{\theta}$, rigid-body acceleration; α, angle-of-attack; M_α, aerodynamic moment effectiveness; $K_{\dot{\alpha}}$, control gain on $\dot{\alpha}$; Z_α, normal acceleration force coefficient; μ, axial velocity; N_a, system-achieved normal acceleration along the negative Z axis; K_{Na}, control gain on N_a (reprinted from Mendel, 1973, p. 33, by courtesy of Marcel Dekker, Inc.).

where $\ddot{\theta}_m(k)$ denotes the measured value of $\ddot{\theta}(k)$, that is corrupted by measurement noise $v_{\ddot{\theta}}(k)$. We shall assume (somewhat unrealistically) that $\alpha(k)$ and $\delta(k)$ can both be measured perfectly. The concatenated measurement equation for N measurements is

$$\begin{pmatrix} \ddot{\theta}_m(k) \\ \ddot{\theta}_m(k-1) \\ \vdots \\ \ddot{\theta}_m(k-N+1) \end{pmatrix} = \begin{pmatrix} \alpha(k) & \delta(k) \\ \alpha(k-1) & \delta(k-1) \\ \vdots & \vdots \\ \alpha(k-N+1) & \delta(k-N+1) \end{pmatrix} \times \begin{pmatrix} M_\alpha \\ M_\delta \end{pmatrix} + \begin{pmatrix} v_{\ddot{\theta}}(k) \\ v_{\ddot{\theta}}(k-1) \\ \vdots \\ v_{\ddot{\theta}}(k-N+1) \end{pmatrix} \tag{3-25}$$

Hence, the least-squares estimates of M_α and M_δ are

$$\begin{pmatrix} \hat{M}_\alpha(k) \\ \hat{M}_\delta(k) \end{pmatrix} = \begin{pmatrix} \sum_{j=0}^{N-1} \alpha^2(k-j) & \sum_{j=0}^{N-1} \alpha(k-j)\delta(k-j) \\ \sum_{j=0}^{N-1} \alpha(k-j)\delta(k-j) & \sum_{j=0}^{N-1} \delta^2(k-j) \end{pmatrix}^{-1} \times \begin{pmatrix} \sum_{j=0}^{N-1} \alpha(k-j)\ddot{\theta}_m(k-j) \\ \sum_{j=0}^{N-1} \delta(k-j)\ddot{\theta}_m(k-j) \end{pmatrix} \quad \square \tag{3-26}$$

FIXED AND EXPANDING MEMORY ESTIMATORS

Estimator $\hat{\boldsymbol{\theta}}_{WLS}(k)$ uses the measurements $z(k-N+1), z(k-N+2), \ldots, z(k)$. When N is fixed ahead of time, $\hat{\boldsymbol{\theta}}_{WLS}(k)$ uses a fixed window of measurements, a window of length N, and, $\hat{\boldsymbol{\theta}}_{WLS}(k)$ is then referred to as a *fixed-memory estimator.* A second approach for choosing N is to set it equal to k; then, $\hat{\boldsymbol{\theta}}_{WLS}(k)$ uses the measurements $z(1), z(2), \ldots, z(k)$. In this case $\hat{\boldsymbol{\theta}}_{WLS}(k)$ uses an expanding window of measurements, a window of length k, and, $\hat{\boldsymbol{\theta}}_{WLS}(k)$ is then referred to as an *expanding-memory estimator*.

SCALE CHANGES

Least-squares (LS) estimates may not be invariant under changes of scale. One way to circumvent this difficulty is to use normalized data.

For example, assume that observers A and B are observing a process; but observer A reads the measurements in one set of units and B in another.

Let $\mathbf{M}$ be a symmetric matrix of scale factors relating A to B; $\mathcal{Z}_A(k)$ and $\mathcal{Z}_B(k)$ denote the total measurement vectors of A and B, respectively. Then

$$\mathcal{Z}_B(k) = \mathcal{H}_B(k)\boldsymbol{\theta} + \mathcal{V}_B(k) = \mathbf{M}\mathcal{Z}_A(k) = \mathbf{M}\mathcal{H}_A(k)\boldsymbol{\theta} + \mathbf{M}\mathcal{V}_A(k) \tag{3-27}$$

which means that

$$\mathcal{H}_B(k) = \mathbf{M}\mathcal{H}_A(k) \tag{3-28}$$

Let $\hat{\boldsymbol{\theta}}_{A,\text{WLS}}(k)$ and $\hat{\boldsymbol{\theta}}_{B,\text{WLS}}(k)$ denote the WLSE's associated with observers A and B, respectively; then $\hat{\boldsymbol{\theta}}_{B,\text{WLS}}(k) = \hat{\boldsymbol{\theta}}_{A,\text{WLS}}(k)$ if

$$\mathbf{W}_B(k) = \mathbf{M}^{-1}\mathbf{W}_A(k)\mathbf{M}^{-1} \tag{3-29}$$

It seems a bit peculiar though to have different weighting matrices for the two WLSE's. In fact, if we begin with $\hat{\boldsymbol{\theta}}_{A,\text{LS}}(k)$ then it is impossible to obtain $\hat{\boldsymbol{\theta}}_{B,\text{LS}}(k)$ such that $\hat{\boldsymbol{\theta}}_{B,\text{LS}}(k) = \hat{\boldsymbol{\theta}}_{A,\text{LS}}(k)$. The reason for this is simple. To obtain $\hat{\boldsymbol{\theta}}_{A,\text{LS}}(k)$, we set $\mathbf{W}_A(k) = \mathbf{I}$, in which case (3-29) reduces to $\mathbf{W}_B(k) = (\mathbf{M}^{-1})^2 \neq \mathbf{I}$.

Next, let $\mathbf{N}_A$ and $\mathbf{N}_B$ denote symmetric normalization matrices for $\mathcal{Z}_A(k)$ and $\mathcal{Z}_B(k)$, respectively. We shall assume that our data is always normalized to the same set of numbers, i.e., that

$$\mathbf{N}_A\mathcal{Z}_A(k) = \mathbf{N}_B\mathcal{Z}_B(k) \tag{3-30}$$

Observe that

$$\mathbf{N}_A\mathcal{Z}_A(k) = \mathbf{N}_A\mathcal{H}_A(k)\boldsymbol{\theta} + \mathbf{N}_A\mathcal{V}_A(k) \tag{3-31}$$

and

$$\mathbf{N}_B\mathcal{Z}_B(k) = \mathbf{N}_B\mathbf{M}\mathcal{Z}_A(k) = \mathbf{N}_B\mathbf{M}\mathcal{H}_A(k)\boldsymbol{\theta} + \mathbf{N}_B\mathbf{M}\mathcal{V}_A(k) \tag{3-32}$$

From (3-30), (3-31), and (3-32), we see that

$$\mathbf{N}_A = \mathbf{N}_B\mathbf{M} \tag{3-33}$$

We now find that

$$\hat{\boldsymbol{\theta}}_{A,\text{WLS}}(k) = (\mathcal{H}_A'\mathbf{N}_A\mathbf{W}_A\mathbf{N}_A\mathcal{H}_A)^{-1}\mathcal{H}_A'\mathbf{N}_A\mathbf{W}_A\mathbf{N}_A\mathcal{Z}_A(k) \tag{3-34}$$

and

$$\hat{\boldsymbol{\theta}}_{B,\text{WLS}}(k) = (\mathcal{H}_A'\mathbf{M}\mathbf{N}_B\mathbf{W}_B\mathbf{N}_B\mathbf{M}\mathcal{H}_A)^{-1}\mathcal{H}_A'\mathbf{M}\mathbf{N}_B\mathbf{W}_B\mathbf{N}_B\mathbf{N}\mathcal{Z}_A(k) \tag{3-35}$$

Substituting (3-33) into (3-35), we then find

$$\hat{\boldsymbol{\theta}}_{B,\text{WLS}}(k) = (\mathcal{H}_A'\mathbf{N}_A\mathbf{W}_B\mathbf{N}_A\mathcal{H}_A')^{-1}\mathcal{H}_A'\mathbf{N}_A\mathbf{W}_B\mathbf{N}_A\mathcal{Z}_A(k) \tag{3-36}$$

Comparing (3-36) and (3-34), we conclude that $\hat{\boldsymbol{\theta}}_{B,\text{WLS}}(k) = \hat{\boldsymbol{\theta}}_{A,\text{WLS}}(k)$ if $\mathbf{W}_B(k) = \mathbf{W}_A(k)$. This is precisely the result we were looking for. It means that, under proper normalization, $\hat{\boldsymbol{\theta}}_{B,\text{WLS}}(k) = \hat{\boldsymbol{\theta}}_{A,\text{WLS}}(k)$, and, as a special case $\hat{\boldsymbol{\theta}}_{B,\text{LS}}(k) = \hat{\boldsymbol{\theta}}_{A,\text{LS}}(k)$.

PROBLEMS

3-1. Derive the formula for $\hat{\boldsymbol{\theta}}_{\text{WLS}}(k)$ by completing the square on the right-hand side of the expression for $J[\hat{\boldsymbol{\theta}}(k)]$ in (3-6).

3-2. Here we explore the estimation of Z_α in (3-23). Assume that N noisy measurements of $N_a(k)$ are available, i.e., $N_{a_m}(k) = Z_\alpha \alpha(k) + \nu_{N_a}(k)$. What is the formula for the least-squares estimator of Z_α?

3-3. Here we explore the simultaneous estimation of M_α, M_δ, and Z_α in (3-22) and (3-23). Assume that N noisy measurements of $\ddot{\theta}(k)$ and $N_a(k)$ are available, i.e., $\ddot{\theta}_m(k) = M_\alpha \alpha(k) + M_\delta \delta(k) + \nu_{\ddot{\theta}}(k)$ and $N_{a_m}(k) = Z_\alpha \alpha(k) + \nu_{N_a}(k)$. Determine the least-squares estimator of M_α, M_δ, and Z_α. Is this estimator different from $\hat{M}_{\alpha_{\text{LS}}}$ and $\hat{M}_{\delta_{\text{LS}}}$ obtained just from $\ddot{\theta}_m(k)$ measurements, and $\hat{Z}_{\alpha_{\text{LS}}}$ obtained just from $N_{a_m}(k)$ measurements?

3-4. In a *curve fitting problem* we wish to fit a given set of data $z(1), z(2), \ldots, z(N)$ by the approximating function

$$\hat{z}(k) = \sum_{j=1}^{n} \hat{\theta}_j \phi_j(k)$$

where $\phi_j(k)(j = 1, 2, \ldots, n)$ are a set of prespecified basis functions.

(a) Obtain a formula for $\hat{\boldsymbol{\theta}}_{\text{LS}}(N)$ that is valid for any set of basis functions.

(b) The simplest approximating function to a set of data is the straight line. In this case $\hat{z}(k) = \hat{\theta}_1 + \hat{\theta}_2 k$, which is known as the least-squares or regression line. Obtain closed-form formulas for $\hat{\theta}_{1,\text{LS}}(N)$ and $\hat{\theta}_{2,\text{LS}}(N)$.

3-5. Suppose $z(k) = \theta_1 + \theta_2 k$, where $z(1) = 3$ miles per hour and $z(2) = 7$ miles per hour. Determine $\hat{\theta}_{1,\text{LS}}$ and $\hat{\theta}_{2,\text{LS}}$ based on these two measurements. Next, redo these calculations by scaling $z(1)$ and $z(2)$ to the units of feet per second. Are the least-squares estimates obtained from these two calculations the same? Use the results developed in the section entitled "Scale Changes" to explain what has happened here.

3-6. **(a)** Under what conditions on scaling matrix $\mathbf{M}$ is scale invariance preserved for a least-squares estimator?

(b) If our original model is nonlinear in the measurements [e.g., $z(k) = \theta z^2(k-1) + \nu(k)$] can anything be done to obtain invariant WLSE's under scaling?

Lesson 4

Least-Squares Estimation: Recursive Processing

INTRODUCTION

In Lesson 3 we assumed that $\mathfrak{X}(k)$ contained N elements, where $N > \dim \boldsymbol{\theta} = n$. Suppose we decide to add more measurements, increasing the total number of them from N to N'. Formula (3-10) in Lesson 3 would not make use of the previously calculated value of $\hat{\boldsymbol{\theta}}$ that is based on N measurements during the calculation of $\hat{\boldsymbol{\theta}}$ that is based on N' measurements. This seems quite wasteful. We intuitively feel that it should be possible to compute the estimate based on N' measurements from the estimate based on N measurements, and a modification of this earlier estimate to account for the $N'-N$ new measurements. In this lesson we shall justify our intuition.

In Lesson 3 we also assumed that $\hat{\boldsymbol{\theta}}$ is determined for a fixed value of n. In many system modeling problems one is interested in a preliminary model in which dimension n is a variable. This is becoming increasingly more important as we begin to model large scale societal, energy, economic, etc. systems in which it may not be clear at the onset what effects are most important. One approach is to recompute $\hat{\boldsymbol{\theta}}$ by means of Formula (3-10) in Lesson 3 for different values of n. This may be very costly, especially for large scale systems, since the number of flops to compute $\hat{\boldsymbol{\theta}}$ is on the order of n^3. A second approach is to obtain $\hat{\boldsymbol{\theta}}$ for $n = n_1$, and to use that estimate in a computationally effective manner to obtain $\hat{\boldsymbol{\theta}}$ for $n = n_2$, where $n_2 > n_1$. These estimators are recursive in the dimension of $\boldsymbol{\theta}$. We shall also examine these estimators.

RECURSIVE LEAST-SQUARES: INFORMATION FORM

To begin, we consider the case when one additional measurement $z(k+1)$, made at t_{k+1}, becomes available:

$$z(k+1) = \mathbf{h}'(k+1)\boldsymbol{\theta} + v(k+1) \tag{4-1}$$

When this equation is combined with our earlier linear model we obtain a new linear model,

$$\mathfrak{Z}(k+1) = \mathcal{H}(k+1)\boldsymbol{\theta} + \mathcal{V}(k+1) \tag{4-2}$$

where

$$\mathfrak{Z}(k+1) = \text{col}\,(z(k+1)|\mathfrak{Z}(k)) \tag{4-3}$$

$$\mathcal{H}(k+1) = \left(\frac{\mathbf{h}'(k+1)}{\mathcal{H}(k)}\right) \tag{4-4}$$

and

$$\mathcal{V}(k+1) = \text{col}\,(v(k+1)|\mathcal{V}(k)) \tag{4-5}$$

Using (3-10) from Lesson 3 and (4-2), it is clear that

$$\hat{\boldsymbol{\theta}}_{WLS}(k+1) = [\mathcal{H}'(k+1)\mathbf{W}(k+1)\mathcal{H}(k+1)]^{-1}\mathcal{H}'(k+1)\mathbf{W}(k+1)\mathfrak{Z}(k+1) \tag{4-6}$$

To proceed further we must assume that $\mathbf{W}$ is diagonal, i.e.,

$$\mathbf{W}(k+1) = \text{diag}\,(w(k+1)|\mathbf{W}(k)) \tag{4-7}$$

We shall now show that it is possible to determine $\hat{\boldsymbol{\theta}}(k+1)$ from $\hat{\boldsymbol{\theta}}(k)$ and $z(k+1)$.

Theorem 4-1. (Information Form of Recursive LSE). *A recursive structure for* $\hat{\boldsymbol{\theta}}_{WLS}(k)$ *is*

$$\hat{\boldsymbol{\theta}}_{WLS}(k+1) = \hat{\boldsymbol{\theta}}_{WLS}(k) + \mathbf{K}_W(k+1)[z(k+1) - \mathbf{h}'(k+1)\hat{\boldsymbol{\theta}}_{WLS}(k)] \tag{4-8}$$

where

$$\mathbf{K}_W(k+1) = \mathbf{P}(k+1)\mathbf{h}(k+1)w(k+1) \tag{4-9}$$

and

$$\mathbf{P}^{-1}(k+1) = \mathbf{P}^{-1}(k) + \mathbf{h}(k+1)w(k+1)\mathbf{h}'(k+1) \tag{4-10}$$

These equations are initialized by $\hat{\boldsymbol{\theta}}_{WLS}(n)$ *and* $\mathbf{P}^{-1}(n)$, *where* $\mathbf{P}(k)$ *is defined below in (4-13), and are used for* $k = n, n+1, \ldots, N-1$.

Proof. Substitute (4-3), (4-4) and (4-7) into (4-6) (sometimes dropping the dependence upon k and $k+1$, for notational simplicity) to see that

$$\hat{\boldsymbol{\theta}}_{\mathrm{WLS}}(k+1) = [\mathcal{H}'(k+1)\mathbf{W}(k+1)\mathcal{H}(k+1)]^{-1}[\mathbf{h}wz + \mathcal{H}'\mathbf{W}\mathfrak{Z}] \tag{4-11}$$

Express $\hat{\boldsymbol{\theta}}_{\mathrm{WLS}}(k)$ as

$$\hat{\boldsymbol{\theta}}_{\mathrm{WLS}}(k) = \mathbf{P}(k)\mathcal{H}'(k)\mathbf{W}(k)\mathfrak{Z}(k) \tag{4-12}$$

where

$$\mathbf{P}(k) = [\mathcal{H}'(k)\mathbf{W}(k)\mathcal{H}(k)]^{-1} \tag{4-13}$$

From (4-12) and (4-13) it is straightforward to show that

$$\mathcal{H}'(k)\mathbf{W}(k)\mathfrak{Z}(k) = \mathbf{P}^{-1}(k)\hat{\boldsymbol{\theta}}_{\mathrm{WLS}}(k) \tag{4-14}$$

and

$$\mathbf{P}^{-1}(k+1) = \mathbf{P}^{-1}(k) + \mathbf{h}(k+1)w(k+1)\mathbf{h}'(k+1) \tag{4-15}$$

It now follows that

$$\begin{aligned}\hat{\boldsymbol{\theta}}_{\mathrm{WLS}}(k+1) &= \mathbf{P}(k+1)[\mathbf{h}wz + \mathbf{P}^{-1}(k)\hat{\boldsymbol{\theta}}_{\mathrm{WLS}}(k)] \\ &= \mathbf{P}(k+1)\{\mathbf{h}wz + [\mathbf{P}^{-1}(k+1) - \mathbf{h}w\mathbf{h}']\hat{\boldsymbol{\theta}}_{\mathrm{WLS}}(k)\} \\ &= \hat{\boldsymbol{\theta}}_{\mathrm{WLS}}(k) + \mathbf{P}(k+1)\mathbf{h}w[z - \mathbf{h}'\hat{\boldsymbol{\theta}}_{\mathrm{WLS}}(k)] \\ &= \hat{\boldsymbol{\theta}}_{\mathrm{WLS}}(k) + \mathbf{K}_{\mathrm{W}}(k+1)[z - \mathbf{h}'\hat{\boldsymbol{\theta}}_{\mathrm{WLS}}(k)]\end{aligned} \tag{4-16}$$

which is (4-8) when gain matrix $\mathbf{K}_{\mathrm{W}}$ is defined as in (4-9).

Based on preceding discussions about dim $\boldsymbol{\theta} = n$ and dim $\mathfrak{Z}(k) = N$, we know that the first value of N for which (3-10) in Lesson 3 can be used is $N = n$; thus, (4-8) must be initialized by $\hat{\boldsymbol{\theta}}_{\mathrm{WLS}}(n)$, which is computed using (3-10) in Lesson 3. Equation (4-10) is also a recursive equation for $\mathbf{P}^{-1}(k+1)$, which is initialized by $\mathbf{P}^{-1}(n) = \mathcal{H}'(n)\mathbf{W}(n)\mathcal{H}(n)$. □

Comments

1. Equation (4-8) can also be expressed as

$$\hat{\boldsymbol{\theta}}_{\mathrm{WLS}}(k+1) = [\mathbf{I} - \mathbf{K}_{\mathrm{W}}(k+1)\mathbf{h}'(k+1)]\hat{\boldsymbol{\theta}}_{\mathrm{WLS}}(k) + \mathbf{K}_{\mathrm{W}}(k+1)z(k+1) \tag{4-17}$$

which demonstrates that *the recursive least-squares estimate (LSE) is a time-varying digital filter that is excited by random inputs* (i.e., the measurements), one whose plant matrix may itself be random, because $\mathbf{K}_{\mathrm{W}}$ and $\mathbf{h}(k+1)$ may be random. The random natures of $\mathbf{K}_{\mathrm{W}}$ and $(\mathbf{I} - \mathbf{K}_{\mathrm{W}}\mathbf{h}')$ make the analysis of this filter exceedingly difficult. If $\mathbf{K}_{\mathrm{W}}$ and $\mathbf{h}$ are deterministic, then stability of this filter can be studied using Lyapunov stability theory.

2. In (4-8), the term $\mathbf{h}'(k+1)\hat{\boldsymbol{\theta}}_{\text{WLS}}(k)$ is a prediction of the actual measurement $z(k+1)$. Because $\hat{\boldsymbol{\theta}}_{\text{WLS}}(k)$ is based on $\mathfrak{Z}(k)$, we express this predicted value as $\hat{z}(k+1|k)$, i.e.,

$$\hat{z}(k+1|k) = \mathbf{h}'(k+1)\hat{\boldsymbol{\theta}}_{\text{WLS}}(k), \tag{4-18}$$

so that

$$\hat{\boldsymbol{\theta}}_{\text{WLS}}(k+1) = \hat{\boldsymbol{\theta}}_{\text{WLS}}(k) + \mathbf{K}_{\text{W}}(k+1)[z(k+1) - \hat{z}(k+1|k)]. \tag{4-19}$$

3. Two recursions are present in our recursive LSE. The first is the *vector recursion* for $\hat{\boldsymbol{\theta}}_{\text{WLS}}$ given by (4-8). Clearly $\hat{\boldsymbol{\theta}}_{\text{WLS}}(k+1)$ cannot be computed from this expression until measurement $z(k+1)$ is available. The second is the *matrix recursion* for $\mathbf{P}^{-1}$ given by (4-10). Observe that values for $\mathbf{P}^{-1}$ (and subsequently $\mathbf{K}_{\text{W}}$) can be precomputed before measurements are made.
4. A digital computer implementation of (4-8)–(4-10) proceeds as follows:

$$\mathbf{P}^{-1}(k+1) \rightarrow \mathbf{P}(k+1) \rightarrow \mathbf{K}_{\text{W}}(k+1) \rightarrow \hat{\boldsymbol{\theta}}_{\text{WLS}}(k+1).$$

5. Equations (4-8)–(4-10) can also be used for $k = 0, 1, \ldots, N-1$ using the following values for $\mathbf{P}^{-1}(0)$ and $\hat{\boldsymbol{\theta}}_{\text{WLS}}(0)$:

$$\mathbf{P}^{-1}(0) = \frac{1}{a^2}\mathbf{I}_n + \mathbf{h}(0)w(0)\mathbf{h}'(0) \tag{4-20}$$

and

$$\hat{\boldsymbol{\theta}}_{\text{WLS}}(0) = \mathbf{P}(0)\left[\frac{1}{a}\boldsymbol{\epsilon} + \mathbf{h}(0)w(0)z(0)\right] \tag{4-21}$$

In these equations (which are derived in Mendel, 1973, pp. 101–106; see, also, Problem 4-1) a is a very large number, ϵ is a very small number, $\boldsymbol{\epsilon}$ is $n \times 1$, and $\boldsymbol{\epsilon} = \text{col}(\epsilon, \epsilon, \ldots, \epsilon)$. When these initial values are used in (4-8)–(4-10) for $k = 0, 1, \ldots, n-1$, then the resulting values obtained for $\hat{\boldsymbol{\theta}}_{\text{WLS}}(n)$ and $\mathbf{P}^{-1}(n)$ are the very same ones that are obtained from the batch formulas for $\hat{\boldsymbol{\theta}}_{\text{WLS}}(n)$ and $\mathbf{P}^{-1}(n)$.

Often $z(0) = 0$, or there is no measurement made at $k = 0$, so that we can set $z(0) = 0$. In this case we can set $w(0) = 0$ so that $\mathbf{P}^{-1}(0) = \mathbf{I}_n/a^2$ and $\hat{\boldsymbol{\theta}}(0) = a\boldsymbol{\epsilon}$. By choosing $\boldsymbol{\epsilon}$ on the order of $1/a^2$, we see that (4-8)–(4-10) can be initialized by setting $\hat{\boldsymbol{\theta}}(0) = \mathbf{0}$ and $\mathbf{P}(0)$ equal to a diagonal matrix of very large numbers.
6. The reason why the results in Theorem 4-1 are referred to as the "information form" of the recursive LSE is deferred until Lesson 11 where connections are made between least-squares and maximum-likelihood estimators (see the section entitled The Linear Model ($\mathcal{H}(k)$ deterministic), Lesson 11).

MATRIX INVERSION LEMMA

Equations (4-10) and (4-9) require the inversion of $n \times n$ matrix $\mathbf{P}$. If n is large than this will be a costly computation. Fortunately, an alternative is available, one that is based on the following *matrix inversion lemma.*

Lemma 4-1. *If the matrices* $\mathbf{A}$, $\mathbf{B}$, $\mathbf{C}$, and $\mathbf{D}$ *satisfy the equation*

$$\mathbf{B}^{-1} = \mathbf{A}^{-1} + \mathbf{C}'\mathbf{D}^{-1}\mathbf{C} \tag{4-22}$$

where all matrix inverses are assumed to exist, then

$$\mathbf{B} = \mathbf{A} - \mathbf{A}\mathbf{C}'(\mathbf{C}\mathbf{A}\mathbf{C}' + \mathbf{D})^{-1}\mathbf{C}\mathbf{A}. \tag{4-23}$$

Proof. Multiply $\mathbf{B}$ by $\mathbf{B}^{-1}$ using (4-23) and (4-22) to show that $\mathbf{B}\mathbf{B}^{-1} = \mathbf{I}$. For a constructive proof of this lemma see Mendel (1973), pp. 96–97. □

Observe that if $\mathbf{A}$ and $\mathbf{B}$ are $n \times n$ matrices, $\mathbf{C}$ is $m \times n$, and $\mathbf{D}$ is $m \times m$, then to compute $\mathbf{B}$ from (4-23) requires the inversion of one $m \times m$ matrix. On the other hand, to compute $\mathbf{B}$ from (4-22) requires the inversion of one $m \times m$ matrix and two $n \times n$ matrices $[\mathbf{A}^{-1}$ and $(\mathbf{B}^{-1})^{-1}]$. When $m < n$ it is definitely advantageous to compute $\mathbf{B}$ using (4-23) instead of (4-22). Observe, also, that in the special case when $m = 1$, matrix inversion in (4-23) is replaced by division.

RECURSIVE LEAST-SQUARES: COVARIANCE FORM

Theorem 4-2. (Covariance Form of Recursive LSE). *Another recursive structure for* $\hat{\theta}_{WLS}(k)$ *is:*

$$\hat{\theta}_{WLS}(k+1) = \hat{\theta}_{WLS}(k) + \mathbf{K}_W(k+1)[z(k+1) - \mathbf{h}'(k+1)\hat{\theta}_{WLS}(k)] \tag{4-24}$$

where

$$\mathbf{K}_W(k+1) = \mathbf{P}(k)\mathbf{h}(k+1)\left[\mathbf{h}'(k+1)\mathbf{P}(k)\mathbf{h}(k+1) + \frac{1}{w(k+1)}\right]^{-1} \tag{4-25}$$

and

$$\mathbf{P}(k+1) = [\mathbf{I} - \mathbf{K}_W(k+1)\mathbf{h}'(k+1)]\mathbf{P}(k) \tag{4-26}$$

These equations are initialized by $\hat{\theta}_{WLS}(n)$ *and* $\mathbf{P}(n)$ *and are used for* $k = n, n+1, \ldots, N-1$.

Proof. We obtain the results in (4-25) and (4-26) by applying the matrix inversion lemma to (4-10), after which our new formula for $\mathbf{P}(k+1)$ is substituted into (4-9). In order to accomplish the first part of this, let $\mathbf{A} = \mathbf{P}(k)$,

$\mathbf{B} = \mathbf{P}(k+1)$, $\mathbf{C} = \mathbf{h}'(k+1)$ and $\mathbf{D} = 1/w(k+1)$. Then (4-10) looks like (4-22), so, using (4-23) we see that

$$\mathbf{P}(k+1) = \mathbf{P}(k) - \mathbf{P}(k)\mathbf{h}(k+1)[\mathbf{h}'(k+1)\mathbf{P}(k)\mathbf{h}(k+1) + w^{-1}(k+1)]^{-1}\mathbf{h}'(k+1)\mathbf{P}(k) \quad (4\text{-}27)$$

Consequently,

$$\begin{aligned}\mathbf{K}_\mathrm{W}(k+1) &= \mathbf{P}(k+1)\mathbf{h}(k+1)\,w(k+1)\\ &= [\mathbf{P} - \mathbf{Ph}(\mathbf{h}'\mathbf{Ph} + w^{-1})^{-1}\mathbf{h}'\mathbf{P}]\mathbf{h}w\\ &= \mathbf{Ph}[\mathbf{I} - (\mathbf{h}'\mathbf{Ph} + w^{-1})^{-1}\mathbf{h}'\mathbf{Ph}]w\\ &= \mathbf{Ph}(\mathbf{h}'\mathbf{Ph} + w^{-1})^{-1}(\mathbf{h}'\mathbf{Ph} + w^{-1} - \mathbf{h}'\mathbf{Ph})w\\ &= \mathbf{Ph}(\mathbf{h}'\mathbf{Ph} + w^{-1})^{-1}\end{aligned}$$

which is (4-25). In order to obtain (4-26), express (4-27) as

$$\begin{aligned}\mathbf{P}(k+1) &= \mathbf{P}(k) - \mathbf{K}_\mathrm{W}(k+1)\mathbf{h}'(k+1)\mathbf{P}(k)\\ &= [\mathbf{I} - \mathbf{K}_\mathrm{W}(k+1)\mathbf{h}'(k+1)]\mathbf{P}(k) \quad \square\end{aligned}$$

Comments

1. The recursive formula for $\hat{\boldsymbol{\theta}}_{\mathrm{WLS}}$, (4-24), is unchanged from (4-8). Only the matrix recursion for $\mathbf{P}$, leading to gain matrix $\mathbf{K}_\mathrm{W}$ has changed. A digital computer implementation of (4-24)–(4-26) proceeds as follows: $\mathbf{P}(k) \rightarrow \mathbf{K}_\mathrm{W}(k+1) \rightarrow \hat{\boldsymbol{\theta}}_{\mathrm{WLS}}(k+1) \rightarrow \mathbf{P}(k+1)$. This order of computations differs from the preceding one.
2. When $z(k)$ is a scalar then the covariance form of the recursive LSE requires no matrix inversions, and only one division.
3. Equations (4-24)–(4-26) can also be used for $k = 0, 1, \ldots, N-1$ using the values for $\mathbf{P}^{-1}(0)$ and $\hat{\boldsymbol{\theta}}_{\mathrm{WLS}}(0)$ given in (4-20) and (4-21).
4. The reason why the results in Theorem 4-2 are referred to as the "covariance form" of the recursive LSE is deferred to Lesson 9 where connections are made between least-squares and best linear unbiased minimum-variance estimators (see p. 79).

WHICH FORM TO USE

We have derived two formulations for a recursive least-squares estimator, the information and covariance forms. In on-line applications, where speed of computation is often the most important consideration, the covariance form is preferable to the information form. This is because a smaller matrix needs to be inverted in the covariance form, namely an $m \times m$ matrix rather than an $m \times n$ matrix (m is often much smaller than n).

The information form is often more useful than the covariance form in analytical studies. For example, it is used to derive the initial conditions for $\mathbf{P}^{-1}(0)$ and $\hat{\boldsymbol{\theta}}_{\text{WLS}}(0)$, which are given in (4-20) and (4-21) (see Mendel, 1973, pp. 101–106). The information form is also to be preferred over the covariance form during the startup of recursive least squares. We demonstrate why this is so next.

We consider the case when

$$\mathbf{P}(0) = a^2\mathbf{I}_n \tag{4-28}$$

where a^2 is a very, very large number. Using the information form, we find that, for $k = 0$, $\mathbf{P}^{-1}(1) = \mathbf{h}(1)w(1)\mathbf{h}'(1) + 1/a^2\mathbf{I}_n$, and, therefore, $\mathbf{K}_{\mathbf{W}}(1) = [\mathbf{h}(1)w(1)\mathbf{h}'(1) + 1/a^2\mathbf{I}_n]^{-1}\mathbf{h}(1)w(1)$. No difficulties are encountered when we compute $\mathbf{K}_{\mathbf{W}}(1)$ using the information form.

Using the covariance form we find, first, that $\mathbf{K}_{\mathbf{W}}(1) \simeq a^2\mathbf{h}(1)[\mathbf{h}'(1)a^2\mathbf{h}(1)]^{-1} = \mathbf{h}(1)[\mathbf{h}'(1)\mathbf{h}(1)]^{-1}$, and then that

$$\mathbf{P}(1) = \{\mathbf{I} - \mathbf{h}(1)[\mathbf{h}'(1)\mathbf{h}(1)]^{-1}\mathbf{h}'(1)\}a^2 \tag{4-29}$$

however, this matrix is singular. To see this, postmultiply both sides of (4-29) by $\mathbf{h}(1)$, to obtain

$$\mathbf{P}(1)\mathbf{h}(1) = \{\mathbf{h}(1) - \mathbf{h}(1)[\mathbf{h}'(1)\mathbf{h}(1)]^{-1}\mathbf{h}'(1)\mathbf{h}(1)\}a^2 = 0 \tag{4-30}$$

Neither $\mathbf{P}(1)$ nor $\mathbf{h}(1)$ equal zero; hence, $\mathbf{P}(1)$ must be a singular matrix for $\mathbf{P}(1)\mathbf{h}(1)$ to equal zero. In fact, once $\mathbf{P}(1)$ becomes singular, all other $\mathbf{P}(j)$, $j \geq 2$, will be singular.

In Lesson 9 we shall show that when $\mathbf{W}^{-1}(k) = \mathbf{E}\{\mathcal{V}(k)\mathcal{V}'(k)\} = \mathcal{R}(k)$, then $\mathbf{P}(k)$ is the covariance matrix of the estimation error, $\tilde{\boldsymbol{\theta}}(k)$. This matrix must be positive definite, and it will be quite difficult to maintain this property if $\mathbf{P}(k)$ is singular; hence, it is advisable to initialize the recursive least-squares estimator using the information form. However, it is also advisable to switch to the covariance formulation as soon after initialization as possible, in order to reduce computing time.

PROBLEMS

4-1. In order to derive the formulas for $\mathbf{P}^{-1}(0)$ and $\hat{\boldsymbol{\theta}}_{\text{WLS}}(0)$, given in (4-20) and (4-21), respectively, one proceeds as follows. Introduce n *artificial measurements* $z^a(-1), z^a(-2), \ldots, z^a(-n)$, where $z^a(-j) \triangleq \epsilon$ in which ϵ is a very small number. Then, assume that the model for $z^a(-j)$ is $z^a(-j) = \frac{1}{a}\theta_j (j = 1, 2, \ldots, n)$ where a is a very large number.

(a) Show that $\hat{\boldsymbol{\theta}}^a_{\text{LS}}(-1) = a\boldsymbol{\epsilon}$, where $n \times 1$ vector $\boldsymbol{\epsilon} = \text{col}(\epsilon, \epsilon, \ldots, \epsilon)$. Additionally, show that $\mathbf{P}^a(-1) = a^2\mathbf{I}_n$.

(b) Show that $\hat{\boldsymbol{\theta}}^a_{\text{WLS}}(0) = \mathbf{P}^a(0)[\boldsymbol{\epsilon}/a + \mathbf{h}(0)w(0)z(0)]$ and $\mathbf{P}^a(0) = [\mathbf{I}_n/a^2 + \mathbf{h}(0)w(0)\mathbf{h}'(0)]^{-1}$.

(c) Show that when the measurements $z^a(-n), \ldots, z^a(-1), z(0), z(1), \ldots, z(l+1)$ are used, then

$$\hat{\boldsymbol{\theta}}^a_{\text{WLS}}(l+1) = \mathbf{P}^a(l+1)\left[\boldsymbol{\epsilon}/a + \sum_{j=0}^{l+1} \mathbf{h}(j)w(j)z(j)\right]$$

and

$$\mathbf{P}^a(l+1) = \left[\mathbf{I}_n/a^2 + \sum_{j=0}^{l+1} \mathbf{h}(j)w(j)\mathbf{h}'(j)\right]^{-1}$$

(d) Show that when the measurements $z(0), z(1), \ldots, z(l+1)$ are used (i.e., the artificial measurements are not used), then

$$\hat{\boldsymbol{\theta}}_{\text{WLS}}(l+1) = \mathbf{P}(l+1)\left[\sum_{j=0}^{l+1} \mathbf{h}(j)w(j)z(j)\right]$$

and

$$\mathbf{P}(l+1) = \left[\sum_{j=0}^{l+1} \mathbf{h}(j)w(j)\mathbf{h}'(j)\right]^{-1}$$

(e) Finally, show that for $a >>>$ and $\boldsymbol{\epsilon} <<<$

$$\mathbf{P}^a(l+1) \rightarrow \mathbf{P}(l+1)$$

and

$$\hat{\boldsymbol{\theta}}^a_{\text{WLS}}(l+1) \rightarrow \hat{\boldsymbol{\theta}}_{\text{WLS}}(l+1)$$

4-2. Prove that once $\mathbf{P}(1)$ becomes singular, all other $\mathbf{P}(j)$, $j \geq 2$, will be singular.

4-3. (Mendel, 1973, Exercise 2–12, pg. 138). The following weighting matrix weights past measurements less heavily than the most recent measurements,

$$\mathbf{W}(k+1) = \text{diag}(w(k+1), w(k), \ldots, w(1)) \triangleq \left(\begin{array}{c|c} \overline{w}(k+1) & \mathbf{0}' \\ \hline 0 & \beta^{t_k+1-t_k}\mathbf{W}(k) \end{array}\right)$$

where

$$0 < \beta < 1$$

(a) Show that $w(j) = \overline{w}(j)\beta^{t_k+1-t_j}$ for $j = 1, \ldots, k+1$.

(b) How must the equations for the recursive weighted least-squares estimator be modified for this weighting matrix? The estimator thus obtained is known as a *fading memory estimator* (Morrison, 1969).

4-4. We showed, in Lesson 3, that it doesn't matter how one chooses the weights, $w(j)$, in the method of least squares, because the weights cancel out in the batch formula for $\hat{\boldsymbol{\theta}}_{\text{LS}}(k)$. On the other hand, $w(k+1)$ appears explicitly in the recursive WLSE, but only in the formula for $\mathbf{K}_{\mathbf{W}}(k+1)$, i.e.,

$$\mathbf{K}_{\mathbf{W}}(k+1) = \mathbf{P}(k)\mathbf{h}(k+1)[\mathbf{h}'(k+1)\mathbf{P}(k)\mathbf{h}(k+1) + w^{-1}(k+1)]^{-1}$$

It would appear that if we set $w(k+1) = w_1$, for all k, we would obtain a $\hat{\boldsymbol{\theta}}_{\text{LS}}(k+1)$ value that would be different from that obtained by setting $w(k+1) = w_2$, for all k. Of course, this cannot be true. Show, *using the formulas for the recursive* WLSE, how they can be made independent of $w(k+1)$, when $w(k+1) = w$, for all k.

4-5. For the data in the accompanying table, do the following:

(a) Obtain the least-squares line $\hat{y}(t) = a + bt$ by means of the batch processing least-squares algorithm;

(b) Obtain the least-squares line by means of the recursive least-squares algorithm, using the recursive startup technique (let $a = 10^8$ and $\epsilon = 10^{-16}$).

t	$y(t)$
0	1
1	5
2	9
3	11

Lesson 5

Least-Squares Estimation: Recursive Processing (continued)

Example 5-1

In order to illustrate some of the Lesson 4 results we shall obtain a recursive algorithm for the least-squares estimator of the scalar θ in the instrument calibration example (Lesson 3). Gain $\mathbf{K}_\mathbf{W}(k+1)$ is computed using (4-9) of Lesson 4, and $\mathbf{P}(k+1)$ is computed using (4-13) of Lesson 4. Generally, we do not compute $\mathbf{P}(k+1)$ using (4-13); but, the simplicity of our example allows us to use this formula to obtain a closed-form expression for $\mathbf{P}(k+1)$ in the most direct way. Recall that $\mathcal{H} = \text{col}\,(1, 1, \ldots, 1)$, which is a $k \times 1$ vector, and $h(k+1) = 1$; thus, setting $\mathbf{W}(k) = \mathbf{I}$ and $w(k+1) = 1$ (in order to obtain the recursive LSE of θ) in the preceding formulas, we find that

$$\mathbf{P}(k+1) = [\mathcal{H}'(k+1)\mathcal{H}(k+1)]^{-1} = \frac{1}{k+1} \tag{5-1}$$

and

$$\mathbf{K}_\mathbf{W}(k+1) = \mathbf{P}(k+1) = \frac{1}{k+1} \tag{5-2}$$

Substituting these results into (4-8) of Lesson 4, we then find

$$\hat{\theta}_{\text{LS}}(k+1) = \hat{\theta}_{\text{LS}}(k) + \frac{1}{k+1}[z(k+1) - \hat{\theta}_{\text{LS}}(k)]$$

or,

$$\hat{\theta}_{\text{LS}}(k+1) = \left(\frac{k}{k+1}\right)\hat{\theta}_{\text{LS}}(k) + \frac{1}{k+1}z(k+1) \tag{5-3}$$

Formula (5-3), which can be used for $k = 0, 1, \ldots, N-1$ by setting $\hat{\theta}_{LS}(0) = 0$, lets us reinterpret the well-known sample mean estimator as a time-varying digital filter [see, also, (1-2) of Lesson 1]. We leave it to the reader to study the stability properties of this first-order filter.

Usually it is in only the simplest of cases that we can obtain closed-form expressions for $\mathbf{K}_W$ and $\mathbf{P}$ and subsequently $\hat{\boldsymbol{\theta}}_{LS}$[or $\hat{\boldsymbol{\theta}}_{WLS}$]; however, we can always obtain values for $\mathbf{K}_W(k+1)$ and $\mathbf{P}(k+1)$ at successive time points using the results in Theorems 4-1 or 4-2. □

GENERALIZATION TO VECTOR MEASUREMENTS

A vector of measurements can occur in any application where it is possible to use more than one sensor; however, it is also possible to obtain a vector of measurements from certain types of individual sensors. In spacecraft applications, it is not unusual to be able to measure attitude, rate, and acceleration. In electrical systems applications, it is not uncommon to be able to measure voltages, currents, and power. Radar measurements often provide information about range, azimuth, and elevation. Radar is an example of a single sensor that provides a vector of measurements.

In the vector measurement case, (4-1) of Lesson 4 is changed from $z(k+1) = \mathbf{h}'(k+1)\boldsymbol{\theta} + v(k+1)$ to

$$\mathbf{z}(k+1) = \mathbf{H}(k+1)\boldsymbol{\theta} + \mathbf{v}(k+1) \tag{5-4}$$

where $\mathbf{z}$ is now an $m \times 1$ vector, $\mathbf{H}$ is $m \times n$ and $\mathbf{v}$ is $m \times 1$.

We leave it to the reader to show that all of the results in Lessons 3 and 4 are unchanged in the vector measurement case; but some notation must be altered (see Table 5-1).

TABLE 5-1 Transformations from Scalar to Vector Measurement Situations, and Vice-Versa

Scalar Measurement	Vector of Measurements
$z(k+1)$	$\mathbf{z}(k+1)$, an $m \times 1$ vector
$v(k+1)$	$\mathbf{v}(k+1)$, an $m \times 1$ vector
$w(k+1)$	$\mathbf{w}(k+1)$, an $m \times m$ matrix
$\mathbf{h}'(k+1)$, a $1 \times n$ matrix	$\mathbf{H}(k+1)$, an $m \times n$ matrix
$\mathfrak{Z}(k)$, an $N \times 1$ vector	$\mathfrak{Z}(k)$, an $Nm \times 1$ vector
$\mathcal{V}(k)$, an $N \times 1$ vector	$\mathcal{V}(k)$, an $Nm \times 1$ vector
$\mathbf{W}(k)$, an $N \times N$ matrix	$\mathbf{W}(k)$, an $Nm \times Nm$ matrix
$\mathcal{H}(k)$, an $N \times n$ matrix	$\mathcal{H}(k)$, an $Nm \times n$ matrix

Source: Reprinted from Mendel, 1973, p. 110. Courtesy of Marcel Dekker, Inc., NY.

CROSS-SECTIONAL PROCESSING

Suppose that at each sampling time t_{k+1} there are q sensors or groups of sensors that provide our vector measurement data. These sensors are corrupted by noise that is uncorrelated from one sensor group to another. The m-dimensional vector $\mathbf{z}(k+1)$ can be represented as

$$\mathbf{z}(k+1) = \operatorname{col}(\mathbf{z}_1(k+1), \mathbf{z}_2(k+1), \ldots, \mathbf{z}_q(k+1)) \tag{5-5}$$

where

$$\mathbf{z}_i(k+1) = \mathbf{H}_i(k+1)\boldsymbol{\theta} + \mathbf{v}_i(k+1) \tag{5-6}$$

$\dim \mathbf{z}_i(k+1) = m_i \times 1$,

$$\sum_{i=1}^{q} m_i = m \tag{5-7}$$

$\mathbf{E}\{\mathbf{v}_i(k+1)\} = \mathbf{0}$, and

$$\mathbf{E}\{\mathbf{v}_i(k+1)\mathbf{v}_j'(k+1)\} = \mathbf{R}_i(k+1)\delta_{ij} \tag{5-8}$$

An alternative to processing all m measurements in one batch (i.e., simultaneously) is available, and is one in which we freeze time at t_{k+1} and recursively process the q batches of measurements one batch at a time. Data $\mathbf{z}_1(k+1)$ are used to obtain an estimate (for notational simplicity, in this section we omit the subscript WLS or LS on $\hat{\boldsymbol{\theta}}$) $\hat{\boldsymbol{\theta}}_1(k+1)$ with $\hat{\boldsymbol{\theta}}_1(k) \triangleq \hat{\boldsymbol{\theta}}(k)$ and $\mathbf{z}(k+1) \triangleq \mathbf{z}_1(k+1)$. When these calculations are completed $\mathbf{z}_2(k+1)$ is processed to obtain the estimate $\hat{\boldsymbol{\theta}}_2(k+1)$. Estimate $\hat{\boldsymbol{\theta}}_1(k+1)$ is used to initialize $\hat{\boldsymbol{\theta}}_2(k+1)$. Each set of data is processed in this manner until the final set $\mathbf{z}_q(k+1)$ has been included. Then time is advanced to t_{k+2} and the cycle is repeated. This type of processing is known as cross-sectional or sequential processing. It is summarized in Figure 5-1 and is contrasted with more usual simultaneous recursive processing in Figure 5-2.

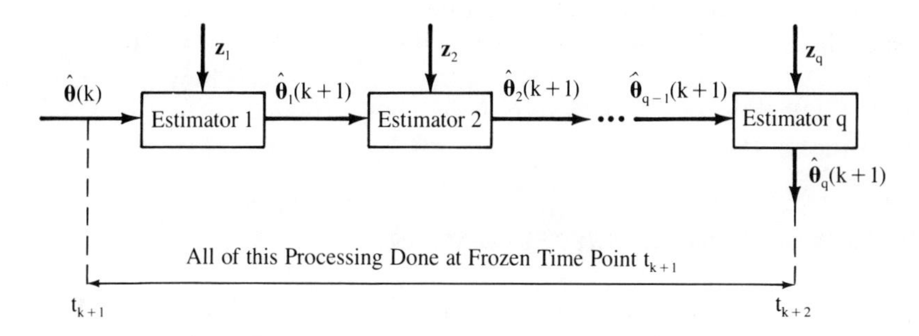

Figure 5-1 Cross-sectional processing of m measurements.

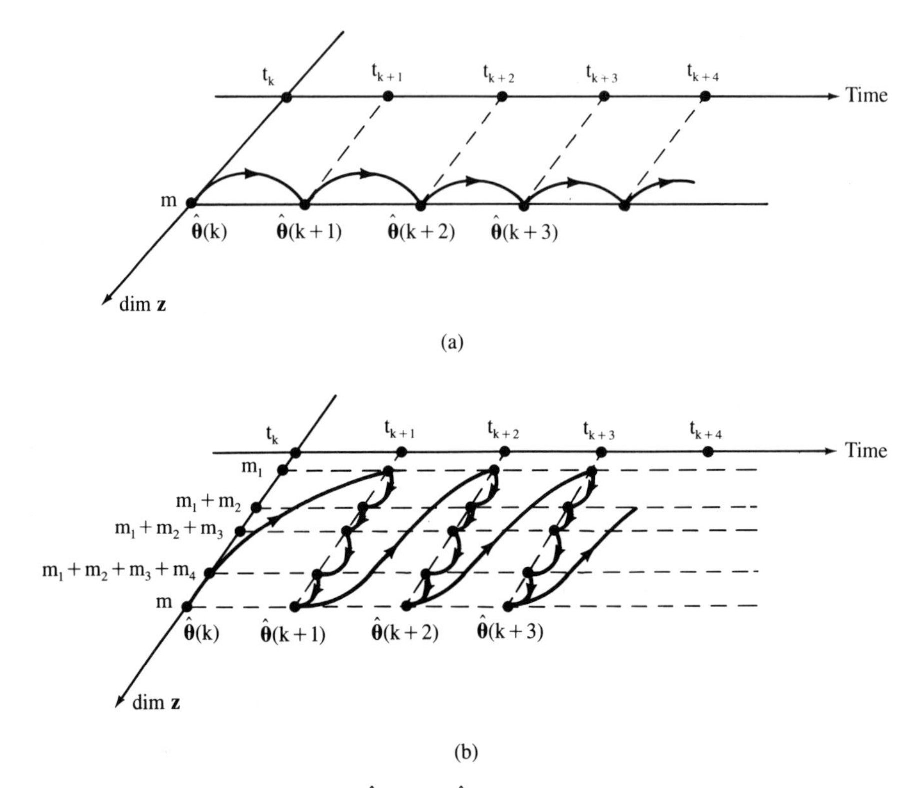

Figure 5-2 Two ways to reach $\hat{\boldsymbol{\theta}}(k+1), \hat{\boldsymbol{\theta}}(k+2), \ldots$. (a) Simultaneous recursive processing performed along the line dim $\mathbf{z} = m$, and (b) cross-sectional recursive processing where, for example, at t_{k+1} the processing is performed along the line TIME $= t_{k+1}$ and stops when that line intersects the line dim $\mathbf{z} = m$.

The remarkable property about cross-sectional processing is that

$$\hat{\boldsymbol{\theta}}_q(k+1) = \hat{\boldsymbol{\theta}}(k+1) \tag{5-9}$$

where $\hat{\boldsymbol{\theta}}(k+1)$ is obtained from simultaneous recursive processing.

A very large computational advantage exists for cross-sectional processing if $m_i = 1$. In this case the matrix inverse $[\mathbf{H}(k+1)\mathbf{P}(k)\mathbf{H}'(k+1) + \mathbf{w}^{-1}(k+1)]^{-1}$ needed in (4-25) of Lesson 4 is replaced by the division $[\mathbf{h}_i'(k+1)\mathbf{P}_i(k)\mathbf{h}_i(k+1) + 1/w_i(k+1)]^{-1}$. See Mendel, 1973, pp. 113–118 for a proof of (5-9)

MULTISTAGE LEAST-SQUARES ESTIMATORS

Suppose we are given a linear model $\mathcal{Z}(k) = \mathcal{H}_1(k)\boldsymbol{\theta}_1 + \mathcal{V}(k)$ with n unknown parameters $\boldsymbol{\theta}_1$, datum $\{\mathcal{Z}(k), \mathcal{H}_1(k)\}$, and the LSE of $\boldsymbol{\theta}_1$, $\hat{\boldsymbol{\theta}}^*_{1,\text{LS}}(k)$, where

$$\hat{\boldsymbol{\theta}}^*_{1,\text{LS}}(k) = [\mathcal{H}_1'(k)\mathcal{H}_1(k)]^{-1}\mathcal{H}_1(k)\mathcal{Z}(k) \tag{5-10}$$

We extend this model to include l additional parameters, $\boldsymbol{\theta}_2$, so that our model is given by

$$\mathcal{Z}(k) = \mathcal{H}_1(k)\boldsymbol{\theta}_1 + \mathcal{H}_2(k)\boldsymbol{\theta}_2 + \mathcal{V}(k) \tag{5-11}$$

For this model, datum $\{\mathcal{Z}(k), \mathcal{H}_1(k), \mathcal{H}_2(k)\}$ is available. We wish to compute the least-squares estimates of $\boldsymbol{\theta}_1$ and $\boldsymbol{\theta}_2$ for the $n + l$ parameter model using the previously computed $\hat{\boldsymbol{\theta}}^*_{1,\mathrm{LS}}(k)$.

Theorem 5-1. *Given the linear model in (5-11), where $\boldsymbol{\theta}_1$ is $n \times 1$ and $\boldsymbol{\theta}_2$ is $\ell \times 1$. The LSEs of $\boldsymbol{\theta}_1$ and $\boldsymbol{\theta}_2$ based on datum $\{\mathcal{Z}(\mathrm{k}), \mathcal{H}_1(\mathrm{k}), \mathcal{H}_2(\mathrm{k})\}$ are found from the following equations:*

$$\hat{\boldsymbol{\theta}}_{1,\mathrm{LS}}(k) = \hat{\boldsymbol{\theta}}^*_{1,\mathrm{LS}}(k) - \mathbf{G}(k)\mathbf{C}(k)\mathcal{H}_2'(k)[\mathcal{Z}(k) - \mathcal{H}_1(k)\hat{\boldsymbol{\theta}}^*_{1,\mathrm{LS}}(k)] \tag{5-12}$$

and

$$\hat{\boldsymbol{\theta}}_{2,\mathrm{LS}}(k) = \mathbf{C}(k)\mathcal{H}_2'(k)[\mathcal{Z}(k) - \mathcal{H}_1(k)\hat{\boldsymbol{\theta}}^*_{1,\mathrm{LS}}(k)] \tag{5-13}$$

where

$$\mathbf{G}(k) = [\mathcal{H}_1'(k)\mathcal{H}_1(k)]^{-1}\mathcal{H}_1'(k)\mathcal{H}_2(k) \tag{5-14}$$

and

$$\mathbf{C}(k) = [\mathcal{H}_2'(k)\mathcal{H}_2(k) - \mathcal{H}_2'(k)\mathcal{H}_1(k)\mathbf{G}(k)]^{-1} \tag{5-15}$$

The results in this theorem were worked out by Åström (1968) and emphasize operations which are performed on the vector of residuals for the n parameter model, namely $\mathcal{Z}(k) - \mathcal{H}_1(k)\hat{\boldsymbol{\theta}}^*_{1,\mathrm{LS}}(k)$. Other forms for $\hat{\boldsymbol{\theta}}_{1,\mathrm{LS}}(k)$ and $\hat{\boldsymbol{\theta}}_{2,\mathrm{LS}}(k)$ appear in Mendel (1975).

Proof. The derivation of (5-12) and (5-13) is based primarily on the block decompositon method for inverting $\mathcal{H}'\mathcal{H}$, where $\mathcal{H} = (\mathcal{H}_1|\mathcal{H}_2)$. See Mendel (1975) for the details. □

Similar results to those in Theorem 5-1 can be developed for the removal of l parameters from a model, for adding or removing parameters one at a time, and for recursive-in-time versions of all these results (see Mendel, 1975). All of these results are referred to as "multistage LSEs."

We conclude this section and lesson with some examples which illustrate problems for which multistage algorithms can be quite useful.

Example 5-2 Identification of a Sampled Impulse Response: Zoom-In Algorithm

A test signal, $u(t)$ is applied at $t = t_0$ to a linear, time-invariant, causal, but unknown system whose output, $y(t)$, and input are measured. The unknown impulse response, $w(t)$, is to be identified using sampled values of $u(t)$ and $y(t)$. For such a system

$$y(t_k) = \int_{t_0}^{t_W} w(\tau)u(t_k - \tau)d\tau \tag{5-16}$$

One approach to identifying $w(t)$ is to discretize (5-16) and to only identify $w(t)$ at discrete values of time. If we assume that (1) $w(t) \approx 0$ for all $t \geq t_w$, (2) $[t_0, t_w]$ is divided into n equal intervals, each of width T, so that $n = (t_w - t_0)/T$ and (3) for $\tau \in [t_{i-1}, t_i]$, $w(\tau) \cong w(t_{i-1})$ and $u(t - \tau) \cong u(t - t_{i-1})$, then

$$y(t_k) \cong \sum_{i=1}^{n} w_1(t_{i-1})u(t_k - t_{i-1}) \tag{5-17}$$

where

$$w_1(t_{i-1}) = Tw(t_{i-1}) \tag{5-18}$$

It is straightforward to identify the n unknown parameters $w_1(t_0)$, $w_1(t_1)$, ..., $w_1(t_{n-1})$ via least squares (see Example 2-1 of Lesson 2); however, for n to be known, t_w must be accurately known and T must be chosen most judiciously. In actual practice t_w is not known that accurately so that n may have to be varied. Multistage LSEs can be used to handle this situation. Sometimes T can be too coarse for certain regions of time, in which case significant features of $w(t)$ such as a ripple may be obscured. In this situation, we would like to "zoom-in" on those intervals of time and rediscretize $y(t)$ just over those intervals, thereby adding more terms to (5-17). Multistage LSEs can also be used to handle this situation, as we demonstrate next.

For illustrative purposes, we present this procedure for the case when the interval of interest equals T, i.e., when $t \in [t_x, t_{x+1}] = T$ which is further divided into q equal intervals, each of width ΔT_x, so that $q = (t_{x+1} - t_x)/\Delta T_x = T/\Delta T_x$. Observe that

$$\begin{aligned}\int_{t_x}^{t_x+1} w(\tau)u(t_k - \tau)d\tau &= \sum_{j=0}^{q-1} \int_{t_x + j\Delta T_x}^{t_x + (j+1)\Delta T_x} w(\tau)u(t_k - \tau)d\tau \\ &= \sum_{j=0}^{q-1} w(t_x + j\Delta T_x)u(t_k - t_x - j\Delta T_x)\Delta T_x \end{aligned} \tag{5-19}$$

hence,

$$\begin{aligned} y(t_k) = \sum_{\substack{i=1 \\ i \neq x}}^{n} w_1(t_{i-1})u(t_k - t_{i-1}) &+ \frac{\Delta T_x}{T} w_1(t_x)u(t_k - t_x) \\ &+ \sum_{j=1}^{q-1} w_2(t_x + j\Delta T_x)u(t_k - t_x - j\Delta T_x) \end{aligned} \tag{5-20}$$

where

$$w_2(t_p) = w(t_p)\Delta T_x \tag{5-21}$$

Equation (5-20) contains $n + q - 1$ parameters.

Let

$$\boldsymbol{\theta}_1 = \text{col}\,(w_1(t_0), w_1(t_1), \ldots, w_1(t_x), \ldots, w_1(t_{n-1})) \tag{5-22}$$

and assume that a least-squares estimate of $\boldsymbol{\theta}_1$, $\hat{\boldsymbol{\theta}}^*_{1,\text{LS}}$, which is based on (5-17), is available. Let

$$\boldsymbol{\theta}_2 = \text{col}\,[w_2(t_x + \Delta T_x), w_2(t_x + 2\Delta T_x), \ldots, w_2(t_x + (q-1)\Delta T_x)] \tag{5-23}$$

To obtain $\hat{\boldsymbol{\theta}}_{\text{LS}} = \text{col}\,(\hat{\boldsymbol{\theta}}_{1,\text{LS}}, \hat{\boldsymbol{\theta}}_{2,\text{LS}})$ proceed as follows: 1) modify $\hat{\boldsymbol{\theta}}^*_{1,\text{LS}}$ by scaling $\hat{w}_{1,\text{LS}}(t_x)$ to $\hat{w}_{1,\text{LS}}(t_x)\Delta T_x/T$ and call the modified result $\hat{\boldsymbol{\theta}}^*_{1,\text{LS}}$; and 2) apply Theorem 5-1 to obtain

TABLE 5-2 Impulse Response Estimates

Number of Parameters in the Model	Average Estimates of Parameters[a,b]										
n	θ_1 (0.39)	θ_2 (0.76)	θ_3 (0.38)	θ_4 (−0.23)	θ_5 (−0.48)	θ_6 (−0.25)	θ_7 (0.14)	θ_8 (0.30)	θ_9 (0.17)	θ_{10} (−0.08)	Average Performance
1	0.34										0.96
2	0.41	0.75									0.75
3	0.39	0.78	0.40								0.63
4	0.39	0.79	0.40	−0.12							0.62
5	0.40	0.79	0.43	−0.17	−0.41						0.44
6	0.39	0.79	0.43	−0.15	−0.44	−0.29					0.40
7	0.39	0.80	0.44	−0.16	−0.44	−0.27	0.06				0.40
8	0.41	0.81	0.43	−0.15	−0.47	−0.30	0.12	0.30			0.29
9	0.43	0.83	0.43	−0.14	−0.47	−0.33	0.10	0.35	0.23		0.18
10	0.43	0.81	0.42	−0.15	−0.46	−0.32	0.10	0.35	0.22	−0.07	0.13

Source: Reprinted from Mendel, 1975, pg. 782, © 1975 IEEE.

[a] Values for parameters shown in parentheses are true values.

[b] Blank spaces in the table denote "no value" for parameters.

$\hat{\theta}_{LS}$. Note that the scaling of $\hat{w}_{1,LS}(t_x)$ in Step 1 is due to the appearance of $w_1(t_x)\Delta T_x / T$ instead of $w_1(t_x)$ in (5-20).

It is straightforward to extend the approach of this example to regions that include more than one sampling interval, T. □

Example 5-3 Impulse Response Identification

An n-stage batch LSE was used to estimate impulse response models having from one to ten parameters [i.e., n in (5-17) ranged from 1–10] for a second-order system (natural frequency of 10.47 rad/sec, damping ratio of 0.12, and, unity gain). The system was forced with an input of randomly chosen ± 1's each of which was equally likely to occur. The system's output was corrupted by zero-mean pseudo-random Gaussian white noise of unity variance. Fifty consecutive samples of the noisy output and noise-free input were then processed by the n-stage batch LSE, and this procedure was repeated ten times from which the average values for the parameter estimates given in Table 5-2 (page 41) were obtained.

The ten models were tested using ten input sequences, each containing 50 samples of randomly chosen ± 1's. The last column in Table 5-2 gives values for the normalized average performance

$$\left[\sum_{i=1}^{50} [\bar{y}_{\text{model}}(t_i) - \bar{y}_{\text{actual}}(t_i)]^2 \Big/ \sum_{i=1}^{50} \bar{y}^2_{\text{actual}}(t_i) \right]^{1/2}$$

in which $\bar{y}(t_i)$ denotes an average over the ten runs. Not too surprisingly, we see that average predicted performance improves as n increases.

All the θ_i results were obtained in one pass through the n-stage LSE using approximately 3150 flops. The same results could have been obtained using 10 LSEs (Lesson 3); but, that would have required approximately 5220 flops. □

PROBLEMS

5-1. Show that, in the vector measurement case, the results given in Lessons 3 and 4 only need to be modified using the transformations listed in Table 5-1.

5-2. Prove that, using cross-sectional processing, $\hat{\theta}_q(k+1) = \hat{\theta}(k+1)$.

5-3. Prove the multistage least-squares estimation Theorem 5-1.

5-4. Extend Example 5-2 to regions of interest equal to mT, where m is a positive integer and T is the original data sampling time.

Lesson 6

Small Sample Properties of Estimators

INTRODUCTION

How do we know whether or not the results obtained from the LSE, or for that matter any estimator, are good? To answer this question, we make use of the fact that all estimators represent transformations of random data. For example, our LSE, $[\mathcal{H}'(k)\mathbf{W}(k)\mathcal{H}(k)]^{-1}\mathcal{H}'(k)\mathbf{W}(k)\mathcal{Z}(k)$, represents a linear transformation on $\mathcal{Z}(k)$. Other estimators may represent nonlinear transformations of $\mathcal{Z}(k)$. *The consequence of this is that $\hat{\boldsymbol{\theta}}(k)$ is itself random.* Its properties must therefore be studied from a statistical viewpoint.

In the estimation literature, it is common to distinguish between small-sample and large-sample properties of estimators. The term "sample" refers to the number of measurements used to obtain $\hat{\boldsymbol{\theta}}$, i.e., the dimension of $\mathcal{Z}$. The phrase "small-sample" means *any number of measurements* (e.g., 1, 2, 100, 10^4, or even an infinite number), whereas the phrase "large-sample" means *an infinite number of measurements*. Large-sample properties are also referred to as asymptotic properties. It should be obvious that *if an estimator possesses a small-sample property it also possesses the associated large-sample property; but, the converse is not always true.*

Why bother studying large-sample properties of estimators if these properties are included in their small-sample properties? Put another way, why not just study small-sample properties of estimators? For many estimators it is relatively easy to study their large-sample properties and virtually impossible to learn about their small-sample properties. An analogous situation occurs in stability theory, where most effort is directed at infinite-time stability behavior rather than at finite-time behavior.

Although "large-sample" means an infinite number of measurements, estimators begin to enjoy their large-sample properties for much fewer than an infinite number of measurements. How few, usually depends on the dimension of $\boldsymbol{\theta}$, n.

A thorough study into $\hat{\boldsymbol{\theta}}$ would mean determining its probability density function $p(\hat{\boldsymbol{\theta}})$. Usually, it is too difficult to obtain $p(\hat{\boldsymbol{\theta}})$ for most estimators (unless $\hat{\boldsymbol{\theta}}$ is multivariate Gaussian); thus, it is customary to emphasize the first- and second-order statistics of $\hat{\boldsymbol{\theta}}$ (or, its associated error $\tilde{\boldsymbol{\theta}} = \boldsymbol{\theta} - \hat{\boldsymbol{\theta}}$), namely the mean and covariance.

We shall examine the following small- and large-sample properties of estimators: unbiasedness and efficiency (small-sample), and asymptotic unbiasedness, consistency, and asymptotic efficiency (large-sample). Small-sample properties are the subject of this lesson, whereas large-sample properties are studied in Lesson 7.

UNBIASEDNESS

Definition 6-1. *Estimator $\hat{\boldsymbol{\theta}}(\mathrm{k})$ is an unbiased estimator of deterministic $\boldsymbol{\theta}$, if*

$$\mathbf{E}\{\hat{\boldsymbol{\theta}}(k)\} = \boldsymbol{\theta} \qquad \text{for } all\ k \tag{6-1}$$

or of random $\boldsymbol{\theta}$, if

$$\mathbf{E}\{\hat{\boldsymbol{\theta}}(k)\} = \mathbf{E}\{\boldsymbol{\theta}\} \qquad \text{for } all\ k \quad \square \tag{6-2}$$

In terms of estimation error, $\tilde{\boldsymbol{\theta}}(k)$, unbiasedness means, that

$$\mathbf{E}\{\tilde{\boldsymbol{\theta}}(k)\} = \mathbf{0} \qquad \text{for } all\ k \tag{6-3}$$

Example 6-1

In the instrument calibration example of Lesson 3, we determined the following LSE of θ:

$$\hat{\theta}_{\mathrm{LS}}(N) = \frac{1}{N}\sum_{i=1}^{N} z(i) \tag{6-4}$$

where

$$z(i) = \theta + v(i) \tag{6-5}$$

Suppose $\mathbf{E}\{v(i)\} = 0$ for $i = 1, 2, \ldots, N$; then,

$$\mathbf{E}\{\hat{\theta}_{\mathrm{LS}}(N)\} = \frac{1}{N}\sum_{i=1}^{N} \mathbf{E}\{z(i)\} = \frac{1}{N}\sum_{i=1}^{N} \theta = \theta$$

which means that $\hat{\theta}_{\mathrm{LS}}(N)$ is an unbiased estimator of θ. $\square$

Many estimators are linear transformations of the measurements, i.e.,

$$\hat{\boldsymbol{\theta}}(k) = \mathbf{F}(k)\mathcal{Z}(k) \tag{6-6}$$

In least-squares, we obtained this linear structure for $\hat{\boldsymbol{\theta}}(k)$ by solving an optimization problem. Sometimes, we begin by assuming that (6-6) is the desired structure for $\hat{\boldsymbol{\theta}}(k)$. We now address the question "when is $\mathbf{F}(k)\mathcal{Z}(k)$ an unbiased estimator of deterministic $\boldsymbol{\theta}$?"

Theorem 6-1. *When* $\mathcal{Z}(k) = \mathcal{H}(k)\boldsymbol{\theta} + \mathcal{V}(k)$, $\mathbf{E}\{\mathcal{V}(k)\} = \mathbf{0}$, *and* $\mathcal{H}(k)$ *is deterministic, then* $\hat{\boldsymbol{\theta}}(k) = \mathbf{F}(k)\mathcal{Z}(k)$ [*where* $\mathbf{F}(k)$ *is deterministic*] *is an unbiased estimator of* $\boldsymbol{\theta}$ *if and only if*

$$\mathbf{F}(k)\mathcal{H}(k) = \mathbf{I} \qquad \textit{for all } k \tag{6-7}$$

Note that this is the first place where we have had to assume any a priori knowledge about noise $\mathcal{V}(k)$.

Proof

a. (*Necessity*). From the model for $\mathcal{Z}(k)$ and the assumed structure for $\hat{\boldsymbol{\theta}}(k)$, we see that

$$\hat{\boldsymbol{\theta}}(k) = \mathbf{F}(k)\mathcal{H}(k)\boldsymbol{\theta} + \mathbf{F}(k)\mathcal{V}(k) \tag{6-8}$$

If $\hat{\boldsymbol{\theta}}(k)$ is an unbiased estimator of $\boldsymbol{\theta}$, $\mathbf{F}(k)$ and $\mathcal{H}(k)$ are deterministic, and $\mathbf{E}\{\mathcal{V}(k)\} = \mathbf{0}$, then

$$\mathbf{E}\{\hat{\boldsymbol{\theta}}(k)\} = \boldsymbol{\theta} = \mathbf{F}(k)\mathcal{H}(k)\boldsymbol{\theta}$$

or

$$[\mathbf{I} - \mathbf{F}(k)\mathcal{H}(k)]\boldsymbol{\theta} = \mathbf{0} \tag{6-9}$$

Obviously, for $\boldsymbol{\theta} \neq \mathbf{0}$, (6-7) is the solution to this equation.

b. (*Sufficiency*). From (6-8) and the nonrandomness of $\mathbf{F}(k)$ and $\mathcal{H}(k)$, we have

$$\mathbf{E}\{\hat{\boldsymbol{\theta}}(k)\} = \mathbf{F}(k)\mathcal{H}(k)\boldsymbol{\theta} \tag{6-10}$$

Assuming the truth of (6-7), it must be that

$$\mathbf{E}\{\hat{\boldsymbol{\theta}}(k)\} = \boldsymbol{\theta}$$

which, of course, means that $\hat{\boldsymbol{\theta}}(k)$ is an unbiased estimator of $\boldsymbol{\theta}$. □

Example 6-2

Matrix $\mathbf{F}(k)$ for the WLSE of $\boldsymbol{\theta}$ is $[\mathcal{H}'(k)\mathbf{W}(k)\mathcal{H}(k)]^{-1}\mathcal{H}'(k)\mathbf{W}(k)$. Observe that this $\mathbf{F}(k)$ matrix satisfies (6-7); thus, when $\mathcal{H}(k)$ is *deterministic the WLSE of* $\boldsymbol{\theta}$ *is unbiased.* Unfortunately, in many interesting applications $\mathcal{H}(k)$ is random, and we cannot apply Theorem 6-1 to study the unbiasedness of the WLSE. We return to this issue in Lesson 8. □

Suppose that we begin by assuming a linear recursive structure for $\hat{\boldsymbol{\theta}}$, namely

$$\hat{\boldsymbol{\theta}}(k+1) = \mathbf{A}(k+1)\hat{\boldsymbol{\theta}}(k) + \mathbf{b}(k+1)z(k+1) \tag{6-11}$$

We then have the following counterpart to Theorem 6-1.

Theorem 6-2. *When* $z(k+1) = \mathbf{h}'(k+1)\boldsymbol{\theta} + v(k+1)$, $E\{v(k+1)\} = 0$, *and* $\mathbf{h}(k+1)$ *is deterministic, then* $\hat{\boldsymbol{\theta}}(k+1)$ *given by (6-11) is an unbiased estimator of* $\boldsymbol{\theta}$ *if*

$$\mathbf{A}(k+1) = \mathbf{I} - \mathbf{b}(k+1)\mathbf{h}'(k+1) \tag{6-12}$$

where $\mathbf{A}(k+1)$ *and* $\mathbf{b}(k+1)$ *are deterministic.* □

We leave the proof of this result to the reader. Unbiasedness means that our recursive estimator does not have two independent design matrices (degrees of freedom), $\mathbf{A}(k+1)$ and $\mathbf{b}(k+1)$. Unbiasedness constraints $\mathbf{A}(k+1)$ to be a function of $\mathbf{b}(k+1)$. When (6-12) is substituted into (6-11), we obtain the following important structure for an unbiased linear recursive estimator of $\boldsymbol{\theta}$,

$$\hat{\boldsymbol{\theta}}(k+1) = \hat{\boldsymbol{\theta}}(k) + \mathbf{b}(k+1)[z(k+1) - \mathbf{h}'(k+1)\hat{\boldsymbol{\theta}}(k)] \tag{6-13}$$

Our recursive WLSE of $\boldsymbol{\theta}$ has this structure; thus, as long as $\mathbf{h}(k+1)$ is deterministic, it produces unbiased estimates of $\boldsymbol{\theta}$. Many other estimators that we shall study will also have this structure.

EFFICIENCY

Did you hear the story about the conventioning statisticians who all drowned in a lake that was on the average 6 in. deep? The point of this rhetorical question is that unbiasedness by itself is not terribly meaningful. We must also study the dispersion about the mean, namely the variance. If the statisticians had known that the variance about the 6 in. average depth was 120 ft, they might not have drowned!

Ideally, we would like our estimator to be unbiased and to have the smallest possible error variance. We consider the case of a scalar parameter first.

Definition 6-2. *An unbiased estimator,* $\hat{\theta}(k)$ *of* θ *is said to be more efficient than any other unbiased estimator,* $\hat{\hat{\theta}}(k)$, *of* θ, *if*

$$\operatorname{Var}(\hat{\theta}(k)) \le \operatorname{Var}(\hat{\hat{\theta}}(k)) \qquad \textit{for all } k \quad \square \tag{6-14}$$

Very often, it is of interest to know if $\hat{\theta}(k)$ satisfies (6-14) for *all* other unbiased estimators, $\hat{\hat{\theta}}(k)$. This can be verified by comparing the variance of

$\theta(k)$ with the smallest error variance that can ever be attained by any unbiased estimator. The following theorem provides a lower bound for $\mathbf{E}\{\tilde{\theta}^2(k)\}$ when θ is a scalar deterministic parameter. Theorem 6-4 generalizes these results to the case of a vector of deterministic parameters.

Theorem 6-3 (Cramer-Rao Inequality). *Let* $\mathbf{z}$ *denote a set of data* [i.e., $\mathbf{z} = \text{col}(z_1, z_2, \ldots, z_k)$; $\mathbf{z}$ *is also short for* $\mathfrak{Z}(k)$]. *If* $\hat{\theta}(k)$ *is an unbiased estimator of deterministic* θ, *then*

$$\mathbf{E}\{\tilde{\theta}^2(k)\} \geq \frac{1}{\mathbf{E}\left\{\left[\frac{\partial}{\partial\theta}\ln p(\mathbf{z})\right]^2\right\}} \quad \textit{for all } k \tag{6-15}$$

Two other ways for expressing (6-15) are

$$\mathbf{E}\{\tilde{\theta}^2(k)\} \geq \frac{1}{\int_{-\infty}^{\infty}\left[\frac{\partial p(z)}{\partial\theta}\right]^2 \frac{1}{p(\mathbf{z})}\,d\mathbf{z}} \quad \textit{for all } k \tag{6-16}$$

and

$$\mathbf{E}\{\tilde{\theta}^2(k)\} \geq \frac{1}{-\mathbf{E}\left\{\frac{\partial^2 \ln p(\mathbf{z})}{\partial\theta^2}\right\}} \quad \textit{for all } k \tag{6-17}$$

where $d\mathbf{z}$ *is short for* $dz_1, dz_2, \ldots, dz_k$. □

Inequalities (6-15), (6-16) and (6-17) are named after Cramer and Rao, who discovered them. They are functions of k because $\mathbf{z}$ is.

Before proving this theorem it is instructive to illustrate its use by means of an example.

Example 6-3

We are given M statistically independent observations of a random variable z that is known to have a Cauchy distribution, i.e.,

$$p(z_i) = \frac{1}{\pi[1 + (z_i - \theta)^2]} \tag{6-18}$$

Parameter θ is unknown and will be estimated using $z_1, z_2, \ldots, z_M$. We shall determine the lower bound for the error-variance of *any* unbiased estimator of θ using (6-15). Observe that we are able to do this without having to specify an estimator structure for $\hat{\theta}$. Without further explanation, we calculate:

$$p(\mathbf{z}) = \prod_{i=1}^{M} p(z_i) = 1\bigg/\left(\pi^M \prod_{i=1}^{M}[1 + (z_i - \theta)^2]\right) \tag{6-19}$$

$$\ln p(\mathbf{z}) = -M \ln \pi - \sum_{i=1}^{M} \ln[1 + (z_i - \theta)^2] \tag{6-20}$$

and

$$\frac{\partial \ln p(\mathbf{z})}{\partial \theta} = \sum_{i=1}^{M} 2(z_i - \theta)/[1 + (z_i - \theta)^2] \tag{6-21}$$

so that

$$\mathbf{E}\left\{\left[\frac{\partial}{\partial \theta} \ln p(\mathbf{z})\right]^2\right\} = \mathbf{E}\left\{\left[\sum_{i=1}^{M} \frac{2(z_i - \theta)}{1 + (z_i - \theta)^2}\right]\left[\sum_{j=1}^{M} \frac{2(z_j - \theta)}{1 + (z_j - \theta)^2}\right]\right\} \tag{6-22}$$

Next, we must evaluate the right-hand side of (6-22). This is tedious to do, but can be accomplished, as follows. Note that

$$\sum_i \sum_j = \sum_{\substack{i \\ i \neq j}} \sum_j + \sum_{\substack{i \\ i = j}} \sum_j = \mathbf{TA} + \mathbf{TB} \tag{6-23}$$

Consider **TA** first, i.e.,

$$\mathbf{TA} = \mathbf{E}\left\{\sum_{i=1}^{M} 2(z_i - \theta)/[1 + (z_i - \theta)^2]\right\} \mathbf{E}\left\{\sum_{j=1}^{M} 2(z_j - \theta)/[1 + (z_j - \theta)^2]\right\} \tag{6-24}$$

where we have made use of statistical independence of the measurements. Observe that

$$\mathbf{E}\left\{\frac{z_i - \theta}{1 + (z_i - \theta)^2}\right\} = \int_{-\infty}^{\infty} \frac{y}{1 + y^2} \cdot \frac{1}{\pi} \frac{1}{1 + y^2}\, dy = 0 \tag{6-25}$$

where $y = z_i - \theta$. The integral is zero because the integrand is an odd function of y. Consequently,

$$\mathbf{TA} = 0 \tag{6-26}$$

Next, consider **TB**, i.e.,

$$\mathbf{TB} = \mathbf{E}\left\{\sum_{i=1}^{M} 4(z_i - \theta)^2/[1 + (z_i - \theta)^2]^2\right\} \tag{6-27}$$

which can also be written as

$$\mathbf{TB} = 4 \sum_{i=1}^{M} \mathbf{TC} \tag{6-28}$$

where

$$\mathbf{TC} = \mathbf{E}\{(z_i - \theta)^2/[1 + (z_i - \theta)^2]^2\} \tag{6-29}$$

or

$$\mathbf{TC} = \int_{-\infty}^{\infty} \frac{y^2}{(1 + y^2)^2} \frac{1}{\pi} \frac{1}{(1 + y^2)}\, dy \tag{6-30}$$

Integrating (6-30) by parts twice, we find that

$$\mathbf{TC} = \frac{1}{8\pi} \int_{-\infty}^{\infty} \frac{dy}{1 + y^2} = \frac{1}{8} \tag{6-31}$$

because the integral in (6-31) is the area under the Cauchy probability density function, which equals π. Substituting (6-31) into (6-28), we determine that

$$\mathbf{TB} = M/2 \tag{6-32}$$

thus, when (6-23) is substituted into (6-22), and that result is substituted into (6-15), we find that

$$\mathbf{E}\{\tilde{\theta}^2(M)\} \geq \frac{2}{M} \qquad \text{for all } M \tag{6-33}$$

Observe that the Cramer-Rao bound depends on the number of measurements used to estimate θ. For large numbers of measurements, this bound equals zero. □

Proof of Theorem 6-3. Because $\hat{\theta}(k)$ is an unbiased estimator of $\hat{\theta}$,

$$\mathbf{E}\{\tilde{\theta}(k)\} = \int_{-\infty}^{\infty} [\hat{\theta}(k) - \theta] p(\mathbf{z}) d\mathbf{z} = 0 \tag{6-34}$$

Differentiating (6-34) with respect to θ, we find that

$$\int_{-\infty}^{\infty} [\hat{\theta}(k) - \theta] \frac{\partial p(\mathbf{z})}{\partial \theta} d\mathbf{z} - \int_{-\infty}^{\infty} p(\mathbf{z})\, d\mathbf{z} = 0$$

which can be rewritten as

$$1 = \int_{-\infty}^{\infty} [\hat{\theta}(k) - \theta] \frac{\partial p(\mathbf{z})}{\partial \theta} d\mathbf{z} \tag{6-35}$$

As an aside, we note that

$$\frac{\partial}{\partial \theta} \ln p(\mathbf{z}) = \frac{\partial p(\mathbf{z})}{\partial \theta} \frac{1}{p(\mathbf{z})} \tag{6-36}$$

so that

$$\frac{\partial p(\mathbf{z})}{\partial \theta} = p(\mathbf{z}) \frac{\partial}{\partial \theta} \ln p(\mathbf{z}) \tag{6-37}$$

Substitute (6-37) into (6-35) to obtain

$$1 = \int_{-\infty}^{\infty} \left[(\hat{\theta}(k) - \theta) \sqrt{p(\mathbf{z})}\right]\left[\sqrt{p(\mathbf{z})} \frac{\partial}{\partial \theta} \ln p(\mathbf{z})\right] d\mathbf{z} \tag{6-38}$$

Recall the Schwarz inequality

$$\left[\int_{-\infty}^{\infty} a(\mathbf{z}) b(\mathbf{z}) d\mathbf{z}\right]^2 \leq \left[\int_{-\infty}^{\infty} a^2(\mathbf{z}) d\mathbf{z}\right]\left[\int_{-\infty}^{\infty} b^2(\mathbf{z}) d\mathbf{z}\right] \tag{6-39}$$

where equality is achieved when $b(\mathbf{z}) = ca(\mathbf{z})$ in which c is an arbitrary constant. Next, square both sides of (6-38) and apply (6-39) to the new right-hand side, to see that

$$1 \leq \left[\int_{-\infty}^{\infty} [\hat{\theta}(k) - \theta]^2 p(\mathbf{z}) d\mathbf{z}\right]\left[\int_{-\infty}^{\infty} \left[\frac{\partial}{\partial \theta} \ln p(\mathbf{z})\right]^2 p(\mathbf{z}) d\mathbf{z}\right]$$

or

$$1 \leq \mathbf{E}\{\tilde{\theta}^2(k)\} \mathbf{E}\left\{\left[\frac{\partial}{\partial \theta} \ln p(\mathbf{z})\right]^2\right\} \tag{6-40}$$

Finally, to obtain (6-15), solve (6-40) for $\mathbf{E}\{\tilde{\theta}^2(k)\}$.

In order to obtain (6-16) from (6-15), observe that

$$\mathbf{E}\left\{\left[\frac{\partial}{\partial\theta}\ln p(\mathbf{z})\right]^2\right\} = \int_{-\infty}^{\infty}\left[\frac{\partial p(\mathbf{z})}{\partial\theta}\frac{1}{p(\mathbf{z})}\right]^2 p(\mathbf{z})\,d\mathbf{z}$$
$$= \int_{-\infty}^{\infty}\left[\frac{\partial p(\mathbf{z})}{\partial\theta}\right]^2 \frac{1}{p(\mathbf{z})}\,d\mathbf{z} \qquad (6\text{-}41)$$

To obtain (6-41) we have also used (6-36).

In order to obtain (6-17), we begin with the identity

$$\int_{-\infty}^{\infty} p(\mathbf{z})\,d\mathbf{z} = 1$$

and differentiate it twice with respect to θ, using (6-37) after each differentiation, to show that

$$\int_{-\infty}^{\infty} p(\mathbf{z})\frac{\partial^2 \ln p(\mathbf{z})}{\partial\theta^2}\,d\mathbf{z} = -\int_{-\infty}^{\infty}\left[\frac{\partial \ln p(\mathbf{z})}{\partial\theta}\right]^2 p(\mathbf{z})\,d\mathbf{z}$$

which can also be expressed as

$$\mathbf{E}\left\{\left[\frac{\partial}{\partial\theta}\ln p(\mathbf{z})\right]^2\right\} = -\mathbf{E}\left\{\frac{\partial^2 \ln p(\mathbf{z})}{\partial\theta^2}\right\} \qquad (6\text{-}42)$$

Substitute (6-42) into (6-15) to obtain (6-17). □

It is sometimes easier to compute the Cramer-Rao bound using one form [i.e., (6-15) or (6-16) or (6-17)] than another. The logarithmic forms are usually used when $p(\mathbf{z})$ is exponential (e.g., Gaussian).

Corollary 6-1. *If the lower bound is achieved in Theorem 6-3, then*

$$\tilde{\theta}(k) = \frac{1}{c}\frac{\partial \ln p(\mathbf{z})}{\partial\theta} \qquad (6\text{-}43)$$

where c *is an arbitrary constant.*

Proof. In deriving the Cramer-Rao bound we used the Schwarz inequality (6-39) for which equality is achieved when $b(\mathbf{z}) = c\,a(\mathbf{z})$. In our case $a(\mathbf{z}) = [\hat{\theta}(k) - \theta]\sqrt{p(\mathbf{z})}$ and $b(\mathbf{z}) = \sqrt{p(\mathbf{z})}\,(\partial/\partial\theta)\ln p(\mathbf{z})$. Setting $b(\mathbf{z}) = c\,a(\mathbf{z})$, we obtain (6-43). □

Equation (6-43) links the structure of an estimator to the property of efficiency, because the left-hand side of (6-43) depends explicitly on $\hat{\theta}(k)$.

We turn next to the general case of a vector of parameters.

Definition 6-3. *An unbiased estimator,* $\hat{\boldsymbol{\theta}}(k)$, *of vector* $\boldsymbol{\theta}$ *is said to be more efficient than any other unbiased estimator,* $\hat{\hat{\boldsymbol{\theta}}}(k)$, *of* $\boldsymbol{\theta}$, *if*

$$\mathbf{E}\{[\boldsymbol{\theta} - \hat{\boldsymbol{\theta}}(k)][\boldsymbol{\theta} - \hat{\boldsymbol{\theta}}(k)]'\} \le \mathbf{E}\{[\boldsymbol{\theta} - \hat{\hat{\boldsymbol{\theta}}}(k)][\boldsymbol{\theta} - \hat{\hat{\boldsymbol{\theta}}}(k)]'\} \quad \square \qquad (6\text{-}44)$$

For a vector of parameters, we see that a more efficient estimator has the smallest error covariance among all unbiased estimators of $\boldsymbol{\theta}$, "smallest" in the sense that $\mathbf{E}\{[\boldsymbol{\theta} - \hat{\boldsymbol{\theta}}(k)][\boldsymbol{\theta} - \hat{\boldsymbol{\theta}}(k)]'\} - \mathbf{E}\{[\boldsymbol{\theta} - \hat{\hat{\boldsymbol{\theta}}}(k)][\boldsymbol{\theta} - \hat{\hat{\boldsymbol{\theta}}}(k)]'\}$ is negative semi-definite.

The generalization of the Cramer-Rao inequality to a vector of parameters is given next.

Theorem 6-4 (Cramer-Rao inequality for a vector of parameters). *Let* $\mathbf{z}$ *denote a set of data as in Theorem 6-3, and* $\hat{\boldsymbol{\theta}}(\mathrm{k})$ *be any unbiased estimator of deterministic* $\boldsymbol{\theta}$ *based on* $\mathbf{z}$. *Then*

$$\mathbf{E}\{\tilde{\boldsymbol{\theta}}(k)\tilde{\boldsymbol{\theta}}'(k)\} \geq \mathbf{J}^{-1} \qquad \textit{for all } \mathrm{k} \tag{6-45}$$

where $\mathbf{J}$ *is the "Fisher information matrix,"*

$$\mathbf{J} = \mathbf{E}\left\{\left[\frac{\partial}{\partial\boldsymbol{\theta}} \ln p(\mathbf{z})\right]\left[\frac{\partial}{\partial\boldsymbol{\theta}} \ln p(\mathbf{z})\right]'\right\} \tag{6-46}$$

which can also be expressed as

$$\mathbf{J} = -\mathbf{E}\left\{\frac{\partial^2}{\partial\boldsymbol{\theta}^2} \ln p(\mathbf{z})\right\} \tag{6-47}$$

Equality holds in (6-45), *if and only if*

$$\left[\frac{\partial}{\partial\boldsymbol{\theta}} \ln p(\mathbf{z})\right]' = c(\boldsymbol{\theta})\tilde{\boldsymbol{\theta}}(k) \quad \square \tag{6-48}$$

A complete proof of this result is given by Sorenson (1980, pp. 94–96). Although the proof is similar to our proof of Theorem 6-3, it is a bit more intricate because of the vector nature of $\boldsymbol{\theta}$.

Inequality (6-45) demonstrates that any unbiased estimator can have a covariance no smaller than $\mathbf{J}^{-1}$. Unfortunately, $\mathbf{J}^{-1}$ is not a greatest lower bound for the error covariance. Other bounds exist which are tighter than (6-45) [e.g., the Bhattacharyya bound (Van Trees, 1968)], but they are even more difficult to compute than $\mathbf{J}^{-1}$.

Corollary 6-2. *Let* $\mathbf{z}$ *denote a set of data as in Theorem 6-3, and* $\hat{\theta}_{\mathrm{i}}(\mathrm{k})$ *be any unbiased estimator of deterministic* θ_{i} *based on* $\mathbf{z}$. *Then*

$$\mathbf{E}\{\tilde{\theta}_i^2(k)\} \geq (\mathbf{J}^{-1})_{ii} \qquad i = 1, 2, \ldots, n \textit{ and all } \mathrm{k} \tag{6-49}$$

where $(\mathbf{J}^{-1})_{ii}$ *is the* i–*ith element in matrix* $\mathbf{J}^{-1}$.

Proof. Inequality (6-45) means that $\mathbf{E}\{\tilde{\boldsymbol{\theta}}(k)\tilde{\boldsymbol{\theta}}'(k)\} - \mathbf{J}^{-1}$ is a positive semi-definite matrix, i.e.,

$$\mathbf{a}'[\mathbf{E}\{\tilde{\boldsymbol{\theta}}(k)\tilde{\boldsymbol{\theta}}'(k)\} - \mathbf{J}^{-1}]\mathbf{a} \geq 0 \tag{6-50}$$

where $\mathbf{a}$ is an arbitrary nonzero vector. Choosing $\mathbf{a} = \mathbf{e}_i$ (the ith unit vector) we obtain (6-49). $\square$

Results similar to those in Theorem 6-4 and Corollary 6-2 are also available for a vector of random parameters (e.g., Sorenson, 1980, pp. 99–100). Let $p(\mathbf{z}, \boldsymbol{\theta})$ denote the joint probability density function between $\mathbf{z}$ and $\boldsymbol{\theta}$. The Cramer-Rao inequality for random parameters is obtained from Theorems 6-3 and 6-4 by replacing $p(\mathbf{z})$ by $p(\mathbf{z}, \boldsymbol{\theta})$. Of course, the expectation is now with respect to $\mathbf{z}$ and $\boldsymbol{\theta}$.

PROBLEMS

6-1. Prove Theorem 6-2, which provides an unbiasedness constraint for the two design matrices that appear in a linear *recursive* estimator.

6-2. Prove Theorem 6-4, which provides the Cramer-Rao bound for a vector of parameters.

6-3. Random variable $X \sim N(x; \mu, \sigma^2)$, and we are given a random sample $\{x_1, x_2, \ldots, x_N\}$. Consider the following estimator for μ,

$$\hat{\mu}(N) = \frac{1}{N+a} \sum_{i=1}^{N} x_i$$

where $a \geq 0$. For what value(s) of a is $\hat{\mu}(N)$ an unbiased estimator of μ?

6-4. Suppose $z_1, z_2, \ldots, z_N$ are random samples from a Gaussian distribution with unknown mean, μ, and variance, σ^2. Reasonable estimators of μ and σ^2 are the sample mean and sample variance,

$$\bar{z} = \frac{1}{N} \sum_{i=1}^{N} z_i$$

and

$$s^2 = \frac{1}{N} \sum_{i=1}^{N} (z_i - \bar{z})^2$$

Is s^2 an unbiased estimator of σ^2? [Hint: Show that $\mathbf{E}\{s^2\} = (N-1)\sigma^2/N$]

6-5. (Mendel, 1973, first part of Exercise 2-9, pg. 137). Show that if $\hat{\boldsymbol{\theta}}$ is an unbiased estimate of $\boldsymbol{\theta}$, $a\,\hat{\boldsymbol{\theta}} + \mathbf{b}$ is an unbiased estimate of $a\,\boldsymbol{\theta} + \mathbf{b}$.

6-6. Suppose that N independent observations $(x_1, x_2, \ldots, x_N)$ are made of a random variable X that is Gaussian, i.e.,

$$p(x_i \,|\, \mu, \sigma^2) = \frac{1}{\sqrt{2\pi}\,\sigma} \exp\left[-(x_i - \mu)^2/2\sigma^2\right]$$

In this problem only μ is unknown. Derive the Cramer-Rao lower bound of $\mathbf{E}\{\tilde{\mu}^2(N)\}$ for an unbiased estimator of μ.

6-7. Repeat Problem 6-6, but in this case assume that only σ^2 is unknown, i.e., derive the Cramer-Rao lower bound of $\mathbf{E}\{[\tilde{\sigma}^2(N)]^2\}$ for an unbiased estimator of σ^2.

6-8. Repeat Problem 6-6, but in this case assume both μ and σ^2 are unknown, i.e., compute $\mathbf{J}^{-1}$ when $\boldsymbol{\theta} = \text{col}\,(\mu, \sigma^2)$.

6-9. Suppose $\hat{\theta}(k)$ is a biased estimator of deterministic θ, with bias $B(\theta)$. Show that

$$\mathbf{E}\{\tilde{\theta}^2(k)\} \geq \frac{\left[1 + \dfrac{\partial B(\theta)}{\partial \theta}\right]^2}{\mathbf{E}\left\{\left[\dfrac{\partial}{\partial \theta} \ln p(\mathbf{z})\right]^2\right\}} \qquad \text{for all } k$$

Lesson 7

Large Sample Properties of Estimators

INTRODUCTION

To begin, we reiterate the fact that "if an estimator possesses a small-sample property it also possesses the associated large-sample property; but, the converse is not always true." In this lesson we shall examine the following large-sample properties of estimators: asymptotic unbiasedness, consistency, and asymptotic efficiency. The first and third properties are natural extensions of the small-sample properties of unbiasedness and efficiency in the limiting situation of an infinite number of measurements. The second property is about convergence of $\hat{\theta}(k)$ to θ.

Before embarking on a discussion of these three large-sample properties, we digress a bit to introduce the concept of asymptotic distribution and its associated asymptotic mean and variance (or covariance, in the vector situation). Doing this will help us better understand these large-sample properties.

ASYMPTOTIC DISTRIBUTIONS

According to Kmenta (1971, pg. 163), ". . . if the distribution of an estimator tends to become more and more similar in form to some specific distribution as the sample size increases, then such a specific distribution is called the *asymptotic distribution* of the estimator in question. . . . What is meant by the asymptotic distribution is not the ultimate form of the distribution, which may be degenerate, but the form that the distribution tends to put on in the last part of its journey to the final collapse (if this occurs)." Consider the situation depicted in Figure 7-1, where $p_i(\hat{\theta})$ denotes the probability density function associated with estimator $\hat{\theta}$ of the scalar parameter θ, based on i measurements. As the number of measurements increases, $p_i(\hat{\theta})$ changes its shape (although, in this example, each one of the density functions is Gaussian). The density function eventually centers itself about the true parameter value θ, and the variance associated with $p_i(\hat{\theta})$ tends to get smaller as i increases. Ultimately, the variance will become so small that in all probability $\hat{\theta} = \theta$. The asymptotic distribution refers to $p_i(\hat{\theta})$ as it evolves from $i = 1, 2, \ldots$, etc., especially for large values of i.

The preceding example illustrates one of the three possible cases that can occur for an asymptotic distribution, namely the case when an estimator has a distribution of the same form regardless of the sample size, and this form is known (e.g., Gaussian). Some estimators have a distribution that, although not necessarily always of the same form, is also known for every sample size. For example, $p_5(\hat{\theta})$ may be uniform, $p_{20}(\hat{\theta})$ may be Rayleigh, and $p_{200}(\hat{\theta})$ may be Gaussian. Finally, for some estimators the distribution is not necessarily known for every sample size, but is known only for $k \to \infty$.

Asymptotic distributions, like other distributions, are characterized by their moments. We are especially interested in their first two moments, namely, the asymptotic mean and variance.

Definition 7-1. *The asymptotic mean is equal to the asymptotic expectation, namely* $\lim_{k\to\infty} \mathbf{E}\{\hat{\theta}(k)\}$. □

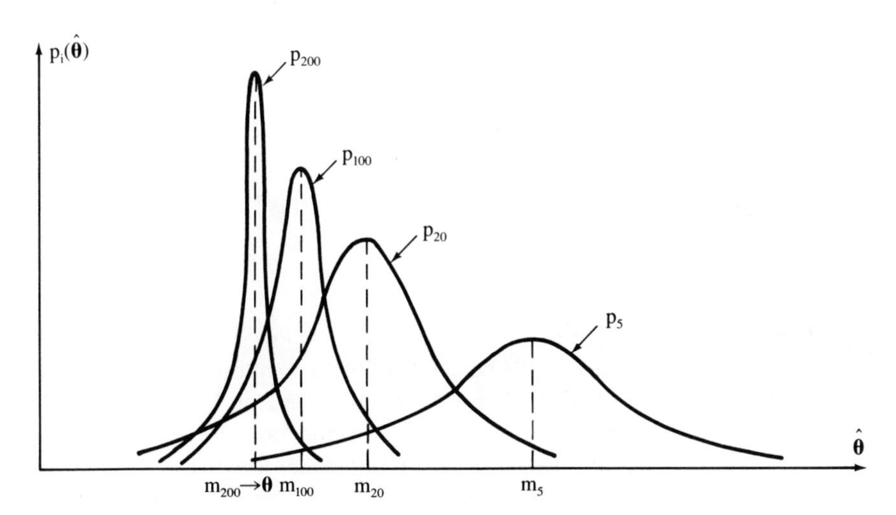

Figure 7-1 Probability density function for estimate of scalar θ, as a function of number of measurements, e.g., p_{20} is the p.d.f. for 20 measurements.

As noted in Goldberger (1964, pg. 116), if $\mathbf{E}\{\hat{\theta}(k)\} = m$ for *all* k, then $\lim_{k\to\infty} \mathbf{E}\{\hat{\theta}(k)\} = \lim_{k\to\infty} m = m$. Alternatively, suppose that

$$\mathbf{E}\{\hat{\theta}(k)\} = m + k^{-1}c_1 + k^{-2}c_2 + \cdots \tag{7-1}$$

where the c's are finite constants; then,

$$\lim_{k\to\infty} \mathbf{E}\{\hat{\theta}(k)\} = \lim_{k\to\infty} \{m + k^{-1}c_1 + k^{-2}c_2 + \cdots\} = m \tag{7-2}$$

Thus if $\mathbf{E}\{\hat{\theta}(k)\}$ is expressible as a power series in $k^0, k^{-1}, k^{-2}, \ldots$, the asymptotic mean of $\hat{\theta}(k)$ is the leading term of this power series; as $k \to \infty$ the terms of "higher order of smallness" in k vanish.

Definition 7-2. *The asymptotic variance, which is short for "variance of the asymptotic distribution" is not equal to* $\lim_{k\to\infty} \text{var}\,[\hat{\theta}(k)]$. *It is defined as*

$$\text{asymptotic var}\,[\hat{\theta}(k)] = \frac{1}{k}\lim_{k\to\infty} \mathbf{E}\left\{k[\hat{\theta}(k) - \lim_{k\to\infty} \mathbf{E}\{\hat{\theta}(k)\}]^2\right\} \quad \square \tag{7-3}$$

Kmenta (1971, pg. 164) states "The asymptotic variance . . . is *not* equal to $\lim_{k\to\infty} \text{var}(\hat{\theta})$. The reason is that in the case of estimators whose variance decreases with an increase in k, the variance will approach zero as $k \to \infty$. This will happen when the distribution collapses on a point. But, as we explained, the asymptotic distribution is not the same as the collapsed (degenerate) distribution, and its variance is *not* zero."

Goldberger (1964, pg. 116) notes that if $\mathbf{E}\{[\hat{\theta}(k) - \lim_{k\to\infty} \mathbf{E}\{\hat{\theta}(k)\}]^2\} = v/k$ for *all* values of k, then asymptotic var $[\hat{\theta}(k)] = v/k$. Alternatively, suppose that

$$\mathbf{E}\{[\hat{\theta}(k) - \lim_{k\to\infty} \mathbf{E}\{\hat{\theta}(k)\}]^2\} = k^{-1}v + k^{-2}c_2 + k^{-3}c_3 + \cdots \tag{7-4}$$

where the the c's are finite constants; then,

$$\begin{aligned}\text{asymptotic var}\,[\hat{\theta}(k)] &= \frac{1}{k}\lim_{k\to\infty} (v + k^{-1}c_2 + k^{-2}c_3 + \cdots) \\ &= \frac{v}{k}\end{aligned} \tag{7-5}$$

Thus if the variance of each $\hat{\theta}(k)$ is expressible as a power series in $k^{-1}, k^{-2}, \ldots$, the asymptotic variance of $\hat{\theta}(k)$ is the leading term of this power series; as k goes to infinity the terms of "higher order of smallness" in k vanish. Observe that if (7-5) is true then the asymptotic variance of $\hat{\theta}(k)$ decreases as $k \to \infty$ . . . which corresponds to the situation depicted in Figure 7-1.

Extensions of our definitions of asymptotic mean and variance to sequences of random *vectors* [e.g., $\hat{\boldsymbol{\theta}}(k)$, $k = 1, 2, \ldots$] are straightforward and can be found in Goldberger (1964, pg. 117).

ASYMPTOTIC UNBIASEDNESS

Definition 7-3. *Estimator* $\hat{\boldsymbol{\theta}}(\mathrm{k})$ *is an asymptotically unbiased estimator of deterministic* $\boldsymbol{\theta}$, *if*

$$\lim_{k\to\infty} \mathbf{E}\{\hat{\boldsymbol{\theta}}(k)\} = \boldsymbol{\theta} \tag{7-6}$$

or of random $\boldsymbol{\theta}$, *if*

$$\lim_{k\to\infty} \mathbf{E}\{\hat{\boldsymbol{\theta}}(k)\} = \mathbf{E}\{\boldsymbol{\theta}\} \quad \square \tag{7-7}$$

Figure 7-1 depicts an example in which the asymptotic mean of $\hat{\theta}$ has converged to θ (note that p_{200} is centered about $m_{200} = \theta$).

Note that for the calculation on the left-hand side of either (7-6) or (7-7) to be performed, the asymptotic distribution of $\hat{\boldsymbol{\theta}}(k)$ must exist, because

$$\mathbf{E}\{\hat{\boldsymbol{\theta}}(k)\} = \int_{-\infty}^{\infty} \cdots \int_{-\infty}^{\infty} \hat{\boldsymbol{\theta}}(k) p(\hat{\boldsymbol{\theta}}(k)) d\hat{\boldsymbol{\theta}}(k) \tag{7-8}$$

Example 7-1

Recall our linear model $\mathscr{Z}(k) = \mathscr{H}(k)\boldsymbol{\theta} + \mathscr{V}(k)$ in which $\mathbf{E}\{\mathscr{V}(k)\} = \mathbf{0}$. Let us assume that each component of $\mathscr{V}(k)$ is uncorrelated and has the same variance σ_v^2. In Lesson 8, we determine an unbiased estimator for σ_v^2. Here, on the other hand, we just assume that

$$\hat{\sigma}_v^2(k) \triangleq \frac{\tilde{\mathscr{Z}}'(k)\tilde{\mathscr{Z}}(k)}{k} \tag{7-9}$$

where $\tilde{\mathscr{Z}}(k) = \mathscr{Z}(k) - \mathscr{H}(k)\hat{\boldsymbol{\theta}}_{LS}(k)$. We leave it to the reader to show that (see Lesson 8)

$$\mathbf{E}\{\hat{\sigma}_v^2(k)\} = \left(\frac{k-n}{k}\right)\sigma_v^2 \tag{7-10}$$

Observe that $\hat{\sigma}_v^2(k)$ is not an unbiased estimator of σ_v^2; but, it is an asymptotically unbiased estimator, because $\lim_{k\to\infty} [(k-n)/k]\,\sigma_v^2 = \sigma_v^2$. $\square$

CONSISTENCY

We now direct our attention to the issue of stochastic convergence. The reader should review the different modes of stochastic convergence, especially convergence in probability and mean-squared convergence (see, for example, Papoulis, 1965).

Definition 7-4. *The probability limit of* $\hat{\theta}(\mathrm{k})$ *is the point* θ^* *on which the distribution of our estimator collapses. We abbreviate "probability limit of* $\hat{\theta}(\mathrm{k})$*" by* plim $\hat{\theta}(\mathrm{k})$. *Mathematically speaking,*

$$\text{plim}\ \hat{\theta}(k) = \theta^* \leftrightarrow \lim_{k\to\infty} Pr\,[|\hat{\theta}(k) - \theta^*| \geq \epsilon] \to 0 \tag{7-11}$$

where ϵ *is a small positive number.* $\square$

Definition 7-5. $\hat{\theta}(\mathrm{k})$ *is a consistent estimator of* θ *if*

$$\text{plim } \hat{\theta}(k) = \theta \quad \square \tag{7-12}$$

Note that "consistency" means the same thing as "convergence in probability." For an estimator to be consistent, its probability limit θ^* must equal its true value θ. Note, also, that a consistent estimator need not be unbiased or asymptotically unbiased.

Why is convergence in probability so popular and widely used in the estimation field? One reason is that plim $(\cdot)$ can be treated as an operator. For example, suppose X_k and Y_k are two random sequences, for which plim $X_k = X$ and plim $Y_k = Y$; then (see Tucker, 1962, for simple proofs of these and other facts),

$$\text{plim } X_k Y_k = (\text{plim } X_k)(\text{plim } Y_k) = XY \tag{7-13}$$

and

$$\text{plim} \left(\frac{X_k}{Y_k}\right) = \frac{\text{plim } X_k}{\text{plim } Y_k} = \frac{X}{Y} \tag{7-14}$$

Additionally, suppose $\mathbf{A}_k$ and $\mathbf{B}_k$ are two commensurate matrix sequences, for which plim $\mathbf{A}_k = \mathbf{A}$ and plim $\mathbf{B}_k = \mathbf{B}$ [note that plim $\mathbf{A}_k$, for example, means $(\text{plim } a_{ij}(k))_{ij}$]; then

$$\text{plim } \mathbf{A}_k \mathbf{B}_k = (\text{plim } \mathbf{A}_k)(\text{plim } \mathbf{B}_k) = \mathbf{AB} \tag{7-15}$$

$$\text{plim } \mathbf{A}_k^{-1} = (\text{plim } \mathbf{A}_k)^{-1} = \mathbf{A}^{-1} \tag{7-16}$$

and

$$\text{plim } \mathbf{A}_k^{-1} \mathbf{B}_k = \mathbf{A}^{-1} \mathbf{B} \tag{7-17}$$

The treatment of plim $(\cdot)$ as an operator often makes the study of consistency quite easy. We shall demonstrate the truth of this in Lesson 8 when we examine the consistency of the least-squares estimator.

A second reason for the importance of consistency is the property that "consistency carries over"; i.e., *any continuous function of a consistent estimator is itself a consistent estimator* [see Tucker, 1967, for a proof of this property, which relies heavily on the preceding treatment of plim $(\cdot)$ as an operator].

Example 7-2

Suppose $\hat{\theta}$ is a consistent estimator of θ. Then $1/\hat{\theta}$ is a consistent estimator of $1/\theta$, $(\hat{\theta})^2$ is a consistent estimator of θ^2, and $\ln \hat{\theta}$ is a consistent estimator of $\ln \theta$. These facts are all due to the consistency carry-over property. $\square$

The reader may be scratching his or her head at this point and wondering about the emphasis placed on these illustrative examples. Isn't, for example, using $\hat{\theta}$ to estimate θ^2 by $(\hat{\theta})^2$ the "natural" thing to do? The answer is "Yes, but only if we know ahead of time that $\hat{\theta}$ is a consistent estimator of θ." If you do not know this to be true, then there is no guarantee that $\widehat{\theta^2} = (\hat{\theta})^2$. In Lesson 11 we show that maximum-likelihood estimators are consistent; thus $\widehat{(\theta^2)_{ML}} = (\hat{\theta}_{ML})^2$. We mention this property about maximum-likelihood estimators here, because one must know whether or not an estimator is consistent before applying the consistency carry-over property. Not all estimators are consistent!

Finally, this carry-over property for consistency does not necessarily apply to other properties. For example, if $\hat{\boldsymbol{\theta}}(k)$ is an unbiased estimator of $\boldsymbol{\theta}$, then $\mathbf{A}\hat{\boldsymbol{\theta}}(k) + \mathbf{b}$ will be an unbiased estimator of $\mathbf{A}\boldsymbol{\theta} + \mathbf{b}$; but, $\hat{\theta^2}(k)$ will not be an unbiased estimator of θ^2.

How do you determine whether or not an estimator is consistent? Often, the direct approach, which makes heavy use of plim $(\cdot)$ operator algebra, is possible. Sometimes, an indirect approach is used, one which examines whether both the bias in $\hat{\theta}(k)$ and variance of $\hat{\theta}(k)$ approach zero as $k \to \infty$. In order to understand the validity of this indirect approach, we digress to discuss mean-squared convergence and its relationship to convergence in probability.

Definition 7-6. *Estimator $\hat{\theta}(k)$ converges to θ in a mean-squared sense, if*

$$\lim_{k\to\infty} \mathbf{E}\{[\hat{\theta}(k) - \theta]^2\} \to 0 \quad \square \tag{7-18}$$

Theorem 7-1. *If $\hat{\theta}(k)$ converges to θ in mean-square, then it converges to θ in probability.*

Proof (Papoulis, 1965, pg. 151). Recall the Inequality of Bienaymé

$$\Pr[|x - a| \geq \epsilon] \leq \mathbf{E}\{|x - a|^2\}/\epsilon^2 \tag{7-19}$$

Let $a = 0$ and $x = \hat{\theta}(k) - \theta$ in (7-19), and take the limit as $k \to \infty$ on both sides of (7-19), to see that

$$\lim_{k\to\infty} \Pr[|\hat{\theta}(k) - \theta| \geq \epsilon] \leq \lim_{k\to\infty} \mathbf{E}\{[\hat{\theta}(k) - \theta]^2\}/\epsilon^2 \tag{7-20}$$

Using the fact that $\hat{\theta}(k)$ converges to θ in mean-square, we see that

$$\lim_{k\to\infty} \Pr[|\hat{\theta}(k) - \theta| \geq \epsilon] \to 0 \tag{7-21}$$

thus, $\hat{\theta}(k)$ converges to θ in probability. $\square$

Recall, from probability theory, that although mean-squared convergence implies convergence in probability, the converse is not true.

Example 7-3 (Kmenta, 1971, pg. 166)

Let $\hat{\theta}(k)$ be an estimator of θ, and let the probability density function of $\hat{\theta}(k)$ be

$\hat{\theta}(k)$	$p(\hat{\theta}(k))$
θ	$1 - \frac{1}{k}$
k	$\frac{1}{k}$

In this example $\hat{\theta}(k)$ can only assume two different values, θ and k. Obviously, $\hat{\theta}(k)$ is consistent, because as $k \to \infty$ the probability that $\hat{\theta}(k)$ equals θ approaches unity, i.e., plim $\hat{\theta}(k) = \theta$. Observe, also, that $\mathbf{E}\{\hat{\theta}(k)\} = 1 + \theta(1 - 1/k)$, which means that $\hat{\theta}(k)$ is biased.

Now let us investigate the mean-squared error between $\hat{\theta}$ and θ; i.e.,

$$\lim_{k\to\infty} \mathbf{E}\{[\hat{\theta}(k) - \theta]^2\} = \lim_{k\to\infty} \left[(\theta - \theta)^2\left(1 - \frac{1}{k}\right) + (k - \theta)^2\frac{1}{k}\right]$$
$$= \lim_{k\to\infty} k \to \infty$$

In this pathological example, the mean-squared error is diverging to infinity; but, $\hat{\theta}(k)$ converges to θ in probability. □

Theorem 7-2. *Let* $\hat{\theta}(k)$ *denote an estimator of* θ. *If* bias $\hat{\theta}(k)$ *and* variance $\hat{\theta}(k)$ *both approach zero as* $k \to \infty$, *then the mean-squared error between* $\hat{\theta}(k)$ *and* θ *approaches zero, and, therefore,* $\hat{\theta}(k)$ *is a consistent estimator of* θ.

Proof. From elementary probability theory, we know that

$$\mathbf{E}\{[\hat{\theta}(k) - \theta]^2\} = [\text{bias } \hat{\theta}(k)]^2 + \text{variance } \hat{\theta}(k) \tag{7-22}$$

If, as assumed, bias $\hat{\theta}(k)$ and variance $\hat{\theta}(k)$ both approach zero as $k \to \infty$, then

$$\lim_{k\to\infty} \mathbf{E}\{[\hat{\theta}(k) - \theta]^2\} \to 0 \tag{7-23}$$

which means that $\hat{\theta}(k)$ converges to θ in mean-square. Thus, by Theorem 7-1 $\hat{\theta}(k)$ also converges to θ in probability. □

The importance of Theorem 7-2 is that it provides a constructive way to test for consistency.

ASYMPTOTIC EFFICIENCY

Definition 7-7. $\hat{\theta}(k)$ *is an asymptotically efficient estimator of scalar parameter* θ, *if*: 1. $\hat{\theta}(k)$ *has an asymptotic distribution with finite mean and variance*, 2. $\hat{\theta}(k)$ *is consistent, and* 3. *the variance of the asymptotic distribution equals*

$$1/\mathbf{E}\left\{\left[\frac{\partial}{\partial\theta} \ln p(\mathbf{z})\right]^2\right\} \quad \square$$

For the case of a vector of parameters conditions 1. and 2. in this definition are unchanged; however, condition 3. is changed to read "if the covariance matrix of the asymptotic distribution equals $\mathbf{J}^{-1}$ (see Theorem 6-4)."

PROBLEMS

7-1. Random variable $X \sim N(x;\ \mu,\ \sigma^2)$, and we are given a random sample $\{x_1, x_2, \ldots, x_N\}$. Consider the following estimator for μ,

$$\hat{\mu}(N) = \frac{1}{N+a}\sum_{i=1}^{N} x_i$$

where $a \geq 0$.

(a) For what value(s) of a is $\hat{\mu}(N)$ an asymptotically unbiased estimator of μ?

(b) Prove that $\hat{\mu}(N)$ is a consistent estimator of μ for all $a \geq 0$.

(c) Compare the results obtained in **(a)** with those obtained in Problem 6-3.

7-2. Suppose $z_1, z_2, \ldots, z_N$ are random samples from a Gaussian distribution with unknown mean, μ, and variance, σ^2. Reasonable estimators of μ and σ^2 are the sample mean and sample variance

$$\bar{z} = \frac{1}{N}\sum_{i=1}^{N} z_i$$

and

$$s^2 = \frac{1}{N}\sum_{i=1}^{N} (z_i - \bar{z})^2$$

(a) Is s^2 an asymptotically unbiased estimator of σ^2? [Hint: Show that $\mathbf{E}\{s^2\} = (N-1)\sigma^2/N$].

(b) Compare the result in **(a)** with that from Problem 6-4.

(c) One can show that the variance of s^2 is

$$\text{var}\,(s^2) = \frac{\mu_4 - \sigma^4}{N} - \frac{2(\mu_4 - 2\sigma^4)}{N^2} + \frac{\mu_4 - 3\sigma^4}{N^3}$$

Explain whether or not s^2 is a consistent estimator of σ^2.

7-3. Random variable $X \sim N(x;\ \mu,\sigma^2)$. Consider the following estimator of the population mean obtained from a random sample of N observations of X,

$$\hat{\mu}(N) = \bar{x} + \frac{a}{N}$$

where a is a finite constant, and $\bar{x}$ is the sample mean.

(a) What are the asymptotic mean and variance of $\hat{\mu}(N)$?

(b) Is $\hat{\mu}(N)$ a consistent estimator of μ?

(c) Is $\hat{\mu}(N)$ asymptotically efficient?

7-4. Let X be a Gaussian variable with mean μ and variance σ^2. Consider the problem of estimating μ from a random sample of observations $x_1, x_2, \ldots, x_N$. Three estimators are proposed:

$$\hat{\mu}_1(N) = \frac{1}{N}\sum_{i=1}^{N} x_i$$

$$\hat{\mu}_2(N) = \frac{1}{N+1}\sum_{i=1}^{N} x_i$$

$$\hat{\mu}_3(N) = \frac{1}{2}x_1 + \frac{1}{2N}\sum_{i=2}^{N} x_i$$

You are to study unbiasedness, efficiency, asymptotic unbiasedness, consistency, and asymptotic efficiency for these estimators. Show the analysis that allows you to complete the following table.

	Estimator		
Properties	$\hat{\mu}_1$	$\hat{\mu}_2$	$\hat{\mu}_3$
Small sample			
Unbiasedness			
Efficiency			
Large sample			
Unbiasedness			
Consistency			
Efficiency			

Entries to this table are *yes* or *no*.

7-5. If plim $X_k = X$ and plim $Y_k = Y$ where X and Y are constants, then plim $(X_k + Y_k) = X + Y$ and plim $c X_k = cX$, where c is a constant. Prove that plim $X_k Y_k = XY$ {Hint: $X_k Y_k = \frac{1}{4}[(X_k + Y_k)^2 - (X_k - Y_k)^2]$}.

Lesson 8

Properties of Least-Squares Estimators

INTRODUCTION

In this lesson we study some small- and large-sample properties of least-squares estimators. Recall that in least-squares we estimate the $n \times 1$ parameter vector $\boldsymbol{\theta}$ of the linear model $\mathcal{Z}(k) = \mathcal{H}(k)\boldsymbol{\theta} + \mathcal{V}(k)$. We will see that most of the results in this lesson require $\mathcal{H}(k)$ to be deterministic, or, $\mathcal{H}(k)$ and $\mathcal{V}(k)$ to be statistically independent. In some applications one or the other of these requirements is met; however, there are many important applications where neither is met.

SMALL SAMPLE PROPERTIES OF LEAST-SQUARES ESTIMATORS

In this section (parts of which are taken from Mendel, 1973, pp. 75–86) we examine the bias and variance of weighted least-squares and least-squares estimators.

To begin, we recall Example 6-2, in which we showed that, when $\mathcal{H}(k)$ *is deterministic, the WLSE of* $\boldsymbol{\theta}$ *is unbiased.* We also showed [after the statement of Theorem 6-2 and Equation (6-13)] that our recursive WLSE of $\boldsymbol{\theta}$ has the requisite structure of an unbiased estimator; but, that unbiasedness of the recursive WLSE of $\boldsymbol{\theta}$ also requires $\mathbf{h}(k+1)$ to be deterministic.

When $\mathcal{H}(k)$ is random, we have the following important result:

Theorem 8-1. *The WLSE of* $\boldsymbol{\theta}$,

$$\hat{\boldsymbol{\theta}}_{\mathrm{WLS}}(k) = [\mathcal{H}'(k)\mathbf{W}(k)\mathcal{H}(k)]^{-1}\mathcal{H}'(k)\mathbf{W}(k)\mathcal{Z}(k) \tag{8-1}$$

is unbiased if $\mathcal{V}$(k) *is zero mean and if* $\mathcal{V}$(k) *and* $\mathcal{H}$(k) *are statistically independent.*

Note that this is the first place where, in connection with least squares, we have had to assume any a priori knowledge about noise $\mathcal{V}(k)$.

Proof. From (8-1) and $\mathfrak{Z}(k) = \mathcal{H}(k)\boldsymbol{\theta} + \mathcal{V}(k)$, we find that

$$\begin{aligned}\hat{\boldsymbol{\theta}}_{\text{WLS}}(k) &= (\mathcal{H}'\mathbf{W}\mathcal{H})^{-1}\mathcal{H}'\mathbf{W}(\mathcal{H}\boldsymbol{\theta} + \mathcal{V}) \\ &= \boldsymbol{\theta} + (\mathcal{H}'\mathbf{W}\mathcal{H})^{-1}\mathcal{H}'\mathbf{W}\mathcal{V} \qquad \text{for all } k\end{aligned} \tag{8-2}$$

where, for notational simplification, we have omitted the functional dependences of $\mathcal{H}$, $\mathbf{W}$, and $\mathcal{V}$ on k. Taking the expectation on both sides of (8-2), it follows that

$$\mathbf{E}\{\hat{\boldsymbol{\theta}}_{\text{WLS}}(k)\} = \boldsymbol{\theta} + \mathbf{E}\{\mathcal{H}'\mathbf{W}\mathcal{H})^{-1}\mathcal{H}'\}\mathbf{W}\mathbf{E}\{\mathcal{V}\} \qquad \text{for all } k \tag{8-3}$$

In deriving (8-3) we have used the fact that $\mathcal{H}(k)$ and $\mathcal{V}(k)$ are statistically independent [recall that if two random variables, a and b, are statistically independent $p(a, b) = p(a)p(b)$; thus, $\mathbf{E}\{ab\} = \mathbf{E}\{a\}\mathbf{E}\{b\}$ and $\mathbf{E}\{g(a)h(b)\} = \mathbf{E}\{g(a)\}\mathbf{E}\{h(b)\}$]. The second term in (8-3) is zero, because $\mathbf{E}\{\mathcal{V}\} = \mathbf{0}$, and therefore

$$\mathbf{E}\{\hat{\boldsymbol{\theta}}_{\text{WLS}}(k)\} = \boldsymbol{\theta} \qquad \text{for all } k \quad \square \tag{8-4}$$

This theorem only states *sufficient conditions* for unbiasedness of $\hat{\boldsymbol{\theta}}_{\text{WLS}}(k)$, which means that, if we do not satisfy these conditions, we *cannot* conclude anything about whether $\hat{\boldsymbol{\theta}}_{\text{WLS}}(k)$ is unbiased or biased. In order to obtain *necessary conditions* for unbiasedness, assume that $\mathbf{E}\{\hat{\boldsymbol{\theta}}_{\text{WLS}}(k)\} = \boldsymbol{\theta}$ and take the expectation on both sides of (8-2). Doing this, we see that

$$\mathbf{E}\{(\mathcal{H}'\mathbf{W}\mathcal{H})^{-1}\mathcal{H}'\mathbf{W}\mathcal{V}\} = \mathbf{0} \tag{8-5}$$

Letting $\mathbf{M} = (\mathcal{H}'\mathbf{W}\mathcal{H})^{-1}\mathcal{H}'\mathbf{W}$ and $\mathbf{m}_i'$ denote the ith row of matrix $\mathbf{M}$, (8-5) can be expressed as the following collection of *orthogonality conditions*,

$$\mathbf{E}\{\mathbf{m}_i'\mathcal{V}\} = 0 \qquad \text{for } i = 1, 2, \ldots, N \tag{8-6}$$

Orthogonality [recall that two random variables a and b are orthogonal if $\mathbf{E}\{ab\} = 0$] is a weaker condition than statistical independence, but is often more difficult to verify ahead of time than independence, especially since $\mathbf{m}_i'$ is a very nonlinear transformation of the random elements of $\mathcal{H}$.

Example 8-1

Recall the impulse response identification Example 2-1, in which $\boldsymbol{\theta} = \text{col}\,[h(1), h(2), \ldots, h(n)]$, where $h(i)$ is the value of the sampled impulse response at time t_i. System input $u(k)$ may be deterministic or random.

If $\{u(k), k = 0, 1, \ldots, N-1\}$ is deterministic, then $\mathcal{H}(N-1)$ [see Equation (2-5)] is deterministic, so that $\hat{\boldsymbol{\theta}}_{\text{WLS}}(N)$ is an unbiased estimator of $\boldsymbol{\theta}$. Often, one uses a random input sequence for $\{u(k), k = 0, 1, \ldots, N-1\}$, such as from a random number generator. This random sequence is in no way related to the measurement noise process, which means that $\mathcal{H}(N-1)$ and $\mathcal{V}(N)$ are statistically independent, and again $\hat{\boldsymbol{\theta}}_{WLS}(N)$ will be an unbiased estimate of the impulse response coefficients. We conclude, therefore, that *WLSEs of impulse response coefficients are unbiased.* □

Example 8-2

As a further illustration of an application of Theorem 8-1, let us take a look at the weighted least-squares estimates of the n α-coefficients in the Example 2-2 AR model. We shall now demonstrate that $\mathcal{H}(N-1)$ and $\mathcal{V}(N-1)$, which are defined in Equation (2-8), are dependent, which means of course that we cannot apply Theorem 8-1 to study the unbiasedness of the WLSEs of the α-coefficients.

We represent the explicit dependences of $\mathcal{H}$ and $\mathcal{V}$ on their elements in the following manner:

$$\mathcal{H} = \mathcal{H}[y(N-1), y(N-2), \ldots, y(0)] \tag{8-7}$$

and

$$\mathcal{V} = \mathcal{V}[u(N-1), u(N-2), \ldots, u(0)] \tag{8-8}$$

Direct iteration of difference equation (7) in Lesson 2 for $k = 1, 2, \ldots, N-1$, reveals that $y(1)$ depends on $u(0)$, $y(2)$ depends on $u(1)$ and $u(0)$, and finally, that $y(N-1)$ depends on $u(N-2), \ldots, u(0)$; thus,

$$\mathcal{H}[y(N-1), y(N-2), \ldots, y(0)] = \mathcal{H}[u(N-2), u(N-3), \ldots, u(0), y(0)] \tag{8-9}$$

Comparing (8-8) and (8-9), we see that $\mathcal{H}$ and $\mathcal{V}$ depend on similar values of random input u; hence, they are statistically dependent. □

Example 8-3

We are interested in estimating the parameter a in the following first-order system:

$$y(k+1) = -ay(k) + u(k) \tag{8-10}$$

where $u(k)$ is a zero-mean white noise sequence. One approach to doing this is to collect $y(k+1), y(k), \ldots, y(1)$ as follows,

$$\underbrace{\begin{bmatrix} y(k+1) \\ y(k) \\ y(k-1) \\ \cdots \\ y(1) \end{bmatrix}}_{\mathcal{Z}(k+1)} = \underbrace{\begin{bmatrix} -y(k) \\ -y(k-1) \\ -y(k-2) \\ \cdots \\ -y(0) \end{bmatrix}}_{\mathcal{H}(k)} a + \underbrace{\begin{bmatrix} u(k) \\ u(k-1) \\ u(k-2) \\ \cdots \\ u(0) \end{bmatrix}}_{\mathcal{V}(k)} \tag{8-11}$$

and, to obtain $\hat{a}_{LS}$. In order to study the bias of $\hat{a}_{LS}$, we use (8-2) in which $\mathbf{W}(k)$ is set equal to $\mathbf{I}$, and $\mathcal{H}(k)$ and $\mathcal{V}(k)$ are defined in (8-11). We also set $\hat{\boldsymbol{\theta}}_{LS}(k) = \hat{a}_{LS}(k+1)$. The argument of $\hat{a}_{LS}$ is $k+1$ instead of k, because the argument of $\mathcal{Z}$ in (8-11) is $k+1$. Doing this, we find that

$$\hat{a}_{LS}(k+1) = a - \frac{\sum_{i=0}^{k} u(i)y(i)}{\sum_{j=0}^{k} y^2(j)} \tag{8-12}$$

thus,

$$\mathbf{E}\{\hat{a}_{LS}(k+1)\} = a - \sum_{i=0}^{k} E\left\{\frac{u(i)y(i)}{\sum_{j=0}^{k} y^2(j)}\right\} \tag{8-13}$$

Note, from (8-10), that $y(j)$ depends at most on $u(j-1)$; therefore,

$$E\left\{\frac{u(k)y(k)}{\sum_{j=0}^{k} y^2(j)}\right\} = E\{u(k)\}E\left\{\frac{y(k)}{\sum_{j=0}^{k} y^2(j)}\right\} = 0 \tag{8-14}$$

because $\mathbf{E}\{u(k)\} = 0$. Unfortunately, all of the remaining terms in (8-13), i.e., for $i = 0, 1, \ldots, k-1$, will not be equal to zero; consequently,

$$\mathbf{E}\{\hat{a}_{LS}(k+1)\} = a + 0 + k \text{ nonzero terms} \tag{8-15}$$

Unless we are very lucky so that the k nonzero terms sum identically to zero, $\mathbf{E}\{\hat{a}_{LS}(k+1)\} \neq a$, which means, of course, that $\hat{a}_{LS}$ is biased.

The results in this example generalize to higher-order difference equations, so that we can conclude that *least-squares estimates of coefficients in an AR model are biased.* □

In the *method of instrumental variables* $\mathcal{H}(k)$ is replaced by $\mathcal{H}^*(k)$ where $\mathcal{H}^*(k)$ is chosen so that it is statistically independent of $\mathcal{V}(k)$. There can be, and in general there will be, many choices of $\mathcal{H}^*(k)$ that may qualify as instrumental variables. It is often difficult to check that $\mathcal{H}^*(k)$ is statistically independent of $\mathcal{V}(k)$.

Next we proceed to compute the covariance matrix of $\tilde{\boldsymbol{\theta}}_{WLS}(k)$, where

$$\tilde{\boldsymbol{\theta}}_{WLS}(k) = \boldsymbol{\theta} - \hat{\boldsymbol{\theta}}_{WLS}(k) \tag{8-16}$$

Theorem 8-2. *If* $\mathbf{E}\{\mathcal{V}(\mathrm{k})\} = \mathbf{0}$, $\mathcal{V}(\mathrm{k})$ *and* $\mathcal{H}(\mathrm{k})$ *are statistically independent, and*

$$\mathbf{E}\{\mathcal{V}(k)\mathcal{V}'(k)\} = \mathcal{R}(k) \tag{8-17}$$

then

$$\text{cov}\,[\tilde{\boldsymbol{\theta}}_{WLS}(k)] = \mathbf{E}_{\mathcal{H}}\{(\mathcal{H}'\mathbf{W}\mathcal{H})^{-1}\mathcal{H}'\mathbf{W}\mathcal{R}\mathbf{W}\mathcal{H}(\mathcal{H}'\mathbf{W}\mathcal{H})^{-1}\} \tag{8-18}$$

Proof. Because $\mathbf{E}\{\mathcal{V}(k)\} = \mathbf{0}$ and $\mathcal{V}(k)$ and $\mathcal{H}(k)$ are statistically independent, $\mathbf{E}\{\tilde{\boldsymbol{\theta}}_{WLS}(k)\} = \mathbf{0}$, so that

$$\text{cov}\,[\tilde{\boldsymbol{\theta}}_{WLS}(k)] = \mathbf{E}\{\tilde{\boldsymbol{\theta}}_{WLS}(k)\tilde{\boldsymbol{\theta}}'_{WLS}(k)\} \tag{8-19}$$

Using (8-2) in (8-16), we see that

$$\tilde{\boldsymbol{\theta}}_{WLS}(k) = -(\mathcal{H}'\mathbf{W}\mathcal{H})^{-1}\mathcal{H}'\mathbf{W}\mathcal{V} \tag{8-20}$$

hence

$$\text{cov}\,[\tilde{\boldsymbol{\theta}}_{WLS}(k)] = \mathbf{E}_{\mathcal{H},\mathcal{V}}\{(\mathcal{H}'\mathbf{W}\mathcal{H})^{-1}\mathcal{H}'\mathbf{W}\mathcal{V}\mathcal{V}'\mathbf{W}\mathcal{H}(\mathcal{H}'\mathbf{W}\mathcal{H})^{-1}\} \tag{8-21}$$

where we have made use of the fact that **W** is a symmetric matrix and the transpose and inverse symbols may be permuted. From probability theory (e.g., Papoulis, 1965), recall that

$$\mathbf{E}_{\mathcal{H},\mathcal{V}}\{\cdot\} = \mathbf{E}_{\mathcal{H}}\{\mathbf{E}_{\mathcal{V}|\mathcal{H}}\{\cdot\,|\mathcal{H}\}\} \tag{8-22}$$

Applying (8-22) to (8-21), we obtain (8-18). □

As it stands, Theorem 8-2 is not too useful because it is virtually impossible to compute the expectation in (8-18), due to the highly nonlinear dependence of $(\mathcal{H}'\mathbf{W}\mathcal{H})^{-1}\mathcal{H}'\mathbf{W}\mathcal{R}\mathbf{W}\mathcal{H}(\mathcal{H}'\mathbf{W}\mathcal{H})^{-1}$ on $\mathcal{H}$. The following special case of Theorem 8-2 is important in practical applications in which $\mathcal{H}(k)$ is deterministic and $\mathcal{R}(k) = \sigma_v^2\,\mathbf{I}$.

Corollary 8-1. *Given the conditions in Theorem 8-2, and that $\mathcal{H}(k)$ is deterministic, and, the components of $\mathcal{V}(k)$ are independent and identically distributed with zero-mean and constant variance σ_v^2, then*

$$\text{cov}\,[\tilde{\boldsymbol{\theta}}_{LS}(k)] = \sigma_v^2\,[\mathcal{H}'(k)\mathcal{H}(k)]^{-1} \tag{8-23}$$

Proof. When $\mathcal{H}$ is deterministic, cov $[\tilde{\boldsymbol{\theta}}_{WLS}(k)]$ is obtained from (8-18) by deleting the expectation on its right-hand side. To obtain cov $[\tilde{\boldsymbol{\theta}}_{LS}(k)]$ when cov $[\mathcal{V}(k)] = \sigma_v^2\,\mathbf{I}$, set $\mathbf{W} = \mathbf{I}$ and $\mathcal{R}(k) = \sigma_v^2\,\mathbf{I}$ in (8-18). The result is (8-23). □

Usually, when we use a least-squares estimation algorithm we do not know the numerical value of σ_v^2. If σ_v^2 is known ahead of time, it can be used directly in the estimate of $\boldsymbol{\theta}$. We show how to do this in Lesson 9. Where do we obtain σ_v^2 in order to compute (8-23)? We can estimate it!

Theorem 8-3. *An unbiased estimator of σ_v^2 is*

$$\hat{\sigma}_v^2(k) = \tilde{\mathcal{Z}}'(k)\tilde{\mathcal{Z}}(k)/(k-n) \tag{8-24}$$

where

$$\tilde{\mathcal{Z}}(k) = \mathcal{Z}(k) - \mathcal{H}(k)\hat{\boldsymbol{\theta}}_{LS}(k) \tag{8-25}$$

Proof. We shall proceed by computing $\mathbf{E}\{\tilde{\mathcal{Z}}'(k)\tilde{\mathcal{Z}}(k)\}$ and then approximating it as $\tilde{\mathcal{Z}}'(k)\tilde{\mathcal{Z}}(k)$, because the latter quantity can be computed from $\mathcal{Z}(k)$ and $\hat{\boldsymbol{\theta}}_{LS}(k)$, as in (8-25).

First, we compute an expression for $\tilde{\mathcal{Z}}(k)$. Substituting both the linear model for $\mathcal{Z}(k)$ and the least-squares formula for $\hat{\boldsymbol{\theta}}_{LS}(k)$ into (8-25), we find that

$$\begin{aligned}\tilde{\mathcal{Z}} &= \mathcal{H}\boldsymbol{\theta} + \mathcal{V} - \mathcal{H}(\mathcal{H}'\mathcal{H})^{-1}\mathcal{H}'(\mathcal{H}\boldsymbol{\theta} + \mathcal{V})\\ &= \mathcal{V} - \mathcal{H}(\mathcal{H}'\mathcal{H})^{-1}\mathcal{H}'\mathcal{V}\\ &= [\mathbf{I}_k - \mathcal{H}(\mathcal{H}'\mathcal{H})^{-1}\mathcal{H}']\mathcal{V}\end{aligned} \tag{8-26}$$

where $\mathbf{I}_k$ is the $k \times k$ identity matrix. Let

$$\mathbf{M} = \mathbf{I}_k - \mathcal{H}(\mathcal{H}'\mathcal{H})^{-1}\mathcal{H}' \tag{8-27}$$

Matrix $\mathbf{M}$ is idempotent, i.e., $\mathbf{M}' = \mathbf{M}$ and $\mathbf{M}^2 = \mathbf{M}$; therefore,

$$\begin{aligned} \mathbf{E}\{\tilde{\mathfrak{Z}}'\tilde{\mathfrak{Z}}\} &= \mathbf{E}\{\mathcal{V}'\mathbf{M}'\mathbf{M}\mathcal{V}\} = \mathbf{E}\{\mathcal{V}'\mathbf{M}\mathcal{V}\} \\ &= \mathbf{E}\{\text{tr } \mathbf{M}\mathcal{V}\mathcal{V}'\} \end{aligned} \tag{8-28}$$

Recall the following well-known facts about the trace of a matrix:

1. $\mathbf{E}\{\text{tr } \mathbf{A}\} = \text{tr } \mathbf{E}\{\mathbf{A}\}$
2. $\text{tr } c\mathbf{A} = c \text{ tr } \mathbf{A}$, where c is a scalar
3. $\text{tr } (\mathbf{A} + \mathbf{B}) = \text{tr } \mathbf{A} + \text{tr } \mathbf{B}$
4. $\text{tr } \mathbf{I}_N = N$
5. $\text{tr } \mathbf{AB} = \text{tr } \mathbf{BA}$

Using these facts, we now continue the development of (8-28), as follows:

$$\begin{aligned} \mathbf{E}\{\tilde{\mathfrak{Z}}'\tilde{\mathfrak{Z}}\} &= \text{tr } [\mathbf{M}\mathbf{E}\{\mathcal{V}\mathcal{V}'\}] = \text{tr } \mathbf{M}\mathcal{R} = \text{tr } \mathbf{M}\sigma_\nu^2 \\ &= \sigma_\nu^2 \text{ tr } \mathbf{M} = \sigma_\nu^2 \text{ tr } [\mathbf{I}_k - \mathcal{H}(\mathcal{H}'\mathcal{H})^{-1}\mathcal{H}'] \\ &= \sigma_\nu^2 k - \sigma_\nu^2 \text{ tr } \mathcal{H}(\mathcal{H}'\mathcal{H})^{-1}\mathcal{H}' \\ &= \sigma_\nu^2 k - \sigma_\nu^2 \text{ tr } (\mathcal{H}'\mathcal{H})(\mathcal{H}'\mathcal{H})^{-1} \\ &= \sigma_\nu^2 k - \sigma_\nu^2 \text{ tr } \mathbf{I}_n = \sigma_\nu^2 (k - n) \end{aligned} \tag{8-29}$$

Solving this equation for σ_ν^2, we find that

$$\sigma_\nu^2 = \frac{\mathbf{E}\{\tilde{\mathfrak{Z}}'\tilde{\mathfrak{Z}}\}}{(k - n)} \tag{8-30}$$

Although this is an exact result for σ_ν^2, it is not one that can be evaluated, because we cannot compute $\mathbf{E}\{\tilde{\mathfrak{Z}}'\tilde{\mathfrak{Z}}\}$.

Using the structure of σ_ν^2 as a starting point, we estimate σ_ν^2 by the simple formula

$$\hat{\sigma}_\nu^2 (k) = \tilde{\mathfrak{Z}}'(k)\tilde{\mathfrak{Z}}(k)/(k - n) \tag{8-31}$$

To show that $\hat{\sigma}_\nu^2 (k)$ is an unbiased estimator of σ_ν^2, we observe that

$$\begin{aligned} \mathbf{E}\{\hat{\sigma}_\nu^2 (k)\} &= \mathbf{E}\{\tilde{\mathfrak{Z}}'\tilde{\mathfrak{Z}}\}/(k - n) \\ &= \sigma_\nu^2 \end{aligned} \tag{8-32}$$

where we have used (8-29) for $\mathbf{E}\{\tilde{\mathfrak{Z}}'\tilde{\mathfrak{Z}}\}$. □

LARGE SAMPLE PROPERTIES OF LEAST-SQUARES ESTIMATORS

Many large sample properties of LSE's are determined by establishing that the LSE is equivalent to another estimator for which it is known that the large sample property holds true. In Lesson 11, for example, we will provide conditions under which the LSE of $\boldsymbol{\theta}$, $\hat{\boldsymbol{\theta}}_{\mathrm{LS}}(k)$, is the same as the maximum-likelihood estimator of $\boldsymbol{\theta}$, $\hat{\boldsymbol{\theta}}_{\mathrm{ML}}(k)$. Because $\hat{\boldsymbol{\theta}}_{\mathrm{ML}}(k)$ is consistent, asymptotically efficient, and asymptotically Gaussian, $\hat{\boldsymbol{\theta}}_{\mathrm{LS}}(k)$ inherits all these properties.

Theorem 8-4. *If*

$$\text{plim}\,[\mathcal{H}'(k)\mathcal{H}(k)/k] = \Sigma_{\mathcal{H}} \tag{8-33}$$

$\Sigma_{\mathcal{H}}^{-1}$ *exists, and*

$$\text{plim}\,[\mathcal{H}'(k)\mathcal{V}(k)/k] = \mathbf{0} \tag{8-34}$$

then

$$\text{plim}\,\hat{\boldsymbol{\theta}}_{\mathrm{LS}}(k) = \boldsymbol{\theta} \tag{8-35}$$

Note that the probability limit of a matrix equals a matrix each of whose elements is the probability limit of the respective matrix element. Assumption (8-33) postulates the existence of a probability limit for the second-order moments of the variables in $\mathcal{H}(k)$, as given by $\Sigma_{\mathcal{H}}$. Assumption (8-34) postulates a zero probability limit for the correlation between $\mathcal{H}(k)$ and $\mathcal{V}(k)$. $\mathcal{H}'(k)\mathcal{V}(k)$ can be thought of as a "filtered" version of noise vector $\mathcal{V}(k)$. For (8-34) to be true "filter $\mathcal{H}'(k)$" must be stable. If, for example, $\mathcal{H}(k)$ is deterministic and $\sigma^2_{\mathcal{V}_i(k)} < \infty$, then (8-34) will be true.

Proof. Beginning with (8-2), but for $\hat{\boldsymbol{\theta}}_{\mathrm{LS}}(k)$ instead of $\hat{\boldsymbol{\theta}}_{\mathrm{WLS}}(k)$, we see that

$$\hat{\boldsymbol{\theta}}_{\mathrm{LS}}(k) = \boldsymbol{\theta} + (\mathcal{H}'\mathcal{H})^{-1}\mathcal{H}'\mathcal{V} \tag{8-36}$$

Operating on both sides of this equation with plim, and using properties (7-15), (7-16), and (7-17), we find that

$$\begin{aligned}\text{plim}\,\hat{\boldsymbol{\theta}}_{\mathrm{LS}}(k) &= \boldsymbol{\theta} + \text{plim}\,[(\mathcal{H}'\mathcal{H}/k)^{-1}(\mathcal{H}'\mathcal{V}/k)]\\ &= \boldsymbol{\theta} + \text{plim}\,(\mathcal{H}'\mathcal{H}/k)^{-1}\,\text{plim}\,(\mathcal{H}'\mathcal{V}/k)\\ &= \boldsymbol{\theta} + \Sigma_{\mathcal{H}}^{-1}\cdot\mathbf{0}\\ &= \boldsymbol{\theta}\end{aligned}$$

which demonstrates that, under the given conditions, $\hat{\boldsymbol{\theta}}_{\mathrm{LS}}(k)$ is a consistent estimator of $\boldsymbol{\theta}$. □

In some important applications Eq. (8-34) does not apply, e.g., Example 8-2. Theorem 8-4 then does not apply, and, the study of consistency is often quite complicated in these cases.

Theorem 8-5. *If (8-33) and (8-34) are true,* $\Sigma_{\mathcal{H}}^{-1}$ *exists, and*

$$\text{plim}\,[\mathcal{V}'(k)\mathcal{V}(k)/k] = \sigma_v^2 \tag{8-37}$$

then

$$\text{plim } \hat{\sigma}_v^2(k) = \sigma_v^2 \tag{8-38}$$

where $\hat{\sigma}_v^2(k)$ *is given by (8-24).*

Proof. From (8-26), we find that

$$\tilde{\mathscr{Z}}'\tilde{\mathscr{Z}} = \mathscr{V}'\mathscr{V} - \mathscr{V}'\mathscr{H}(\mathscr{H}'\mathscr{H})^{-1}\mathscr{H}'\mathscr{V} \tag{8-39}$$

Consequently,

$$\begin{aligned}
\text{plim } \hat{\sigma}_v^2(k) &= \text{plim } \tilde{\mathscr{Z}}'\tilde{\mathscr{Z}}/(k-n) \\
&= \text{plim } \mathscr{V}'\mathscr{V}/(k-n) - \text{plim } \mathscr{V}'\mathscr{H}(\mathscr{H}'\mathscr{H})^{-1}\mathscr{H}'\mathscr{V}/(k-n) \\
&= \sigma_v^2 - \text{plim } \mathscr{V}'\mathscr{H}/(k-n) \cdot \text{plim } [\mathscr{H}'\mathscr{H}/(k-n)]^{-1} \\
&\quad \cdot \text{plim } \mathscr{H}'\mathscr{V}/(k-n) \\
&= \sigma_v^2 - \mathbf{0}' \cdot \Sigma_{\mathscr{H}}^{-1} \cdot \mathbf{0} \\
&= \sigma_v^2 \quad \square
\end{aligned}$$

PROBLEMS

8-1. Suppose that $\hat{\theta}_{LS}$ is an unbiased estimator of θ. Is $\hat{\theta}_{LS}^2$ an unbiased estimator of θ^2? (Hint: Use the least-squares batch algorithm to study this question.)

8-2. For $\hat{\boldsymbol{\theta}}_{WLS}(k)$ to be an unbiased estimator of $\boldsymbol{\theta}$ we required $\mathbf{E}\{\mathscr{V}(k)\} = \mathbf{0}$. This problem considers the case when $\mathbf{E}\{\mathscr{V}(k)\} \neq \mathbf{0}$.

(a) Assume that $\mathbf{E}\{\mathscr{V}(k)\} = \mathscr{V}_0$ where $\mathscr{V}_0$ is known to us. How is the concatenated measurement equation $\mathscr{Z}(k) = \mathscr{H}(k)\boldsymbol{\theta} + \mathscr{V}(k)$ modified in this case so we can use the results derived in this lesson to obtain $\hat{\boldsymbol{\theta}}_{WLS}(k)$ or $\hat{\boldsymbol{\theta}}_{LS}(k)$?

(b) Assume that $\mathbf{E}\{\mathscr{V}(k)\} = m_{\mathscr{V}}\mathbf{1}$ where $m_{\mathscr{V}}$ is constant but is unknown. How is the concatenated measurement equation $\mathscr{Z}(k) = \mathscr{H}(k)\boldsymbol{\theta} + \mathscr{V}(k)$ modified in this case so that we can obtain least-squares estimates of both $\boldsymbol{\theta}$ and $m_{\mathscr{V}}$?

8-3. Consider the stable autoregressive model $y(k) = \theta_1 y(k-1) + \cdots + \theta_K y(k-K) + \epsilon(k)$ in which the $\epsilon(k)$ are identically distributed random variables with mean zero and finite variance σ^2. Prove that the least-squares estimates of $\theta_1, \ldots, \theta_K$ are consistent (see also, Ljung, 1976).

8-4. In this lesson we have assumed that the $\mathscr{H}(k)$ variables have been measured without error. Here we examine the situation when $\mathscr{H}_m(k) = \mathscr{H}(k) + \mathscr{N}(k)$ in which $\mathscr{H}(k)$ denotes a matrix of true values and $\mathscr{N}(k)$ a matrix of measurement errors. The basic linear model is now

$$\mathscr{Z}(k) = \mathscr{H}(k)\boldsymbol{\theta} + \mathscr{V}(k) = \mathscr{H}_m(k)\boldsymbol{\theta} + [\mathscr{V}(k) - \mathscr{N}(k)\boldsymbol{\theta}].$$

Prove that $\hat{\boldsymbol{\theta}}_{LS}(k)$ is not a consistent estimator of $\boldsymbol{\theta}$.

Lesson 9

Best Linear Unbiased Estimation

INTRODUCTION

Least-squares estimation, as described in Lessons 3, 4 and 5, is for the linear model

$$\mathscr{Z}(k) = \mathscr{H}(k)\boldsymbol{\theta} + \mathscr{V}(k) \tag{9-1}$$

where $\boldsymbol{\theta}$ is a deterministic, but unknown vector of parameters, $\mathscr{H}(k)$ can be deterministic or random, and we do not know anything about $\mathscr{V}(k)$ ahead of time. By minimizing $\tilde{\mathscr{Z}}'(k)\mathbf{W}(k)\tilde{\mathscr{Z}}(k)$, where $\tilde{\mathscr{Z}}(k) = \mathscr{Z}(k) - \mathscr{H}(k)\hat{\boldsymbol{\theta}}_{\text{WLS}}(k)$, we determined that $\hat{\boldsymbol{\theta}}_{\text{WLS}}(k)$ is a linear transformation of $\mathscr{Z}(k)$, i.e., $\hat{\boldsymbol{\theta}}_{\text{WLS}}(k) = \mathbf{F}_{\text{WLS}}(k)\mathscr{Z}(k)$. After establishing the structure of $\hat{\boldsymbol{\theta}}_{\text{WLS}}(k)$, we studied its small- and large-sample properties. Unfortunately, $\hat{\boldsymbol{\theta}}_{\text{WLS}}(k)$ is not always unbiased or efficient. These properties were not built into $\hat{\boldsymbol{\theta}}_{\text{WLS}}(k)$ during its design.

In this lesson we develop our second estimator. It will be both unbiased and efficient, *by design*. In addition, we want the estimator to be a linear function of the measurements $\mathscr{Z}(k)$. This estimator is called a best linear unbiased estimator (BLUE) or an unbiased minimum-variance estimator (UMVE). To keep notation relatively simple, we will use $\hat{\boldsymbol{\theta}}_{\text{BLU}}(k)$ to denote the BLUE of $\boldsymbol{\theta}$.

As in least-squares, we begin with the linear model in (9-1), where $\boldsymbol{\theta}$ is deterministic. Now, however, *$\mathscr{H}(k)$ must be deterministic and $\mathscr{V}(k)$ is assumed to be zero mean with positive definite known covariance matrix $\mathscr{R}(k)$.* An example of such a covariance matrix occurs for white noise. In the case of scalar measurements, $z(k)$, this means that scalar noise $v(k)$ is white, i.e.,

$$\mathbf{E}\{v(k)v(j)\} = \sigma_i^2(k)\delta_{kj} \tag{9-2}$$

where δ_{kj} is the Kronecker δ (i.e., $\delta_{kj} = 0$ for $k \neq j$ and $\delta_{kj} = 1$ for $k = j$) thus,

$$\mathcal{R}(k) = \mathbf{E}\{\mathcal{V}(k)\mathcal{V}'(k)\} = \text{diag}\,[\sigma_v^2(k), \sigma_v^2(k-1), \ldots, \sigma_v^2(k-N+1)] \tag{9-3}$$

In the case of vector measurements, $\mathbf{z}(k)$, this means that vector noise, $\mathbf{v}(k)$, is white, i.e.,

$$\mathbf{E}\{\mathbf{v}(k)\mathbf{v}'(j)\} = \mathbf{R}(k)\delta_{kj} \tag{9-4}$$

thus,

$$\mathcal{R}(k) = \text{diag}\,[\mathbf{R}(k), \mathbf{R}(k-1), \ldots, \mathbf{R}(k-N+1)] \tag{9-5}$$

PROBLEM STATEMENT AND OBJECTIVE FUNCTION

We begin by assuming the following linear structure for $\hat{\boldsymbol{\theta}}_{\text{BLU}}(k)$,

$$\hat{\boldsymbol{\theta}}_{\text{BLU}}(k) = \mathbf{F}(k)\mathcal{Z}(k) \tag{9-6}$$

where, for notational simplicity, we have omitted subscripting $\mathbf{F}(k)$ as $\mathbf{F}_{\text{BLU}}(k)$. We shall design $\mathbf{F}(k)$ such that

a. $\hat{\boldsymbol{\theta}}_{\text{BLU}}(k)$ is an unbiased estimator of $\boldsymbol{\theta}$, and
b. the error variance for each one of the n parameters is minimized. In this way, $\hat{\boldsymbol{\theta}}_{\text{BLU}}(k)$ *will be unbiased and efficient, by design.*

Recall, from Theorem 6-1, that *unbiasedness constrains design matrix* $\mathbf{F}(k)$, *such that*

$$\mathbf{F}(k)\mathcal{H}(k) = \mathbf{I} \qquad \text{for all } k \tag{9-7}$$

Our objective now is to choose the elements of $\mathbf{F}(k)$, subject to the constraint of (9-7), in such a way that the error variance for each one of the n parameters is minimized.

In solving for $\mathbf{F}_{\text{BLU}}(k)$, it will be convenient to partition matrix $\mathbf{F}(k)$, as

$$\mathbf{F}(k) = \begin{pmatrix} \mathbf{f}_1'(k) \\ \hline \mathbf{f}_2'(k) \\ \hline \vdots \\ \hline \mathbf{f}_n'(k) \end{pmatrix} \tag{9-8}$$

Equation (9-7) can now be expressed in terms of the vector components of $\mathbf{F}(k)$. For our purposes, it is easier to work with the transpose of (9-7), $\mathcal{H}'\mathbf{F}' = \mathbf{I}$, which can be expressed as

$$\mathcal{H}'[\mathbf{f}_1 | \mathbf{f}_2 | \ldots | \mathbf{f}_n] = [\mathbf{e}_1 | \mathbf{e}_2 | \ldots | \mathbf{e}_n] \tag{9-9}$$

where $\mathbf{e}_i$ is the ith unit vector,

$$\mathbf{e}_i = \text{col}\,(0, 0, \ldots, 0, 1, 0, \ldots, 0) \tag{9-10}$$

in which the nonzero element occurs in the ith position. Equating respective elements on both sides of (9-9), we find that

$$\mathcal{H}'(k)\mathbf{f}_i(k) = \mathbf{e}_i \qquad i = 1, 2, \ldots, n \tag{9-11}$$

Our single unbiasedness constraint on matrix $\mathbf{F}(k)$ is now a set of n constraints on the rows of $\mathbf{F}(k)$.

Next, we express $\mathbf{E}\{[\boldsymbol{\theta}_i - \hat{\boldsymbol{\theta}}_{i,\text{BLU}}(k)]^2\}$ in terms of $\mathbf{f}_i$ $(i = 1, 2, \ldots, N)$. We shall make use of (9-11), (9-1), and the following equivalent representation of (9-6)

$$\hat{\theta}_{\text{i,BLU}}(k) = \mathbf{f}_i'(k)\mathcal{Z}(k) \qquad i = 1, 2, \ldots, n \tag{9-12}$$

Proceeding, we find that

$$\begin{aligned}
\mathbf{E}\{[\theta_i - \hat{\theta}_{i,\text{BLU}}(k)]^2\} &= \mathbf{E}\{(\theta_i - \mathbf{f}_i'\mathcal{Z})^2\} = \mathbf{E}\{(\theta_i - \mathcal{Z}'\mathbf{f}_i)^2\} \\
&= \mathbf{E}\{\theta_i^2 - 2\theta_i\mathcal{Z}'\mathbf{f}_i + (\mathcal{Z}'\mathbf{f}_i)^2\} \\
&= \mathbf{E}\{\theta_i^2 - 2\theta_i(\mathcal{H}\boldsymbol{\theta} + \mathcal{V})'\mathbf{f}_i + [(\mathcal{H}\boldsymbol{\theta} + \mathcal{V})'\mathbf{f}_i]^2\} \\
&= \mathbf{E}\{\theta_i^2 - 2\theta_i\boldsymbol{\theta}'\mathcal{H}'\mathbf{f}_i - 2\theta_i\mathcal{V}'\mathbf{f}_i + [\boldsymbol{\theta}'\mathcal{H}'\mathbf{f}_i + \mathcal{V}'\mathbf{f}_i]^2\} \\
&= \mathbf{E}\{\theta_i^2 - 2\theta_i\boldsymbol{\theta}'\mathbf{e}_i - 2\theta_i\mathcal{V}'\mathbf{f}_\text{i} + [\boldsymbol{\theta}'\mathbf{e}_i + \mathcal{V}'\mathbf{f}_i]^2\} \\
&= \mathbf{E}\{\mathbf{f}_i'\mathcal{V}\mathcal{V}'\mathbf{f}_i\} = \mathbf{f}_i'\mathcal{R}\mathbf{f}_i
\end{aligned} \tag{9-13}$$

Observe that the error-variance for the ith parameter depends only on the ith row of design matrix $\mathbf{F}(k)$. We, therefore, establish the following objective function:

$$\begin{aligned}
J_i(\mathbf{f}_i, \boldsymbol{\lambda}_i) &= \mathbf{E}\{[\theta_i - \hat{\theta}_{i,\text{BLU}}(k)]^2\} + \boldsymbol{\lambda}_i'(\mathcal{H}'\mathbf{f}_i - \mathbf{e}_i) \\
&= \mathbf{f}_i'\mathcal{R}\mathbf{f}_i + \boldsymbol{\lambda}_i'(\mathcal{H}'\mathbf{f}_i - \mathbf{e}_i)
\end{aligned} \tag{9-14}$$

where $\boldsymbol{\lambda}_i$ is the ith vector of the Lagrange multipliers, that is associated with the ith unbiasedness constraint. Our objective now is to minimize J_i with respect to $\mathbf{f}_i$ and $\boldsymbol{\lambda}_i$ $(i = 1, 2, \ldots, N)$.

DERIVATION OF ESTIMATOR

A necessary condition for minimizing $J_i(\mathbf{f}_i, \boldsymbol{\lambda}_i)$ is $\partial J_i(\mathbf{f}_i, \boldsymbol{\lambda}_i)/\partial \mathbf{f}_i = \mathbf{0}$ $(i = 1, 2, \ldots, n)$; hence,

$$2\mathcal{R}\mathbf{f}_i + \mathcal{H}\boldsymbol{\lambda}_i = \mathbf{0} \tag{9-15}$$

from which we determine $\mathbf{f}_i$, as

$$\mathbf{f}_i = -\frac{1}{2}\mathcal{R}^{-1}\mathcal{H}\boldsymbol{\lambda}_i \tag{9-16}$$

For (9-16) to be valid, $\mathcal{R}^{-1}$ must exist. Any noise $\mathcal{V}(k)$ whose covariance matrix $\mathcal{R}$ is positive definite qualifies. Of course, if $\mathcal{V}(k)$ is white, then $\mathcal{R}$ is diagonal (or block diagonal) and $\mathcal{R}^{-1}$ exists. This may also be true if $\mathcal{V}(k)$ is not white. A second necessary condition for minimizing J_i $(\mathbf{f}_i, \boldsymbol{\lambda}_i)$ is $\partial J_i (\mathbf{f}_i, \boldsymbol{\lambda}_i)/\partial \boldsymbol{\lambda}_i = \mathbf{0}$ $(i = 1, 2, \ldots, n)$, which gives us the unbiasedness constraints

$$\mathcal{H}'\mathbf{f}_i = \mathbf{e}_i \qquad i = 1, 2, \ldots, n \tag{9-17}$$

To determine $\boldsymbol{\lambda}_i$, substitute (9-16) into (9-17). Doing this, we find that

$$\boldsymbol{\lambda}_i = -2(\mathcal{H}'\mathcal{R}^{-1}\mathcal{H})^{-1}\mathbf{e}_i \tag{9-18}$$

whereupon

$$\mathbf{f}_i = \mathcal{R}^{-1}\mathcal{H}(\mathcal{H}'\mathcal{R}^{-1}\mathcal{H})^{-1}\ \mathbf{e}_i \tag{9-19}$$

$(i = 1, 2, \ldots, n)$. Matrix $\mathbf{F}(k)$ is reconstructed from $\mathbf{f}_i(k)$, as follows:

$$\begin{aligned} \mathbf{F}'(k) &= (\mathbf{f}_1|\mathbf{f}_2|\ldots|\mathbf{f}_n) \\ &= \mathcal{R}^{-1}\mathcal{H}(\mathcal{H}'\mathcal{R}^{-1}\mathcal{H})^{-1}(\mathbf{e}_1|\mathbf{e}_2|\ldots|\mathbf{e}_n) \\ &= \mathcal{R}^{-1}\mathcal{H}(\mathcal{H}'\mathcal{R}^{-1}\mathcal{H})^{-1} \end{aligned} \tag{9-20}$$

Hence

$$\mathbf{F}_{\text{BLU}}(k) = [\mathcal{H}'(k)\mathcal{R}^{-1}(k)\mathcal{H}(k)]^{-1}\mathcal{H}'(k)\mathcal{R}^{-1}(k) \tag{9-21}$$

which means that

$$\boxed{\hat{\boldsymbol{\theta}}_{\text{BLU}}(k) = [\mathcal{H}'(k)\mathcal{R}^{-1}(k)\mathcal{H}(k)]^{-1}\mathcal{H}'(k)\mathcal{R}^{-1}(k)\mathcal{Z}(k)} \tag{9-22}$$

COMPARISON OF $\hat{\theta}_{\text{BLU}}(k)$ AND $\hat{\theta}_{\text{WLS}}(k)$

We are struck by the close similarity between $\hat{\boldsymbol{\theta}}_{\text{BLU}}(k)$ and $\hat{\boldsymbol{\theta}}_{\text{WLS}}(k)$.

Theorem 9-1. *The BLUE of* $\boldsymbol{\theta}$ *is the special case of the WLSE of* $\boldsymbol{\theta}$ *when*

$$\mathbf{W}(k) = \mathcal{R}^{-1}(k) \tag{9-23}$$

If $\mathbf{W}(\mathrm{k})$ *is diagonal, then (9-23) requires* $\mathcal{V}(\mathrm{k})$ *to be white.*

Proof. Compare the formulas for $\hat{\boldsymbol{\theta}}_{\text{BLU}}(k)$ in (9-22) and $\hat{\boldsymbol{\theta}}_{\text{WLS}}(k)$ in (10) of Lesson 3. If $\mathbf{W}(k)$ is a diagonal matrix, then $\mathcal{R}(k)$ is a diagonal matrix only if $\mathcal{V}(k)$ is white. □

Matrix $\mathcal{R}^{-1}(k)$ weights the contributions of precise measurements heavily and deemphasizes the contributions of imprecise measurements. The best linear unbiased estimation design technique has led to a weighting matrix that is quite sensible. See Problem 9-2.

Corollary 9-1. *All results obtained in Lessons 3, 4, and 5 for* $\hat{\boldsymbol{\theta}}_{WLS}(k)$ *can be applied to* $\hat{\boldsymbol{\theta}}_{BLU}(k)$ *by setting* $\mathbf{W}(k) = \mathcal{R}^{-1}(k)$. □

We leave it to the reader to explore the full implications of this important corollary, by reexamining the wide range of topics, which were discussed in Lessons 3, 4 and 5.

Theorem 9-2 (Gauss-Markov Theorem). *If* $\mathcal{R}(k) = \sigma_v^2\,\mathbf{I}$, *then* $\hat{\boldsymbol{\theta}}_{BLU}(k) = \hat{\boldsymbol{\theta}}_{LS}(k)$.

Proof. Using (9-22) and the fact that $\mathcal{R}(k) = \sigma_v^2\,\mathbf{I}$, we find that

$$\hat{\boldsymbol{\theta}}_{BLU}(k) = \left(\frac{1}{\sigma_v^2}\mathcal{H}'\mathcal{H}\right)^{-1}\mathcal{H}'\frac{1}{\sigma_v^2}\mathcal{Z} = (\mathcal{H}'\mathcal{H})^{-1}\mathcal{H}'\mathcal{Z} = \hat{\boldsymbol{\theta}}_{LS}(k) \quad \square$$

Why is this a very important result? We have connected two seemingly different estimators, one of which—$\hat{\boldsymbol{\theta}}_{BLU}(k)$—has the properties of unbiased and minimum variance by design; hence, in this case $\hat{\boldsymbol{\theta}}_{LS}(k)$ inherits these properties. Remember though that the derivation of $\hat{\boldsymbol{\theta}}_{BLU}(k)$ required $\mathcal{H}(k)$ to be deterministic; thus, *Theorem 9-2 is "conditioned" on $\mathcal{H}(k)$ being deterministic.*

SOME PROPERTIES OF $\hat{\theta}_{BLU}(k)$

To begin, we direct our attention at the covariance matrix of parameter estimation error $\tilde{\theta}_{BLU}(k)$.

Theorem 9-3. *If* $\mathcal{V}(k)$ *is zero mean, then*

$$\operatorname{cov}[\tilde{\boldsymbol{\theta}}_{BLU}(k)] = [\mathcal{H}'(k)\mathcal{R}^{-1}(k)\mathcal{H}(k)]^{-1} \tag{9-24}$$

Proof. We apply Corollary 9-1 to $\operatorname{cov}[\tilde{\boldsymbol{\theta}}_{WLS}(k)]$ [given in (18) of Lesson 8] for the case when $\mathcal{H}(k)$ is deterministic, to see that

$$\begin{aligned}\operatorname{cov}[\tilde{\boldsymbol{\theta}}_{BLU}(k)] &= \operatorname{cov}[\tilde{\boldsymbol{\theta}}_{WLS}(k)]\Big|_{\mathbf{W}(k) = \mathcal{R}^{-1}(k)}\\ &= (\mathcal{H}'\mathcal{R}^{-1}\mathcal{H})^{-1}\mathcal{H}'\mathcal{R}^{-1}\mathcal{R}\mathcal{R}^{-1}\mathcal{H}(\mathcal{H}'\mathcal{R}^{-1}\mathcal{H})^{-1}\\ &= (\mathcal{H}'\mathcal{R}^{-1}\mathcal{H})^{-1} \quad \square\end{aligned}$$

Observe the great simplification of the expression for $\operatorname{cov}[\tilde{\boldsymbol{\theta}}_{WLS}(k)]$, when $\mathbf{W}(k) = \mathcal{R}^{-1}(k)$. Note, also, that the error variance of $\hat{\theta}_{i,BLU}(k)$ is given by the ith diagonal element of $\operatorname{cov}[\tilde{\boldsymbol{\theta}}_{BLU}(k)]$.

Corollary 9-2. *When* $\mathbf{W}(k) = \mathcal{R}^{-1}(k)$ *then matrix* $\mathbf{P}(k)$, *which appears in the recursive WLSE of* $\boldsymbol{\theta}$ *equals* $\operatorname{cov}[\tilde{\boldsymbol{\theta}}_{BLU}(k)]$, *i.e.*,

$$\mathbf{P}(k) = \operatorname{cov}[\tilde{\boldsymbol{\theta}}_{BLU}(k)] \tag{9-25}$$

Proof. Recall Equation (13) of Lesson 4, that

$$\mathbf{P}(k) = [\mathcal{H}'(k)\mathbf{W}(k)\mathcal{H}(k)]^{-1} \tag{9-26}$$

When $\mathbf{W}(k) = \mathcal{R}^{-1}(k)$, then

$$\mathbf{P}(k) = [\mathcal{H}'(k)\mathcal{R}^{-1}(k)\mathcal{H}(k)]^{-1} \tag{9-27}$$

hence, $\mathbf{P}(k) = \text{cov}[\tilde{\boldsymbol{\theta}}_{\text{BLU}}(k)]$ because of (9-24) . □

Soon we will examine a recursive BLUE. Matrix $\mathbf{P}(k)$ will have to be calculated, just as it has to be calculated for the recursive WLSE. Every time $\mathbf{P}(k)$ is calculated in our recursive BLUE, we obtain a quantitative measure of how well we are estimating $\boldsymbol{\theta}$. Just look at the diagonal elements of $\mathbf{P}(k)$, $k = 1, 2, \ldots$. The same statement cannot be made for the meaning of $\mathbf{P}(k)$ in the recursive WLSE. In the recursive WLSE, $\mathbf{P}(k)$ has no special meaning.

Next, we examine $\text{cov}[\tilde{\boldsymbol{\theta}}_{\text{BLU}}(k)]$ in more detail.

Theorem 9-4. $\hat{\boldsymbol{\theta}}_{\text{BLU}}(\text{k})$ *is a most efficient estimator of* $\boldsymbol{\theta}$ *within the class of all unbiased estimators that are linearly related to the measurements* $\mathcal{Z}(\text{k})$.

Proof (Mendel, 1973, pp. 155–156). According to Definition 6-3, we must show that

$$\boldsymbol{\Sigma} = \text{cov}[\tilde{\boldsymbol{\theta}}_a(k)] - \text{cov}[\tilde{\boldsymbol{\theta}}_{\text{BLU}}(k)] \tag{9-28}$$

is positive semidefinite. In (9-28), $\tilde{\boldsymbol{\theta}}_a(k)$ is the error associated with an arbitrary linear unbiased estimate of $\boldsymbol{\theta}$. For convenience, we write $\boldsymbol{\Sigma}$ as

$$\boldsymbol{\Sigma} = \boldsymbol{\Sigma}_a - \boldsymbol{\Sigma}_{\text{BLU}} \tag{9-29}$$

In order to compute $\boldsymbol{\Sigma}_a$ we use the facts that

$$\hat{\boldsymbol{\theta}}_a(k) = \mathbf{F}_a(k)\mathcal{Z}(k) \tag{9-30}$$

and

$$\mathbf{F}_a(k)\mathcal{H}(k) = \mathbf{I} \tag{9-31}$$

thus,

$$\begin{aligned} \boldsymbol{\Sigma}_a &= \mathbf{E}\{(\boldsymbol{\theta} - \mathbf{F}_a\mathcal{Z})(\boldsymbol{\theta} - \mathbf{F}_a\mathcal{Z})'\} \\ &= \mathbf{E}\{(\boldsymbol{\theta} - \mathbf{F}_a\mathcal{H}\boldsymbol{\theta} - \mathbf{F}_a\mathcal{V})(\boldsymbol{\theta} - \mathbf{F}_a\mathcal{H}\boldsymbol{\theta} - \mathbf{F}_a\mathcal{V})'\} \\ &= \mathbf{E}\{(\mathbf{F}_a\mathcal{V})(\mathbf{F}_a\mathcal{V})'\} \\ &= \mathbf{F}_a\mathcal{R}\mathbf{F}_a' \end{aligned} \tag{9-32}$$

Because $\hat{\boldsymbol{\theta}}_{\text{BLU}}(k) = \mathbf{F}(k)\mathcal{Z}(k)$ and $\mathbf{F}(k)\mathcal{H}(k) = \mathbf{I}$,

$$\boldsymbol{\Sigma}_{\text{BLU}} = \mathbf{F}\mathcal{R}\mathbf{F}' \tag{9-33}$$

Substituting (9-32) and (9-33) into (9-29), and making repeated use of the unbiasedness constraints $\mathcal{H}'\mathbf{F}' = \mathcal{H}'\mathbf{F}_a' = \mathbf{F}_a\mathcal{H}_a = \mathbf{I}$, we find that

$$\begin{aligned}\mathbf{\Sigma} &= \mathbf{F}_a\mathcal{R}\mathbf{F}_a' - \mathbf{F}\mathcal{R}\mathbf{F}' \\ &= \mathbf{F}_a\mathcal{R}\mathbf{F}_a' - \mathbf{F}\mathcal{R}\mathbf{F}' + 2(\mathcal{H}'\mathcal{R}^{-1}\mathcal{H})^{-1} - (\mathcal{H}'\mathcal{R}^{-1}\mathcal{H})^{-1} \\ &\quad - (\mathcal{H}'\mathcal{R}^{-1}\mathcal{H})^{-1} \\ &= \mathbf{F}_a\mathcal{R}\mathbf{F}_a' - \mathbf{F}\mathcal{R}\mathbf{F}' + 2(\mathcal{H}'\mathcal{R}^{-1}\mathcal{H})^{-1}(\mathcal{H}'\mathcal{R}^{-1}\mathcal{R}\mathbf{F}') \\ &\quad - (\mathcal{H}'\mathcal{R}^{-1}\mathcal{H})^{-1}(\mathcal{H}'\mathcal{R}^{-1}\mathcal{R}\mathbf{F}_a') \\ &\quad - (\mathbf{F}_a\mathcal{R}\mathcal{R}^{-1}\mathcal{H})(\mathcal{H}'\mathcal{R}^{-1}\mathcal{H})^{-1}\end{aligned} \tag{9-34}$$

Making use of the structure of $\mathbf{F}(k)$, given in (9-21), we see that $\mathbf{\Sigma}$ can also be written as

$$\begin{aligned}\mathbf{\Sigma} &= \mathbf{F}_a\mathcal{R}\mathbf{F}_a' - \mathbf{F}\mathcal{R}\mathbf{F}' + 2\mathbf{F}\mathcal{R}\mathbf{F}' - \mathbf{F}\mathcal{R}\mathbf{F}_a' - \mathbf{F}_a\mathcal{R}\mathbf{F}' \\ &= (\mathbf{F}_a - \mathbf{F})\mathcal{R}(\mathbf{F}_a - \mathbf{F})'\end{aligned} \tag{9-35}$$

In order to investigate the definiteness of $\mathbf{\Sigma}$, consider the definiteness of $\mathbf{a}'\mathbf{\Sigma}\mathbf{a}$, where $\mathbf{a}$ is an arbitrary nonzero vector,

$$\mathbf{a}'\mathbf{\Sigma}\mathbf{a} = [(\mathbf{F}_a - \mathbf{F})'\mathbf{a}]'\mathcal{R}[(\mathbf{F}_a - \mathbf{F})'\mathbf{a}] \tag{9-36}$$

Matrix $\mathbf{F}$ (i.e., $\mathbf{F}_{BLU}$) is unique; therefore $(\mathbf{F}_a - \mathbf{F})'\mathbf{a}$ is a nonzero vector, unless $\mathbf{F}_a - \mathbf{F}$ and $\mathbf{a}$ are orthogonal, which is a possibility that cannot be excluded. Because matrix $\mathcal{R}$ is positive definite, $\mathbf{a}'\mathbf{\Sigma}\mathbf{a} \geq 0$, which means that $\mathbf{\Sigma}$ is positive semidefinite. □

These results serve as further confirmation that designing $\mathbf{F}(k)$ as we have done, by minimizing only the diagonal elements of $\text{cov}[\tilde{\theta}_{BLU}(k)]$, is sound.

Corollary 9-3. *If* $\mathcal{R}(\mathrm{k}) = \Sigma_v^2\,\mathbf{I}$, *then* $\hat{\theta}_{LS}(\mathrm{k})$ *is a most efficient estimator of* θ.

The proof of this result is a direct consequence of Theorems 9-2 and 9-4. □

At the end of Lesson 3 we noted that $\hat{\theta}_{WLS}(k)$ may not be invariant under scale changes. We demonstrate next that $\hat{\theta}_{BLU}(k)$ is invariant to such changes.

Theorem 9-5. $\hat{\theta}_{BLU}(\mathrm{k})$ *is invariant under changes of scale.*

Proof (Mendel, 1973, pp. 156–157). Assume that observers A and B are observing a process; but, observer A reads the measurements in one set of units and B in another. Let $\mathbf{M}$ be a symmetric matrix of scale factors relating A to B (e.g., 5,280 ft/mile, 454 g/lb, etc.), and $\mathfrak{Z}_A(k)$ and $\mathfrak{Z}_B(k)$ denote the total measurement vectors of A and B, respectively. Then

$$\mathfrak{Z}_B(k) = \mathcal{H}_B(k)\theta + \mathcal{V}_B(k) = \mathbf{M}\mathfrak{Z}_A(k) = \mathbf{M}\mathcal{H}_A(k)\theta + \mathbf{M}\mathcal{V}_A(k) \tag{9-37}$$

which means that

$$\mathcal{H}_B(k) = \mathbf{M}\mathcal{H}_A(k) \tag{9-38}$$

$$\mathcal{V}_B(k) = \mathbf{M}\mathcal{V}_A(k) \tag{9-39}$$

and

$$\mathcal{R}_B(k) = \mathbf{M}\mathcal{R}_A(k)\mathbf{M}' = \mathbf{M}\mathcal{R}_A(k)\mathbf{M} \tag{9-40}$$

Let $\hat{\boldsymbol{\theta}}_{A,\text{BLU}}(k)$ and $\hat{\boldsymbol{\theta}}_{B,\text{BLU}}(k)$ denote the BLUE's associated with observers A and B, respectively; then,

$$\begin{aligned}\hat{\boldsymbol{\theta}}_{B,\text{BLU}}(k) &= (\mathcal{H}_B'\mathcal{R}_B^{-1}\mathcal{H}_B)^{-1}\mathcal{H}_B'\mathcal{R}_B^{-1}\mathcal{Z}_B \\ &= [\mathcal{H}_A'\mathbf{M}(\mathbf{M}\mathcal{R}_A\mathbf{M})^{-1}\mathbf{M}\mathcal{H}_A]^{-1}\mathcal{H}_A'\mathbf{M}(\mathbf{M}\mathcal{R}_A\mathbf{M})^{-1}\mathbf{M}\mathcal{Z}_A \\ &= (\mathcal{H}_A'\mathbf{M}\mathbf{M}^{-1}\mathcal{R}_A^{-1}\mathbf{M}^{-1}\mathbf{M}\mathcal{H}_A)^{-1}\mathcal{H}_A'\mathbf{M}\mathbf{M}^{-1}\mathcal{R}_A^{-1}\mathbf{M}^{-1}\mathbf{M}\mathcal{Z}_A \\ &= (\mathcal{H}_A'\mathcal{R}_A^{-1}\mathcal{H}_A)^{-1}\mathcal{H}_A'\mathcal{R}_A^{-1}\mathcal{Z}_A \\ &= \hat{\boldsymbol{\theta}}_{A,\text{BLU}}(k) \quad \square\end{aligned} \tag{9-41}$$

RECURSIVE BLUEs

Because of Corollary 9-1, we obtain recursive formulas for the BLUE of $\boldsymbol{\theta}$ by setting $1/w(k+1) = r(k+1)$ in the recursive formulas for the WLSEs of $\boldsymbol{\theta}$ which are given in Lesson 4. In the case of a vector of measurements, we set (see Table 5-1) $\mathbf{w}^{-1}(k+1) = \mathbf{R}(k+1)$.

Theorem 9-6 (Information Form of Recursive BLUE). *A recursive structure for $\hat{\boldsymbol{\theta}}_{\text{BLU}}(\text{k})$ is:*

$$\hat{\boldsymbol{\theta}}_{\text{BLU}}(k+1) = \hat{\boldsymbol{\theta}}_{\text{BLU}}(k) + \mathbf{K}_{\mathbf{B}}(k+1)[z(k+1) - \mathbf{h}'(k+1)\hat{\boldsymbol{\theta}}_{\text{BLU}}(k)] \tag{9-42}$$

where

$$\mathbf{K}_{\mathbf{B}}(k+1) = \mathbf{P}(k+1)\mathbf{h}(k+1)r^{-1}(k+1) \tag{9-43}$$

and

$$\mathbf{P}^{-1}(k+1) = \mathbf{P}^{-1}(k) + \mathbf{h}(k+1)r^{-1}(k+1)\mathbf{h}'(k+1) \tag{9-44}$$

These equations are initialized by $\hat{\boldsymbol{\theta}}_{\text{BLU}}(\text{n})$ and $\mathbf{P}^{-1}(\text{n})$ [where $\mathbf{P}(\text{k})$ is cov $[\tilde{\boldsymbol{\theta}}_{\text{BLU}}(\text{k})]$, *given in (9-31)] and, are used for* $\text{k} = \text{n}, \text{n}+1, \ldots, \text{N}-1$. *These equations can also be used for* $\text{k} = 0, 1, \ldots, \text{N}-1$ *as long as* $\hat{\boldsymbol{\theta}}_{\text{BLU}}(0)$ *and* $\mathbf{P}^{-1}(0)$ *are chosen using Equations (21) and (20) in Lesson 4, respectively, in which* $\text{w}(0)$ *is replaced by* $\text{r}^{-1}(0)$. $\square$

Theorem 9-7 (Covariance Form of Recursive BLUE). *Another recursive structure for $\hat{\boldsymbol{\theta}}_{\text{BLU}}(\text{k})$ is (9-42) in which*

$$\mathbf{K}_{\mathbf{B}}(k+1) = \mathbf{P}(k)\mathbf{h}(k+1)\,[\mathbf{h}'(k+1)\mathbf{P}(k)\mathbf{h}(k+1) + r(k+1)]^{-1} \tag{9-45}$$

and

$$\mathbf{P}(k+1) = [\mathbf{I} - \mathbf{K}_B(k+1)\mathbf{h}'(k+1)]\mathbf{P}(k) \tag{9-46}$$

These equations are initialized by $\hat{\theta}_{BLU}(n)$ *and* $\mathbf{P}(n)$, *and are used for* $k = n, n+1, \ldots, N-1$. *They can also be used for* $k = 0, 1, \ldots, N-1$ *as long as* $\hat{\theta}_{BLU}(0)$ *and* $\mathbf{P}(0)$ *are chosen using Equations (4-21) and (4-20), respectively, in which* $w(0)$ *is replaced by* $r^{-1}(0)$. □

Recall that, in best-linear unbiased estimation, $\mathbf{P}(k) = \text{cov}[\tilde{\theta}_{BLU}(k)]$. Observe, in Theorem 9-7, that we compute $\mathbf{P}(k)$ recursively, and not $\mathbf{P}^{-1}(k)$. This is why the results in Theorem 9-7 (and, subsequently, Theorem 4-2) are referred to as the *covariance form* of recursive BLUE.

PROBLEMS

9-1. (Mendel, 1973, Exercise 3-2, pg. 175). Assume $\mathcal{H}(k)$ is random, and that $\hat{\theta}_{BLU}(k) = \mathbf{F}(k)\mathcal{Z}(k)$.

(a) Show that unbiasedness of the estimate is attained when $\mathbf{E}\{\mathbf{F}(k)\mathcal{H}(k)\} = \mathbf{I}$.

(b) At what point in the derivation of $\hat{\theta}_{BLU}(k)$ do the computations break down because $\mathcal{H}(k)$ is random?

9-2. Here we examine the situation when $\mathcal{V}(k)$ is colored noise and how to use a model to compute $\mathcal{R}(k)$. Now our linear model is

$$\mathbf{z}(k+1) = \mathcal{H}(k+1)\theta + \mathbf{v}(k+1)$$

where $\mathbf{v}(k)$ is colored noise modeled as

$$\mathbf{v}(k+1) = \mathbf{A}_v\mathbf{v}(k) + \xi(k)$$

We assume that deterministic matrix $\mathbf{A}_v$ is known and that $\xi(k)$ is zero-mean white noise with covariance $\mathcal{R}_\xi(k)$. Working with the *measurement difference* $\mathbf{z}^*(k+1) = \mathbf{z}(k+1) - \mathbf{A}_v\mathbf{z}(k)$ write down the formula for $\hat{\theta}_{BLU}(k)$ in batch form. Be sure to define all concatenated quantities.

9-3. (Sorenson, 1980, Exercise 3-15, pg. 130). Suppose $\hat{\theta}_1$ and $\hat{\theta}_2$ are unbiased estimators of θ with var $(\hat{\theta}_1) = \sigma_1^2$ and var $(\hat{\theta}_2) = \sigma_2^2$. Let $\hat{\theta}_3 = \alpha\hat{\theta}_1 + (1-\alpha)\hat{\theta}_2$.

(a) Prove that $\hat{\theta}_1$ is unbiased.

(b) Assume that $\hat{\theta}_1$ and $\hat{\theta}_2$ are statistically independent, and find the mean-squared error of $\hat{\theta}_3$.

(c) What choice of α minimizes the mean-squared error?

9-4. (Mendel, 1973, Exercise 3-12, pp. 176–177). A series of measurements $\mathbf{z}(k)$ are made, where $\mathbf{z}(k) = \mathbf{H}\theta + \mathbf{v}(k)$, $\mathbf{H}$ is an $m \times n$ *constant* matrix, $\mathbf{E}\{\mathbf{v}(k)\} = \mathbf{0}$, and $\text{cov}[\mathbf{v}(k)] = \mathcal{R}$ is a constant matrix.

(a) Using the two formulations of the recursive BLUE show that (Ho, 1963, pp. 152–154):

(i) $\mathbf{P}(k+1)\mathbf{H}' = \mathbf{P}(k)\mathbf{H}'[\mathbf{H}\mathbf{P}(k)\mathbf{H}' + \mathbf{R}]^{-1}\mathbf{R}$, and

(ii) $\mathbf{H}\mathbf{P}(k) = \mathbf{R}[\mathbf{H}\mathbf{P}(k-1)\mathbf{H}' + \mathbf{R}]^{-1}\mathbf{H}\mathbf{P}(k-1)$.

(b) Next, show that

(i) $\mathbf{P}(k)\mathbf{H}' = \mathbf{P}(k-2)\mathbf{H}'[2\mathbf{H}\mathbf{P}(k-2)\mathbf{H}' + \mathbf{R}]^{-1}\mathbf{R}$;

(ii) $\mathbf{P}(k)\mathbf{H}' = \mathbf{P}(k-3)\mathbf{H}'[3\mathbf{H}\mathbf{P}(k-3)\mathbf{H}' + \mathbf{R}]^{-1}\mathbf{R}$; and

(iii) $\mathbf{P}(k)\mathbf{H}' = \mathbf{P}(0)\mathbf{H}'[k\mathbf{H}\mathbf{P}(0)\mathbf{H}' + \mathbf{R}]^{-1}\mathbf{R}$.

(c) Finally, show that the asymptotic form ($k \rightarrow \infty$) for the BLUE of $\boldsymbol{\theta}$ is (Ho, 1963, pp. 152–154)

$$\hat{\boldsymbol{\theta}}_{\text{BLU}}(k+1) = \hat{\boldsymbol{\theta}}_{\text{BLU}}(k) + \frac{1}{(k+1)}\mathbf{P}(0)\mathbf{H}'[\mathbf{H}\mathbf{P}(0)\mathbf{H}']^{-1}[\mathbf{z}(k+1) - \mathbf{H}(k+1)\hat{\boldsymbol{\theta}}_{\text{BLU}}(k)]$$

This equation, with its $1/(k+1)$ weighting function, represents a form of multidimensional stochastic approximation.

Lesson 10

Likelihood

INTRODUCTION

This lesson provides background material for the method of maximum-likelihood. It explains the relationship of *likelihood* to *probability*, and when the terms *likelihood* and *likelihood ratio* can be used interchangeably. The major reference for this lesson is Edwards (1972), a most delightful book.

LIKELIHOOD DEFINED

To begin, we define what is meant by an hypothesis, H, and results (of an experiment), R. Suppose scalar parameter θ can assume only two values, 0 or 1; then, we say that there are two hypotheses associated with θ, namely H_0 and H_1, where, for H_0, $\theta = 0$, and for H_1, $\theta = 1$. This is the situation of a *binary hypothesis*. Suppose next that scalar parameter θ can assume ten values, a, b, c, d, e, f, g, h, i, j; then, we say there are ten hypotheses associated with θ, namely $H_1, H_2, H_3, \ldots, H_{10}$, where, for H_1, $\theta = a$, for H_2, $\theta = b, \ldots$, and for H_{10}, $\theta = j$. Parameter θ may also assume values from an interval, i.e., $a \leq \theta \leq b$. In this case, we have an infinite, uncountable number of hypotheses about θ, each one associated with a real number in the interval $[a, b]$. Finally, we may have a vector of parameters each one of whose elements has a collection of hypotheses associated with it. For example, suppose that each one of the n elements of $\boldsymbol{\theta}$ is either 0 or 1. Vector $\boldsymbol{\theta}$ is then characterized by 2^n hypotheses.

Results, R, are the outputs of an experiment. In our work on parameter estimation for the linear model $\mathcal{Z}(k) = \mathcal{H}(k)\boldsymbol{\theta} + \mathcal{V}(k)$, the "results" are the data in $\mathcal{Z}(k)$ and $\mathcal{H}(k)$.

We let $P(R|H)$ denote the probability of obtaining results R given hypothesis H according to some probability model, e.g., $p[z(k)|\boldsymbol{\theta}]$. In probability, $P(R|H)$ is always viewed as a function of R for *fixed* values of H. Usually the explicit dependence of P on H is not shown. In order to understand the differences between probability and likelihood, it is important to show the explicit dependence of P on H.

Example 10-1

Random number generators are often used to generate a sequence of random numbers that can then be used as the input sequence to a dynamical system, or as an additive measurement noise sequence. To run a random number generator, you must choose a probability model. The Gaussian model is often used; however, it is characterized by two parameters, mean μ and variance σ^2. In order to obtain a stream of Gaussian random numbers from the random number generator, you must fix μ and σ^2. Let μ_T and σ_T^2 denote (true) values chosen for μ and σ^2. The Gaussian probability density function for the generator is $p[z(k)|\mu_T, \sigma_T^2]$, and the numbers we obtain at its output, $z(1), z(2), \ldots$, are of course quite dependent on the hypothesis $H_T = (\mu_T, \sigma_T^2)$. □

For fixed H we can apply the axioms of probability (e.g., see Papoulis, 1965). If, for example, results R_1 and R_2 are mutually exclusive, then $P(R_1 \text{ or } R_2|H) = P(R_1|H) + P(R_2|H)$.

Definition 10-1 (Edwards, 1972, pg. 9). *Likelihood*, L(H|R), *of the hypothesis* H *given the results* R *and a specific probability model is proportional to* P(R|H), *the constant of proportionality being arbitrary, i.e.,*

$$L(H|R) = cP(R|H) \quad \square \tag{10-1}$$

For likelihood R is fixed (i.e., given ahead of time) and H is variable. *There are no axioms of likelihood.* Likelihood cannot be compared using different data sets (i.e., different results, say, R_1 and R_2) unless the data sets are statistically independent.

Example 10-2

Suppose we are given a sequence of Gaussian random numbers, using the random number generator that was described in Example 10-1, but, we do not know μ_T and σ_T^2. Is it possible to infer (i.e., estimate) what the values of μ and σ^2 were that most likely generated the given sequence? The method of maximum-likelihood, which we study in Lesson 11, will show us how to do this. The starting point for the estimation of μ and σ^2 will be $p[z(k)|\mu, \sigma^2]$, where now $z(k)$ is fixed and μ and σ^2 are treated as variables. □

Example 10-3 (Edwards, 1972, pg. 10)

To further illustrate the difference between $P(R|H)$ and $L(H|R)$, we consider the following binomial model which we assume describes the occurrence of boys and girls in a family of two children:

$$P(R|p) = \frac{(m+f)!}{m!\,f!} p^m (1-p)^f \tag{10-2}$$

where p denotes the probability of a male child, m equals the number of male children, f equals the number of female children, and, in this example,

$$m + f = 2 \tag{10-3}$$

Our objective is to determine p; but, to do this we need some results. Knocking on neighbor's doors and conducting a simple survey, we establish two data sets:

$$R_1 = \{1 \text{ boy and } 1 \text{ girl}\} \Rightarrow m = 1 \quad \text{and} \quad f = 1 \tag{10-4}$$

$$R_2 = \{2 \text{ boys}\} \Rightarrow m = 2 \quad \text{and} \quad f = 0 \tag{10-5}$$

In order to keep the determination of p simple for this meager collection of data, we shall only consider two values for p, i.e., two hypotheses,

$$\left.\begin{aligned} H_1 : p = 1/4 \\ H_2 : p = 1/2 \end{aligned}\right\} \tag{10-6}$$

To begin, we create a *table of probabilities*, in which the entries are $P(R_i|H_j$ fixed), where this is computed using (10-2). For H_1 (i.e., $p = 1/4$), $P(R_1|1/4) = 3/8$ and $P(R_2|1/4) = 1/16$; for H_2 (i.e., $p = 1/2$), $P(R_1|1/2) = 1/2$ and $P(R_2|1/2) = 1/4$. These results are collected together in Table 10-1.

TABLE 10-1 $P(R_i|H_j$ **fixed)**

	R_1	R_2
H_1	3/8	1/16
H_2	1/2	1/4

Next, we create a *table of likelihoods*, using (10-1). In this table (Table 10-2) the entries are $L(H_i|R_j$ fixed). Constants c_1 and c_2 are arbitrary and c_1, for example, appears in each one of the table entries in the R_1 column.

TABLE 10-2 $L(H_i|R_j$ **fixed)**

	R_1	R_2
H_1	$3/8\ c_1$	$1/16\ c_2$
H_2	$1/2\ c_1$	$1/4\ c_2$

What can we conclude from the table of likelihoods? First, for data R_1, the likelihood of H_1 is 3/4 the likelihood of H_2. The number 3/4 was obtained by taking the ratio of likelihoods $L(H_1|R_1)$ and $L(H_2|R_1)$. Second, on data R_2, the likelihood of H_1 is 1/4 the likelihood of H_2 [note that $1/4 = 1/16\, c_2/1/4\, c_2$]. Finally, we conclude that, even from our two meager results, the value $p = 1/2$ appears to be more likely than the value $p = 1/4$, which, of course, agrees with our intuition. □

LIKELIHOOD RATIO

In the preceding example we were able to draw conclusions about the likelihood of one hypothesis versus a second hypothesis by comparing ratios of likelihood, defined, of course, on the same set of data. Forming the ratios of likelihood, we obtain the *likelihood ratio.*

Definition 10-2 (Edwards, 1972, pg. 10). *The likelihood ratio of two hypotheses on the same data is the ratio of the likelihoods on the data. Let* $\mathrm{L}(\mathrm{H}_1, \mathrm{H}_2|\mathrm{R})$ *denote likelihood ratio; then,*

$$L(H_1, H_2|R) = \frac{L(H_1|R)}{L(H_2|\mathrm{R})} = \frac{P(R|H_1)}{P(R|H_2)} \quad \square \tag{10-7}$$

Observe that likelihood ratio statements do not depend on the arbitrary constant c which appears in the definition of likelihood, because c cancels out of the ratio $cP(R|H_1)/cP(R|H_2)$.

Theorem 10-1 (Edwards, 1972, pg. 11). *Likelihood ratios of two hypotheses on statistically independent sets of data may be multiplied together to form the likelihood ratio of the combined data.*

Proof. Let $L(H_1, H_2|R_1 \;\&\; R_2)$ denote the likelihood ratio of H_1 and H_2 on the combined data $R_1 \;\&\; R_2$, i.e.,

$$L(H_1, H_2|R_1 \;\&\; R_2) = \frac{L(H_1|R_1 \;\&\; R_2)}{L(H_2|R_1 \;\&\; R_2)} = \frac{P(R_1 \;\&\; R_2|H_1)}{P(R_1 \;\&\; R_2|H_2)} \tag{10-8}$$

Because R_1 & R_2 are statistically independent data, $P(R_1 \;\&\; R_2|H_i) = P(R_1|H_i)P(R_2|H_i)$; hence,

$$\begin{aligned} L(H_1, H_2 \,|\, R_1 \;\&\; R_2) &= \frac{P(R_1|H_1)}{P(R_1|H_2)} \times \frac{P(R_2|H_1)}{P(R_2|H_2)} \\ &= L(H_1, H_2|R_1) \times L(H_1, H_2|R_2) \quad \square \end{aligned} \tag{10-9}$$

Example 10-4

We can now state the conclusions that are given at the end of Example 10-3 more formally. From Table 10-2, we see that $L(1/4,1/2|R_1) = 3/4$ and $L(1/4,1/2|R_2) = 1/4$. Additionally, because R_1 and R_2 are data from independent experiments, $L(1/4, 1/2|$

R_1 & R_2) = 3/4 × 1/4 = 3/16. This reinforces our intuition that $p = 1/2$ is much more likely than $p = 1/4$. □

RESULTS DESCRIBED BY CONTINUOUS DISTRIBUTIONS

Suppose the results R have a continuous distribution; then, we know from probability theory that the probabilty of obtaining a result that lies in the interval $(R, R + dR)$ is $P(R|H)dR$, as $dR \to 0$. $P(R|H)$ is then a probability density. In this case, $L(H|R) = cP(R|H)dR$; but, cdR can be defined as a new arbitrary constant, c_1, so that $L(H|R) = c_1P(R|H)$. In likelihood ratio statements $c_1 = cdR$ disappears entirely; thus, likelihood and likelihood ratio are unaffected by the nature of the distribution of R.

Recall, that a transformation of variables greatly affects probability because of the dR which appears in the probability formula. *Likelihood and likelihood ratio, on the other hand, are unaffected by transformations of variables*, because of the absorption of dR into c_1.

MULTIPLE HYPOTHESES

Thus far, all of our attention has been directed at the case of two hypotheses. In order to apply likelihood and likelihood ratio concepts to parameter estimation problems, where parameters take on more than two values, we must extend our preceding results to the case of multiple hypotheses. As stated by Edwards (1972, pg. 11), "Instead of forming all the pairwise likelihood ratios it is simpler to present the same information in terms of the likelihood ratios for the several hypotheses versus one of their number, which may be chosen quite arbitrarily for this purpose."

Our extensions rely very heavily on the results for the two hypotheses case, because of the convenient introduction of an arbitrary comparison hypothesis, H^*. Let H_i denote the ith hypothesis; then,

$$L(H_i, H^*|R) = \frac{L(H_i|R)}{L(H^*|R)} \tag{10-10}$$

Observe, also, that

$$\frac{L(H_i, H^*|R)}{L(H_j, H^*|R)} = \frac{L(H_i|R)}{L(H_j|R)} = L(H_i, H_j|R) \tag{10-11}$$

which means that we can compute the likelihood-ratio between *any* two hypotheses H_i and H_j if we can compute the likelihood-ratio function $L(H_k, H^*|R)$.

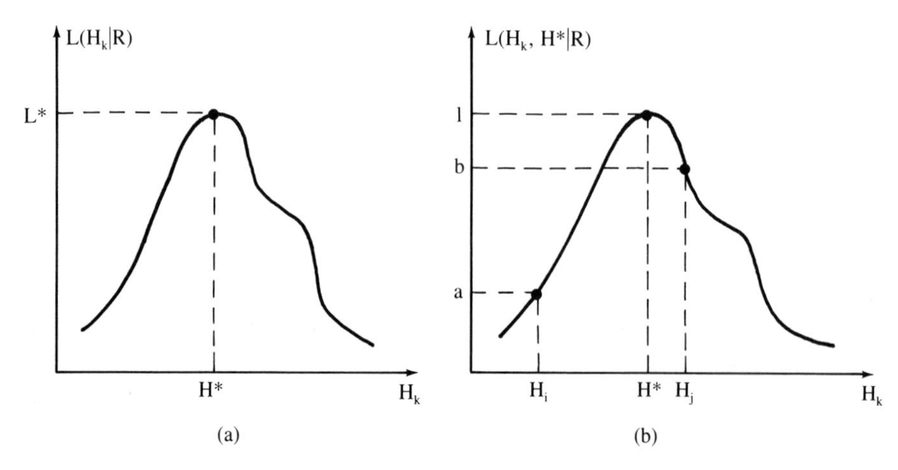

Figure 10-1 Multiple hypotheses case: (a) likelihood $L(H_k|R)$ versus H_k, and (b) likelihood ratio $L(H_k,H^*|R)$ versus H_k. Comparison hypothesis H^* has been chosen to be the hypothesis associated with the maximum value of $L(H_k|R)$, L^*.

Figure 10-1(a) depicts a likelihood function $L(H_k|R)$. Any value of H_k can be chosen as the comparison hypothesis. We choose H^* as the hypothesis associated with the maximum value of $L(H_k|R)$, so that the maximum value of $L(H_k,H^*|R)$ will be normalized to unity. The likelihood ratio function $L(H_k,H^*|R)$, depicted in Figure 10-1(b), was obtained from Figure 10-1(a) and (10-10). In order to compute $L(H_i,H_j|R)$, we determine from Figure 10-1(b), that $L(H_i,H^*|R) = a$ and $L(H_j,H^*|R) = b$, so that $L(H_i,H_j|R) = a/b$.

Is it really necessary to know H^* in order to carry out the normalization of $L(H_k,H^*|R)$ depicted in Figure 10-1(b)? No, because H^* can be eliminated by a clever "conceptual" choice of constant c. We can choose c, such that

$$L(H^*|R) = L^* \triangleq 1 \tag{10-12}$$

According to (10-1), this means that

$$c = 1/P(R|H^*) \tag{10-13}$$

If c is "chosen" in this manner, then

$$L(H_k,H^*|R) = \frac{L(H_k|R)}{L(H^*|R)} = L(H_k|R) \tag{10-14}$$

which means that, *in the case of multiple hypotheses, likelihood and likelihood ratio can be used interchangeably*. This helps to explain why authors use different names for the function that is the starting point in the method of maximum likelihood, including *likelihood function* and *likelihood-ratio function*.

PROBLEMS

10-1. A test is said to be a *likelihood-ratio test* if there is a number c such that this test leads to (d_i refers to the ith decision)

$$\begin{aligned} d_1 &\quad \text{if } L(H_1,H_2|R) > c \\ d_2 &\quad \text{if } L(H_1,H_2|R) < c \\ d_1 \text{ or } d_2 &\quad \text{if } L(H_1,H_2|R) = c \end{aligned}$$

Consider a sequence of n tosses of a coin of which m are heads, i.e., $P(R|p) = p^m(1-p)^{n-m}$. Let $H_1 = p_1$ and $H_2 = p_2$ where $p_1 > p_2$ and p represents the probability of a head. Show that $L(H_1,H_2|R)$ increases as m increases. Then show that $L(H_1,H_2|R)$ increases when $\hat{p} = m/n$ increases, and that the likelihood-ratio test which consists of accepting H_1 if $L(H_1,H_2|R)$ is larger than some constant is equivalent to accepting H_1 if $\hat{p}$ is larger than some related constant.

10-2. Consider n independent observations $x_1, x_2, \ldots, x_n$ which are normally distributed with mean μ and known standard deviation σ. Our two hypotheses are: $H_1: \mu = \mu_1$ and $H_2: \mu = \mu_2$. Using the sample mean as an estimator of μ, show that the likelihood-ratio test (defined in Problem 10-1) consists of accepting H_1 if the sample mean exceeds some constant.

10-3. Suppose that the data consists of a single observation z with Cauchy density given by

$$p(z|\theta) = \frac{1}{\pi[1 + (z - \theta)^2]}$$

Test (see Problem 10-1) $H_1: \theta = \theta_1 = 1$ versus $H_2: \theta = \theta_2 = -1$ when $c = 1/2$, i.e., show that for $c = 1/2$, we accept H_1 when $z > -0.35$ *or* when $z < -5.65$.

Lesson 11

Maximum-Likelihood Estimation

LIKELIHOOD*

Let us consider a vector of unknown parameters $\boldsymbol{\theta}$ that describes a collection of N independent identically distributed observations $z(k)$, $k = 1, 2, \ldots, N$. We collect these measurements into an $N \times 1$ vector $\mathfrak{Z}(N)$, ($\mathfrak{Z}$ for short),

$$\mathfrak{Z} = \text{col}\,(z(1), z(2), \ldots, z(N) \tag{11-1}$$

The likelihood of $\boldsymbol{\theta}$, given the observations $\mathfrak{Z}$, is defined to be proportional to the value of the probability density function of the observations given the parameters

$$l(\boldsymbol{\theta}|\mathfrak{Z}) \propto p(\mathfrak{Z}|\boldsymbol{\theta}) \tag{11-2}$$

where l is the likelihood function and p is the conditional joint probability density function. Because $z(i)$ are independent and identically distributed,

$$l(\boldsymbol{\theta}|\mathfrak{Z}) \propto p(z(1)|\boldsymbol{\theta})p(z(2)|\boldsymbol{\theta}) \cdots p(z(N)|\boldsymbol{\theta}) \tag{11-3}$$

In many applications $p(\mathfrak{Z}|\boldsymbol{\theta})$ is exponential (e.g., Gaussian). It is easier then to work with the natural logarithm of $l(\boldsymbol{\theta}|\mathfrak{Z})$ than with $l(\boldsymbol{\theta}|\mathfrak{Z})$. Let

$$\text{L}(\boldsymbol{\theta}|\mathfrak{Z}) = \ln l(\boldsymbol{\theta}|\mathfrak{Z}) \tag{11-4}$$

Quantity L is sometimes referred to as the log-likelihood function, the support function (Kmenta, 1971), the likelihood function [Mehra (1971), Schweppe

* The material in this section is taken from Mendel (1983b, pp. 94–95).

(1965), and Stepner and Mehra (1973), for example], or the conditional likelihood function (Nahi, 1969). We shall use these terms interchangeably.

MAXIMUM-LIKELIHOOD METHOD AND ESTIMATES*

The *maximum-likelihood method* is based on the relatively simple idea that different populations generate different samples and that any given sample is more likely to have come from some populations than from others. The maximum-likelihood estimate (MLE) $\hat{\boldsymbol{\theta}}_{\text{ML}}$ is the value of $\boldsymbol{\theta}$ that maximizes l or L for a particular set of measurements $\mathfrak{X}$. The logarithm of l is a monotonic transformation of l (i.e., whenever l is decreasing or increasing, $\ln l$ is also decreasing or increasing); therefore, the point corresponding to the maximum of l is also the point corresponding to the maximum of $\ln l = \text{L}$.

Obtaining an MLE involves specifying the likelihood function and finding those values of the parameters that give this function its maximum value. It is required that, if L is differentiable, the partial derivative of L (or l) with respect to each of the unknown parameters $\theta_1, \theta_2, \ldots, \theta_n$ equal zero:

$$\left.\frac{\partial \text{L}(\boldsymbol{\theta}|\mathfrak{X})}{\partial \theta_i}\right|_{\boldsymbol{\theta} = \hat{\boldsymbol{\theta}}_{\text{ML}}} = 0 \qquad \text{for all } i = 1, 2, \ldots, n \tag{11-5}$$

To be sure that the solution of (11-5) gives, in fact, a maximum value of $\text{L}(\boldsymbol{\theta}|\mathfrak{X})$, certain second-order conditions must be fulfilled. Consider a Taylor series expansion of $\text{L}(\boldsymbol{\theta}|\mathfrak{X})$ about $\hat{\boldsymbol{\theta}}_{\text{ML}}$, i.e.,

$$\begin{aligned}\text{L}(\boldsymbol{\theta}|\mathfrak{X}) &= \text{L}(\hat{\boldsymbol{\theta}}_{\text{ML}}|\mathfrak{X}) + \sum_{i=1}^{n} \frac{\partial \text{L}(\hat{\boldsymbol{\theta}}_{\text{ML}}|\mathfrak{X})}{\partial \theta_i}(\theta_i - \hat{\theta}_{i,\text{ML}}) \\ &\quad + \frac{1}{2}\sum_{i=1}^{n}\sum_{j=1}^{n} \frac{\partial^2 \text{L}(\hat{\boldsymbol{\theta}}_{\text{ML}}|\mathfrak{X})}{\partial \theta_i \partial \theta_j}(\theta_i - \hat{\theta}_{i,\text{ML}})(\theta_j - \hat{\theta}_{j,\text{ML}}) + \cdots \end{aligned} \tag{11-6}$$

where $\partial \text{L}(\hat{\boldsymbol{\theta}}_{\text{ML}}|\mathfrak{X})/\partial\theta_i$, for example, is short for $[\partial \text{L}(\boldsymbol{\theta}|\mathfrak{X})/\partial\theta_i]|_{\boldsymbol{\theta} = \hat{\boldsymbol{\theta}}_{\text{ML}}}$. Because $\hat{\boldsymbol{\theta}}_{\text{ML}}$ is the MLE of $\boldsymbol{\theta}$, the second term on the right-hand side of (11-6) is zero [by virtue of (11-5)]; hence,

$$\begin{aligned}\text{L}(\boldsymbol{\theta}|\mathfrak{X}) &= \text{L}(\hat{\boldsymbol{\theta}}_{\text{ML}}|\mathfrak{X}) \\ &\quad + \frac{1}{2}\sum_{i=1}^{n}\sum_{j=1}^{n} \frac{\partial^2 \text{L}(\hat{\boldsymbol{\theta}}_{\text{ML}}|\mathfrak{X})}{\partial \theta_i \partial \theta_j}(\theta_i - \hat{\theta}_{i,\text{ML}})(\theta_j - \hat{\theta}_{j,\text{ML}}) + \cdots \end{aligned} \tag{11-7}$$

Recognizing that the second term in (11-7) is a quadratic form, we can write it in a more compact notation as $1/2(\boldsymbol{\theta} - \hat{\boldsymbol{\theta}}_{\text{ML}})'\mathbf{J}_o(\hat{\boldsymbol{\theta}}_{\text{ML}}|\mathfrak{X})(\boldsymbol{\theta} - \hat{\boldsymbol{\theta}}_{\text{ML}})$, where $\mathbf{J}_o(\hat{\boldsymbol{\theta}}_{\text{ML}}|\mathfrak{X})$, the *observed Fisher information matrix* [see Equation (6-46)], is

$$\mathbf{J}_o(\hat{\boldsymbol{\theta}}_{\text{ML}}|\mathfrak{X}) = \left.\left(\frac{\partial^2 L(\boldsymbol{\theta}|\mathfrak{X})}{\partial\theta_i\partial\theta_j}\right)\right|_{\boldsymbol{\theta} = \hat{\boldsymbol{\theta}}_{\text{ML}}} \qquad i, j = 1, 2, \ldots, n \tag{11-8}$$

* The material in this section is taken from Mendel (1983b, pp. 95–98).

hence,

$$L(\boldsymbol{\theta}|\mathfrak{Z}) = L(\hat{\boldsymbol{\theta}}_{ML}|\mathfrak{Z}) + \frac{1}{2}(\boldsymbol{\theta} - \hat{\boldsymbol{\theta}}_{ML})'\mathbf{J}_o(\hat{\boldsymbol{\theta}}_{ML}|\mathfrak{Z})(\boldsymbol{\theta} - \hat{\boldsymbol{\theta}}_{ML}) + \cdots \tag{11-9}$$

Now let us examine sufficient conditions for the likelihood function to be maximum. We assume that, close to $\boldsymbol{\theta}$, $L(\boldsymbol{\theta}|\mathfrak{Z})$ is approximately quadratic, in which case

$$L(\boldsymbol{\theta}|\mathfrak{Z}) \approx L(\hat{\boldsymbol{\theta}}_{ML}|\mathfrak{Z}) + \frac{1}{2}(\boldsymbol{\theta} - \hat{\boldsymbol{\theta}}_{ML})'\mathbf{J}_o(\hat{\boldsymbol{\theta}}_{ML}|\mathfrak{Z})(\boldsymbol{\theta} - \hat{\boldsymbol{\theta}}_{ML}) \tag{11-10}$$

From vector calculus, it is well known that a sufficient condition for a function of n variables to be maximum is that the matrix of second partial derivatives of that function, evaluated at the extremum, must be negative definite. For $L(\boldsymbol{\theta}|\mathfrak{Z})$ in (11-10), this means that *a sufficient condition for* $L(\boldsymbol{\theta}|\mathfrak{Z})$ *to be maximized is*

$$\mathbf{J}_o(\hat{\boldsymbol{\theta}}_{ML}|\mathfrak{Z}) < 0 \tag{11-11}$$

Example 11-1

This is a continuation of Example 10-2. We observe a random sample $\{z(1), z(2), \ldots, z(N)\}$ at the output of a Gaussian random number generator and wish to find the maximum-likelihood estimators of μ and σ^2.

The Gaussian density function $p(z|\mu,\sigma^2)$ is

$$p(z|\mu,\sigma^2) = (2\pi\sigma^2)^{-1/2}\exp\left\{-\frac{1}{2}[(z-\mu)/\sigma]^2\right\} \tag{11-12}$$

Its natural logarithm is

$$\ln p(z|\mu,\sigma^2) = -\frac{1}{2}\ln(2\pi\sigma^2) - \frac{1}{2}[(z-\mu)/\sigma]^2 \tag{11-13}$$

The likelihood function is

$$l(\mu,\sigma^2) = p(z(1)|\mu,\sigma^2)p(z(2)|\mu,\sigma^2)\cdots p(z(N)|\mu,\sigma^2) \tag{11-14}$$

and its logarithm is

$$L(\mu,\sigma^2) = \sum_{i=1}^{N} \ln p(z(i)|\mu,\sigma^2) \tag{11-15}$$

Substituting for $\ln p(z(i)|\mu,\sigma^2)$ gives

$$\begin{aligned} L(\mu,\sigma^2) &= \sum_{i=1}^{N}\left[-\frac{1}{2}\ln(2\pi\sigma^2) - \frac{1}{2}\left(\frac{z(i)-\mu}{\sigma}\right)^2\right] \\ &= -\frac{N}{2}\ln(2\pi\sigma^2) - \frac{1}{2\sigma^2}\sum_{i=1}^{N}[z(i)-\mu]^2 \end{aligned} \tag{11-16}$$

There are two unknown parameters in L, μ and σ^2. Differentiating L with respect to each of them gives

$$\frac{\partial L}{\partial \mu} = \frac{1}{\sigma^2} \sum_{i=1}^{N} [z(i) - \mu] \tag{11-17a}$$

$$\frac{\partial L}{\partial (\sigma^2)} = -\frac{N}{2}\frac{1}{\sigma^2} + \frac{1}{2\sigma^4} \sum_{i=1}^{N} [z(i) - \mu]^2 \tag{11-17b}$$

Equating these partial derivatives to zero, we obtain

$$\frac{1}{\hat{\sigma}^2_{ML}} \sum_{i=1}^{N} [z(i) - \hat{\mu}_{ML}] = 0 \tag{11-18a}$$

$$-\frac{N}{2}\frac{1}{\hat{\sigma}^2_{ML}} + \frac{1}{2\hat{\sigma}^4_{ML}} \sum_{i=1}^{N} [z(i) - \hat{\mu}_{ML}]^2 = 0 \tag{11-18b}$$

For $\hat{\sigma}^2_{ML}$ different from zero (11-18a) reduces to

$$\sum_{i=1}^{N} [z(i) - \hat{\mu}_{ML}] = 0$$

giving

$$\hat{\mu}_{ML} = \frac{1}{N} \sum_{i=1}^{N} z(i) = \bar{z} \tag{11-19}$$

Thus *the MLE of the mean of a Gaussian population is equal to the sample mean* $\bar{z}$. Once again we see that the sample mean is an optimal estimator.

Observe that (11-18a) and (11-18b) can be solved for $\hat{\sigma}^2_{ML}$, the MLE of σ^2. Multiplying Eq. (11-18b) by $2\hat{\sigma}^4_{ML}$ leads to

$$-N\hat{\sigma}^2_{ML} + \sum_{i=1}^{N} [z(i) - \hat{\mu}_{ML}]^2 = 0$$

Substituting $\bar{z}$ for $\hat{\mu}_{ML}$ and solving for $\hat{\sigma}^2_{ML}$ gives

$$\hat{\sigma}^2_{ML} = \frac{1}{N} \sum_{i=1}^{N} [z(i) - \bar{z}]^2 \tag{11-20}$$

Thus, *the MLE of the variance of a Gaussian population is simply equal to the sample variance.* □

PROPERTIES OF MAXIMUM-LIKELIHOOD ESTIMATES

The importance of maximum-likelihood estimation is that it produces estimates that have very desirable properties.

Theorem 11-1. *Maximum-likelihood estimates are:* (1) *consistent,* (2) *asymptotically Gaussian with mean* $\boldsymbol{\theta}$ *and covariance matrix* $\frac{1}{N}\mathbf{J}^{-1}$, *in which* $\mathbf{J}$

is the Fisher Information Matrix [*Equation* (6-45)], *and*, (3) *asymptotically efficient.*

Proof. For proofs of consistency, asymptotic normality and asymptotic efficiency, see Sorenson (1980, pp. 187–190; 190–192; and 192–193, respectively). These proofs, though somewhat heuristic, convey the ideas needed to prove the three parts of this theorem. More detailed analyses can be found in Cramer (1946) and Zacks (1971). See also, Problems 11-13, 11-14, and 11-15. □

Theorem 11-2 (Invariance Property of MLE's). *Let* $\mathbf{g}(\boldsymbol{\theta})$ *be a vector function mapping* $\boldsymbol{\theta}$ *into an interval in r-dimensional Euclidean space. Let* $\hat{\boldsymbol{\theta}}_{ML}$ *be a MLE of* $\boldsymbol{\theta}$; *then* $\mathbf{g}(\hat{\boldsymbol{\theta}}_{ML})$ *is a MLE of* $\mathbf{g}(\boldsymbol{\theta})$; *i.e.*,

$$[\widehat{\mathbf{g}(\boldsymbol{\theta})}]_{ML} = \mathbf{g}(\hat{\boldsymbol{\theta}}_{ML}) \tag{11-21}$$

Proof (See Zacks, 1971). Note that Zacks points out that in many books, this theorem is cited only for the case of one-to-one mappings, $\mathbf{g}(\boldsymbol{\theta})$. His proof does not require $\mathbf{g}(\boldsymbol{\theta})$ to be one-to-one. Note, also, that the proof of this theorem is related to the "consistency carry-over" property of a consistent estimator, which was discussed in Lesson 7. □

Example 11-2

We wish to obtain a MLE of the variance σ_ν^2 in our linear model $z(k) = \mathbf{h}'(k)\boldsymbol{\theta} + \nu(k)$. One approach is to let $\theta_1 = \sigma_\nu^2$, establish the log-likelihood function for θ_1 and maximize it, to determine $\hat{\theta}_{1,ML}$. Usually, mathematical programming (i.e., search techniques) must be used to determine $\hat{\theta}_{1,ML}$. Here is where a difficulty can occur, because θ_1 (a variance) is known to be positive; thus, $\hat{\theta}_{1,ML}$ must be constrained to be positive. Unfortunately, constrained mathematical programming techniques are more difficult than unconstrained ones.

A second approach is to let $\theta_2 = \sigma_\nu$, establish the log-likelihood function for θ_2 (it will be the same as the one for θ_1, except that θ_1 will be replaced by θ_2^2), and maximize it to determine $\hat{\theta}_{2,ML}$. Because θ_2 is a standard deviation, which can be positive or negative, unconstrained mathematical programming can be used to determine $\hat{\theta}_{2,ML}$. Finally, we use the Invariance Property of MLE's to compute $\hat{\theta}_{1,ML}$, as

$$\hat{\theta}_{1,ML} = (\hat{\theta}_{2,ML})^2 \quad \square \tag{11-22}$$

THE LINEAR MODEL ($\mathcal{H}(k)$ deterministic)

We return now to the linear model

$$\mathcal{Z}(k) = \mathcal{H}(k)\boldsymbol{\theta} + \mathcal{V}(k) \tag{11-23}$$

in which $\boldsymbol{\theta}$ is an $n \times 1$ vector of deterministic parameters, $\mathcal{H}(k)$ *is deterministic*, and $\mathcal{V}(k)$ is zero mean white noise, with covariance matrix $\mathcal{R}(k)$. This is precisely the same model that was used to derive the BLUE of $\boldsymbol{\theta}$, $\hat{\boldsymbol{\theta}}_{BLU}(k)$. Our objectives in this paragraph are twofold: (1) to derive the MLE of $\boldsymbol{\theta}$, $\hat{\boldsymbol{\theta}}_{ML}(k)$, and (2) to relate $\hat{\boldsymbol{\theta}}_{ML}(k)$ to $\hat{\boldsymbol{\theta}}_{BLU}(k)$ and $\hat{\boldsymbol{\theta}}_{LS}(k)$.

In order to derive the MLE of $\boldsymbol{\theta}$, we need to determine a formula for $p(\mathcal{Z}(k)|\boldsymbol{\theta})$. To proceed, *we assume that $\mathcal{V}(k)$ is Gaussian*, with multivariate density function $p(\mathcal{V}(k))$, where

$$p(\mathcal{V}(k)) = \frac{1}{\sqrt{(2\pi)^N|\mathcal{R}(k)|}}\exp\left[-\frac{1}{2}\mathcal{V}'(k)\mathcal{R}^{-1}(k)\mathcal{V}(k)\right] \tag{11-24}$$

Recall (e.g., Papoulis, 1965) that linear transformations on, and linear combinations of, Gaussian random vectors are themselves Gaussian random vectors. For this reason, it is clear that when $\mathcal{V}(k)$ is Gaussian, $\mathcal{Z}(k)$ is as well. The multivariate Gaussian density function of $\mathcal{Z}(k)$, derived from $p(\mathcal{V}(k))$, is

$$p(\mathcal{Z}(k)|\boldsymbol{\theta}) = \frac{1}{\sqrt{(2\pi)^N|\mathcal{R}(k)|}}\exp\left\{-\frac{1}{2}[\mathcal{Z}(k) - \mathcal{H}(k)\boldsymbol{\theta}]'\mathcal{R}^{-1}(k)[\mathcal{Z}(k) - \mathcal{H}(k)\boldsymbol{\theta}]\right\} \tag{11-25}$$

Theorem 11-3. *When* $p(\mathcal{Z}|\boldsymbol{\theta})$ *is multivariate Gaussian and* $\mathcal{H}(k)$ *is deterministic, then the principle of ML leads to the BLUE of* $\boldsymbol{\theta}$, *i.e.*,

$$\hat{\boldsymbol{\theta}}_{ML}(k) = \hat{\boldsymbol{\theta}}_{BLU}(k) \tag{11-26}$$

Proof. We must maximize $p(\mathcal{Z}|\boldsymbol{\theta})$ with respect to $\boldsymbol{\theta}$. This can be accomplished by minimizing the argument of the exponential in (11-25); hence, $\hat{\boldsymbol{\theta}}_{ML}(k)$ is the solution of

$$\left.\frac{d}{d\boldsymbol{\theta}}(\mathcal{Z} - \mathcal{H}\boldsymbol{\theta})'\mathcal{R}^{-1}(\mathcal{Z} - \mathcal{H}\boldsymbol{\theta})\right|_{\boldsymbol{\theta} = \hat{\boldsymbol{\theta}}_{ML}} = \mathbf{0} \tag{11-27}$$

This equation can also be expressed as $dJ[\hat{\boldsymbol{\theta}}_{ML}]/d\hat{\boldsymbol{\theta}} = \mathbf{0}$ where $J[\hat{\boldsymbol{\theta}}_{ML}] = \tilde{\mathcal{Z}}'(k)\mathcal{R}^{-1}(k)\tilde{\mathcal{Z}}(k)$ and $\tilde{\mathcal{Z}}(k) = \mathcal{Z}(k) - \mathcal{H}\hat{\boldsymbol{\theta}}_{ML}(k)$. Comparing this version of (11-27) with Equation (3-4) and the subsequent derivation of the WLSE of $\boldsymbol{\theta}$, we conclude that

$$\hat{\boldsymbol{\theta}}_{ML}(k) = \left.\hat{\boldsymbol{\theta}}_{WLS}(k)\right|_{\mathbf{W}(k) = \mathcal{R}^{-1}(k)} \tag{11-28}$$

but, we also know, from Lesson 9, that

$$\left.\hat{\boldsymbol{\theta}}_{WLS}(k)\right|_{\mathbf{W}(k) = \mathcal{R}^{-1}(k)} = \hat{\boldsymbol{\theta}}_{BLU}(k) \tag{11-29}$$

From (11-28) and (11-29), we conclude that

$$\hat{\boldsymbol{\theta}}_{ML}(k) = \hat{\boldsymbol{\theta}}_{BLU}(k) \quad \square \tag{11-30}$$

We now suggest a reason why Theorem 9-6 (and, subsequently, Theorem 4-1) is referred to as the "information form" of recursive BLUE. From

Theorems 11-1 and 11-3 we know that, when $\mathcal{H}(k)$ is deterministic, $\hat{\theta}_{BLU} \sim N(\hat{\theta}; \theta, 1/N\,\mathbf{J}^{-1})$. This means that $\mathbf{P}(k) = \text{cov}[\tilde{\theta}_{BLU}(k)]$ is proportional to $\mathbf{J}^{-1}$. Observe, in Theorem 9-6, that we compute $\mathbf{P}^{-1}$ recursively, and not $\mathbf{P}$. Because $\mathbf{P}^{-1}$ is proportional to the Fisher information matrix $\mathbf{J}$, the results in Theorem 9-6 (and 4-1) are therefore referred to as the "information form" of recursive BLUE.

A second more pragmatic reason is due to the fact that the inverse of any covariance matrix is known as an information matrix (e.g., Anderson and Moore, 1979, pg. 138). Consequently, any algorithm that is in terms of information matrices is known as an "information form" algorithm.

Corollary 11-1. *If* $p[\mathcal{Z}(k)|\theta]$ *is multivariate Gaussian,* $\mathcal{H}(k)$ *is deterministic, and* $\mathcal{R}(k) = \sigma_v^2\mathbf{I}$, *then*

$$\hat{\theta}_{ML}(k) = \hat{\theta}_{BLU}(k) = \hat{\theta}_{LS}(k) \tag{11-31}$$

These estimators are: unbiased, most efficient (within the class of linear estimators), consistent, and Gaussian.

Proof. To obtain (11-31), combine the results in Theorems 11-3 and 9-2. The estimators are:

1. unbiased, because $\hat{\theta}_{BLU}(k)$ is unbiased;
2. most efficient, because $\hat{\theta}_{BLU}(k)$ is most efficient;
3. consistent, because $\hat{\theta}_{ML}(k)$ is consistent; and
4. Gaussian, because they depend linearly upon $\mathcal{Z}(k)$ which is Gaussian. □

Observe that, when $\mathcal{Z}(k)$ is Gaussian, this corollary permits us to make statements about small-sample properties of MLE's. Usually, we cannot make such statements.

A LOG-LIKELIHOOD FUNCTION FOR AN IMPORTANT DYNAMICAL SYSTEM

In practice there are two major problems in obtaining MLE's of parameters in models of dynamical systems

1. obtaining an expression for $L\{\theta|\mathcal{Z}\}$, and
2. maximizing $L\{\theta|\mathcal{Z}\}$ with respect to θ.

In this section we direct our attention to the first problem for a linear, time-invariant dynamical system that is excited by a known forcing function, has deterministic initial conditions, and has measurements that are corrupted by additive white Gaussian noise. This system is described by the following state-equation model:

$$\mathbf{x}(k + 1) = \mathbf{\Phi x}(k) + \mathbf{\Psi u}(k) \tag{11-32}$$

and

$$\mathbf{z}(k + 1) = \mathbf{Hx}(k + 1) + \mathbf{v}(k + 1) \qquad k = 0, 1, \ldots, N - 1 \tag{11-33}$$

In this model, $\mathbf{u}(k)$ is known ahead of time, $\mathbf{x}(0)$ is deterministic, $\mathbf{E}\{\mathbf{v}(k)\} = \mathbf{0}$, $\mathbf{v}(k)$ is Gaussian, and $\mathbf{E}\{\mathbf{v}(k)\mathbf{v}'(j)\} = \mathbf{R}\delta_{kj}$.

To begin, we must establish the parameters that constitute $\boldsymbol{\theta}$. In theory $\boldsymbol{\theta}$ could contain all of the elements in $\mathbf{\Phi}$, $\mathbf{\Psi}$, $\mathbf{H}$ and $\mathbf{R}$. In practice, however, these matrices are never completely unknown. State equations are either derived from physical principles (e.g., Newton's laws) or associated with a canonical model (e.g., controllability canonical form); hence, we usually know that certain elements in $\mathbf{\Phi}$, $\mathbf{\Psi}$ and $\mathbf{H}$ are identically zero or are known constants.

Even though all of the elements in $\mathbf{\Phi}$, $\mathbf{\Psi}$, $\mathbf{H}$ and $\mathbf{R}$ will not be unknown in an application, there still may be more unknowns present than can possibly be identified. How many parameters can be identified, and which parameters can be identified by maximum-likelihood estimation (or, for that matter, by any type of estimation) is the subject of *identifiability of systems* (Stepner and Mehra, 1973). *We shall assume that* $\boldsymbol{\theta}$ *is identifiable.* Identifiability is akin to "existence." When we assume $\boldsymbol{\theta}$ is identifiable, we assume that it is possible to identify $\boldsymbol{\theta}$ by ML methods. This means that all of our statements are predicated by the statement: If $\boldsymbol{\theta}$ is identifiable, then . . . .

We let

$$\boldsymbol{\theta} = \text{col}\,(\text{elements of } \mathbf{\Phi}, \mathbf{\Psi}, \mathbf{H}, \text{ and } \mathbf{R}) \tag{11-34}$$

Example 11-3

The "controllable canonical form" state-variable representation for the discrete-time autoregressive moving average (ARMA) model

$$H(z) = \frac{\beta_1 z^{n-1} + \beta_2 z^{n-2} + \cdots + \beta_{n-1} z + \beta_n}{z_n + \alpha_1 z^{n-1} + \cdots + \alpha_{n-1} z + \alpha_n} \tag{11-35}$$

which implies the ARMA difference equation (z denotes the unit advance operator)

$$\begin{aligned} y(k + n) + \alpha_1 y(k + n - 1) + \cdots + \alpha_n y(k) \\ = \beta_1 u(k + n - 1) + \cdots + \beta_n u(k) \end{aligned} \tag{11-36}$$

is

$$\begin{pmatrix} x_1(k+1) \\ x_2(k+1) \\ \vdots \\ x_n(k+1) \end{pmatrix} = \underbrace{\begin{pmatrix} 0 & 1 & 0 & \cdots & 0 \\ 0 & 0 & 1 & \cdots & 0 \\ \vdots & \vdots & \vdots & \ddots & \vdots \\ -\alpha_n & -\alpha_{n-1} & -\alpha_{n-2} & \cdots & -\alpha_1 \end{pmatrix}}_{\mathbf{\Phi}} \begin{pmatrix} x_1(k) \\ x_2(k) \\ \vdots \\ x_n(k) \end{pmatrix} + \underbrace{\begin{pmatrix} 0 \\ 0 \\ \vdots \\ 1 \end{pmatrix}}_{\mathbf{\Psi}} u(k) \tag{11-37}$$

and

$$y(k) = \underbrace{(\beta_n, \beta_{n-1}, \ldots, \beta_1)}_{\mathbf{H}}\mathbf{x}(k) \tag{11-38}$$

For this model there are no unknown parameters in matrix $\boldsymbol{\Psi}$, and, $\boldsymbol{\Phi}$ and $\mathbf{H}$ each contain exactly n (unknown) parameters. Matrix $\boldsymbol{\Phi}$ contains the n α-parameters which are associated with the poles of $H(z)$, whereas matrix $\mathbf{H}$ contains the n β-parameters which are associated with the zero of $H(z)$. *In general, an* n*th-order, single-input single-output system is completely characterized by* $2n$ *parameters.* □

Our objective now is to determine $L(\boldsymbol{\theta}|\mathfrak{X})$ for the system in (11-32) and (11-33). To begin, we must determine $p(\mathfrak{X}|\boldsymbol{\theta}) = p(\mathbf{z}(1), \mathbf{z}(2), \ldots, \mathbf{z}(N)|\boldsymbol{\theta})$. This is easy to do, because of the whiteness of noise $\mathbf{v}(k)$, i.e., $p(\mathbf{z}(1), \mathbf{z}(2), \ldots, \mathbf{z}(N)) = p(\mathbf{z}(1))p(\mathbf{z}(2)) \ldots p(\mathbf{z}(N))$; thus,

$$L(\boldsymbol{\theta}|\mathfrak{X}) = \ln\left[\prod_{i=1}^{N} p(\mathbf{z}(i)|\boldsymbol{\theta})\right] \tag{11-39}$$

From the Gaussian nature of $\mathbf{v}(k)$ and the linear measurement model in (11-33), we know that

$$p(\mathbf{z}(i)|\boldsymbol{\theta}) = \frac{1}{\sqrt{(2\pi)^m|\mathbf{R}|}} \exp\left\{-\frac{1}{2}[\mathbf{z}(i) - \mathbf{H}\mathbf{x}(i)]'\mathbf{R}^{-1}[\mathbf{z}(i) - \mathbf{H}\mathbf{x}(i)]\right\} \tag{11-40}$$

thus,

$$L(\boldsymbol{\theta}|\mathfrak{X}) = -\frac{1}{2}\sum_{i=1}^{N}[\mathbf{z}(i) - \mathbf{H}\mathbf{x}(i)]'\mathbf{R}^{-1}[\mathbf{z}(i) - \mathbf{H}\mathbf{x}(i)] - \frac{N}{2}\ln|\mathbf{R}| - \frac{N}{2}m\ln 2\pi \tag{11-41}$$

The log-likelihood function $L(\boldsymbol{\theta}|\mathfrak{X})$ is a function of $\boldsymbol{\theta}$. To indicate which quantities on the right-hand side of (11-41) may depend on $\boldsymbol{\theta}$, we subscript all such quantities with $\boldsymbol{\theta}$. Additionally, because $\frac{N}{2}m \ln 2\pi$ does not depend on $\boldsymbol{\theta}$ we neglect it in subsequent discussions. Our final log-likelihood function is:

$$L(\boldsymbol{\theta}|\mathfrak{X}) = -\frac{1}{2}\sum_{i=1}^{N}[\mathbf{z}(i) - \mathbf{H}_{\theta}\mathbf{x}_{\theta}(i)]'\mathbf{R}_{\theta}^{-1}[\mathbf{z}(i) - \mathbf{H}_{\theta}\mathbf{x}_{\theta}(i)] - \frac{N}{2}\ln|\mathbf{R}_{\theta}| \tag{11-42}$$

Observe that $\boldsymbol{\theta}$ occurs explicitly and implicitly in $L(\boldsymbol{\theta}|\mathfrak{X})$. Matrices $\mathbf{H}_{\theta}$ and $\mathbf{R}_{\theta}$ contain the explicit dependence of $L(\boldsymbol{\theta}|\mathfrak{X})$ on $\boldsymbol{\theta}$, whereas state vector $\mathbf{x}_{\theta}(i)$ contains the implicit dependence of $L(\boldsymbol{\theta}|\mathfrak{X})$ on $\boldsymbol{\theta}$. In order to numerically calculate the right-hand side of (11-42), we must solve state equation (11-32). This can be done when values of the unknown parameters which appear in $\boldsymbol{\Phi}$ and $\boldsymbol{\Psi}$ are given specific values; for, then (11-32) becomes

$$\mathbf{x}_{\theta}(k+1) = \boldsymbol{\Phi}_{\theta}\mathbf{x}_{\theta}(k) + \boldsymbol{\Psi}_{\theta}\mathbf{u}(k) \qquad \mathbf{x}_{\theta}(0) \text{ known} \tag{11-43}$$

In essence, then, *state equation (11-43) is a constraint that is associated with the computation of the log-likelihood function.*

How do we determine $\hat{\boldsymbol{\theta}}_{ML}$ for $L(\boldsymbol{\theta}|\mathfrak{Z})$ given in (11-42) [subject to its constraint in (11-43)]? No simple closed-form solution is possible, because $\boldsymbol{\theta}$ enters into $L(\boldsymbol{\theta}|\mathfrak{Z})$ in a complicated *nonlinear* manner. The only way presently known for obtaining $\hat{\boldsymbol{\theta}}_{ML}$ is by means of mathematical programming, which is beyond the scope of this course (e.g., see Mendel, 1983, pp. 141–142).

Comment. This completes our studies of methods for estimating unknown deterministic parameters. Prior to studying methods for estimating unknown random parameters, we pause to review an important body of material about multivariate Gaussian random variables.

PROBLEMS

11-1. If $\hat{a}$ is the MLE of a, what is the MLE of $a^{1/a}$? Explain your answer.

11-2. (Sorenson, 1980, Theorem 5.1, pg. 185). If an estimator exists such that equality is satisfied in the Cramer-Rao inequality, prove that it can be determined as the solution of the likelihood equation.

11-3. Consider a sequence of independently distributed random variables $x_1, x_2, \ldots, x_N$, having the probability density function $\theta^2 x_i e^{-\theta x_i}$, where $\theta > 0$.
- **(a)** Derive $\hat{\theta}_{ML}(N)$.
- **(b)** You want to study whether or not $\hat{\theta}_{ML}(N)$ is an unbiased estimator of θ. Explain (without working out all the details) how you would do this.

11-4. Consider a random variable z which can take on the values $z = 0, 1, 2, \ldots$. This variable is Poisson distributed, i.e., its probability function $P(z)$ is $P(z) = \mu^z e^{-\mu}/z!$ Let $z(1)$, $z(2), \ldots, z(N)$ denote N independent observations of z. Find $\hat{\mu}_{ML}(N)$.

11-5. If $p(z) = \theta e^{-\theta z}$, $z > 0$ and $p(z) = 0$, otherwise, find $\hat{\theta}_{ML}$ given a sample of N independent observations.

11-6. Find the maximum-likelihood estimator of θ from a sample of N independent observations that are uniformly distributed over the interval $(0, \theta)$.

11-7. Derive the maximum-likelihood estimate of the signal power, defined as ζ^2, in the signal $z(k) = \zeta s(k)$, $k = 0, 1, \ldots, N$, where $s(k)$ is a scalar, stationary, zero-mean Gaussian random sequence with autocorrelation $\mathbf{E}\{s(i)s(j)\} = \phi(j - i)$.

11-8. Suppose x is a binary variable that assumes a value of 1 with probability a and a value of 0 with probability $(1 - a)$. The probability distribution of x can be described by

$$P(x) = (1 - a)^{(1 - x)} a^x$$

Suppose we draw a random sample of N values $\{x_1, x_2, \ldots, x_N\}$. Find the MLE of a.

11-9. (Mendel, 1973, Exercise 3-15, pg. 177). Consider the linear model for which $\mathcal{V}(k)$ is Gaussian with zero mean and $\mathcal{R}(k) = \sigma^2\mathbf{I}$.

(a) Show that the maximum-likelihood estimator of σ^2, denoted $\hat{\sigma}^2_{\text{ML}}$, is

$$\hat{\sigma}^2_{\text{ML}} = \frac{(\mathscr{Z} - \mathscr{H}\hat{\boldsymbol{\theta}}_{\text{ML}})'(\mathscr{Z} - \mathscr{H}\hat{\boldsymbol{\theta}}_{\text{ML}})}{N}$$

where $\hat{\boldsymbol{\theta}}_{\text{ML}}$ is the maximum-likelihood estimator of $\boldsymbol{\theta}$.

(b) Show that $\hat{\sigma}^2_{\text{ML}}$ is biased, but that it is asymptotically unbiased.

11-10. We are given N independent samples $\mathbf{z}(1), \mathbf{z}(2), \ldots, \mathbf{z}(N)$ of the identically distributed two-dimensional random vector $\mathbf{z}(i) = \text{col}[z_1(i), z_2(i)]$, with the Gaussian density function

$$p(z_1(i), z_2(i)|\rho) = \frac{1}{2\pi\sqrt{1-\rho^2}} \exp\left\{-\frac{z_1^2(i) - 2\rho z_1(i)z_2(i) + z_2^2(i)}{2(1-\rho^2)}\right\}$$

where ρ is the correlation coefficient between z_1 and z_2.

(a) Determine $\hat{\rho}_{\text{ML}}$ (Hint: You will obtain a cubic equation for $\hat{\rho}_{\text{ML}}$ and must show that $\hat{\rho}_{\text{ML}} = r$, where r equals the sample correlation coefficient).

(b) Determine the Cramer-Rao bound for $\hat{\rho}_{\text{ML}}$.

11-11. It is well known, from linear system theory, that the choice of state variables used to describe a dynamical system is not unique. Consider the nth-order difference equation

$$\begin{aligned} y(k+n) + a_1 y(k+n-1) + \cdots + a_n y(k) \\ = b_n^0 m(k+n-1) + b_{n-1}^0 m(k+n-2) + \cdots + b_1^0 m(k). \end{aligned}$$

A state-variable model for this system is

$$\begin{pmatrix} x_1(k+1) \\ x_2(k+1) \\ \vdots \\ x_n(k+1) \end{pmatrix} = \begin{pmatrix} 0 & 1 & 0 & \cdots & 0 \\ 0 & 0 & 1 & \cdots & 0 \\ \vdots & \vdots & \vdots & \ddots & \vdots \\ -a_n & -a_{n-1} & -a_{n-2} & \cdots & -a_1 \end{pmatrix} \begin{pmatrix} x_1(k) \\ x_2(k) \\ \vdots \\ x_n(k) \end{pmatrix} + \begin{pmatrix} b_1 \\ b_2 \\ \vdots \\ b_n \end{pmatrix} m(k)$$

and

$$y(k) = (1\ 0 \ldots 0)x(k)$$

where

$$\begin{pmatrix} b_1 \\ b_2 \\ \vdots \\ b_n \end{pmatrix} = \begin{pmatrix} 1 & 0 & 0 & \cdots & 0 & 0 \\ a_1 & 1 & 0 & \cdots & 0 & 0 \\ a_2 & a_1 & 1 & \cdots & 0 & 0 \\ \vdots & \vdots & \vdots & \ddots & \vdots & \vdots \\ a_{n-2} & a_{n-3} & a_{n-4} & \cdots & 1 & 0 \\ a_{n-1} & a_{n-2} & a_{n-3} & \cdots & a_1 & 1 \end{pmatrix} \begin{pmatrix} b_1^0 \\ b_2^0 \\ \vdots \\ b_n^0 \end{pmatrix}$$

Suppose maximum-likelihood estimates have been determined for the a_i- and b_i-parameters. How does one compute the maximum-likelihood estimates of the b_i^0-parameters?

11-12. Prove that, for a random sample of measurements, a maximum-likelihood estimator is a consistent estimator [Hints: (1) Show that $\mathbf{E}\{\partial \ln p(\mathscr{Z}|\boldsymbol{\theta})/\partial\boldsymbol{\theta}|\boldsymbol{\theta}\} = \mathbf{0}$; (2) Expand $\partial \ln p(\mathscr{Z}|\hat{\boldsymbol{\theta}}_{\text{ML}})/\partial\boldsymbol{\theta}$ in a Taylor series about $\boldsymbol{\theta}$, and show that $\partial \ln p(\mathscr{Z}|\boldsymbol{\theta})/\partial\boldsymbol{\theta} = -(\hat{\boldsymbol{\theta}}_{\text{ML}} - \boldsymbol{\theta})'\partial^2 \ln p(\mathscr{Z}|\boldsymbol{\theta}^*)/\partial\boldsymbol{\theta}^2$, where $\boldsymbol{\theta}^* = \lambda\boldsymbol{\theta} + (1-\lambda)\hat{\boldsymbol{\theta}}_{\text{ML}}$ and $0 \le \lambda \le 1$; (3) Show that $\partial \ln p(\mathscr{Z}|\boldsymbol{\theta})/\partial\boldsymbol{\theta} = \sum_{i=1}^{N} \partial \ln p(z(i)|\boldsymbol{\theta})/\partial\boldsymbol{\theta}$ and ∂^2 ln

$p(\mathcal{Z}|\boldsymbol{\theta})/\partial\boldsymbol{\theta}^2 = \sum_{i=1}^{N} \partial^2 \ln p(z(i)|\boldsymbol{\theta})/\partial\boldsymbol{\theta}^2$; (4) Using the strong law of large numbers to assert that, with probability one, sample averages converge to ensemble averages, and assuming that $\mathbf{E}\{\partial^2 \ln p(z(i)|\boldsymbol{\theta})/\partial\boldsymbol{\theta}^2\}$ is negative definite, show that $\hat{\boldsymbol{\theta}}_{ML} \rightarrow \boldsymbol{\theta}$ with probability one; thus, $\hat{\boldsymbol{\theta}}_{ML}$ is a consistent estimator of θ. The steps in this proof have been taken from Sorenson, 1980, pp. 187–191.]

11-13. Prove that, for a random sample of measurements, a maximum-likelihood estimator is asymptotically Gaussian with mean value $\boldsymbol{\theta}$ and covariance matrix $(N\mathbf{J})^{-1}$ where $\mathbf{J}$ is the Fisher information matrix for a single measurement $z(i)$. [Hints: (1) Expand $\partial \ln p(\mathcal{Z}|\hat{\boldsymbol{\theta}}_{ML})/\partial\boldsymbol{\theta}$ in a Taylor series about $\boldsymbol{\theta}$ and neglect second- and higher-order terms; (2) Show that

$$\frac{1}{N}\frac{\partial}{\partial\boldsymbol{\theta}} \ln p(\mathcal{Z}|\boldsymbol{\theta}) = (\hat{\boldsymbol{\theta}}_{ML} - \boldsymbol{\theta})' \left[-\frac{1}{N}\frac{\partial^2}{\partial\boldsymbol{\theta}^2} \ln p(\mathcal{Z}|\boldsymbol{\theta}) \right]$$

(3) Let $\mathbf{s}(\boldsymbol{\theta}) = \partial \ln p(\mathcal{Z}|\boldsymbol{\theta})/\partial\boldsymbol{\theta}$ and show that $\mathbf{s}(\boldsymbol{\theta}) = \sum_{i=1}^{N} \mathbf{s}_i(\boldsymbol{\theta})$, where $\mathbf{s}_i(\boldsymbol{\theta}) = \partial \ln p(z(i)|\boldsymbol{\theta})/\partial\boldsymbol{\theta}$; (4) Let $\bar{\mathbf{s}}$ denote the sample mean of $\mathbf{s}_i(\boldsymbol{\theta})$, and show that the distribution of $\bar{\mathbf{s}}$ asymptotically converges to a Gaussian distribution having mean zero and covariance $\mathbf{J}/N$; (5) Using the strong law of large numbers we know that $\frac{1}{N}\,\partial^2 \ln p(\mathcal{Z}|\boldsymbol{\theta})/\partial\boldsymbol{\theta}^2 \rightarrow \mathbf{E}\{\partial^2 \ln p(z(i)|\boldsymbol{\theta})/\partial\boldsymbol{\theta}^2|\boldsymbol{\theta}\}$; consequently, show that $(\hat{\boldsymbol{\theta}}_{ML} - \boldsymbol{\theta})'\mathbf{J}$ is asymptotically Gaussian with zero mean and covariance $\mathbf{J}/N$; (6) Complete the proof of the theorem. The steps in this proof have been taken from Sorenson, 1980, pg. 192.]

11-14. Prove that, for a random sample of measurements, a maximum-likelihood estimator is asymptotically efficient [Hints: (1) Let $\mathbf{S}(\boldsymbol{\theta}) = -\mathbf{E}\{\partial^2 \ln p(\mathcal{Z}|\boldsymbol{\theta})/\partial\boldsymbol{\theta}^2|\boldsymbol{\theta}\}$ and show that $\mathbf{S}(\boldsymbol{\theta}) = N\mathbf{J}$ where $\mathbf{J}$ is the Fisher information matrix for a single measurement $z(i)$; (2) Use the result stated in Problem 11-13 to complete the proof. The steps in this proof have been taken from Sorenson, 1980, pg. 193.]

Lesson 12

Elements of Multivariate Gaussian Random Variables

INTRODUCTION

Gaussian random variables are important and widely used for at least two reasons. First, they often provide a model that is a reasonable approximation to observed random behavior. Second, if the random phenomenon that we observe at the macroscopic level is the superposition of an arbitrarily large number of independent random phenomena, which occur at the microscopic level, the macroscopic description is justifiably Gaussian.

Most (if not all) of the material in this lesson should be a review for a reader who has had a course in probability theory. We collect a wide range of facts about multivariate Gaussian random variables here, in one place, because they are often needed in the remaining lessons.

UNIVARIATE GAUSSIAN DENSITY FUNCTION

A random variable y is said to be distributed as the *univariate Gaussian distribution* with mean m_y and variance σ_y^2 [i.e., $y \sim N(y; m_y, \sigma_y^2)$] if the density function of y is given as

$$p(y) = \frac{1}{\sqrt{2\pi\sigma_y^2}} \exp[-(y - m_y)^2/2\sigma_y^2] \qquad -\infty < y < \infty \tag{12-1}$$

For notational simplicity throughout this chapter, we do not condition density functions on their parameters; this conditioning is understood. Density $p(y)$ is the familiar bell-shaped curve, centered at $y = m_y$.

MULTIVARIATE GAUSSIAN DENSITY FUNCTION

Let $y_1, y_2, \ldots, y_m$ be random variables, and $\mathbf{y} = \text{col}\,(y_1, y_2, \ldots, y_m)$. The function

$$p(y_1, y_2, \ldots, y_m) = p(\mathbf{y}) = \frac{1}{\sqrt{(2\pi)^m|\mathbf{P_y}|}} \exp\left[-\frac{1}{2}(\mathbf{y} - \mathbf{m_y})'\mathbf{P_y}^{-1}(\mathbf{y} - \mathbf{m_y})\right] \tag{12-2}$$

is said to be a *multivariate* (m-*variate*) *Gaussian density function* [i.e., $\mathbf{y} \sim N(\mathbf{y}; \mathbf{m_y}, \mathbf{P_y})$]. In (12-2),

$$\mathbf{m_y} = \mathbf{E}\{\mathbf{y}\} \tag{12-3}$$

and

$$\mathbf{P_y} = \mathbf{E}\{(\mathbf{y} - \mathbf{m_y})(\mathbf{y} - \mathbf{m_y})'\} \tag{12-4}$$

Note that, although we refer to $p(\mathbf{y})$ as a density function, it is actually a *joint density function* between the random variables $y_1, y_2, \ldots, y_m$. If $\mathbf{P_y}$ is not positive definite, it is more convenient to define the multivariate Gaussian distribution by its characteristic function. We will not need to do this.

JOINTLY GAUSSIAN RANDOM VECTORS

Let $\mathbf{x}$ and $\mathbf{y}$ individually be n- and m-dimensional Gaussian random vectors, i.e., $\mathbf{x} \sim N(\mathbf{x}; \mathbf{m_x}, \mathbf{P_x})$ and $\mathbf{y} \sim N(\mathbf{y}; \mathbf{m_y}, \mathbf{P_y})$. Let $\mathbf{P_{xy}}$ and $\mathbf{P_{yx}}$ denote the cross-covariance matrices between $\mathbf{x}$ and $\mathbf{y}$, i.e.,

$$\mathbf{P_{xy}} = \mathbf{E}\{(\mathbf{x} - \mathbf{m_x})(\mathbf{y} - \mathbf{m_y})'\} \tag{12-5}$$

and

$$\mathbf{P_{yx}} = \mathbf{E}\{(\mathbf{y} - \mathbf{m_y})(\mathbf{x} - \mathbf{m_x})'\} \tag{12-6}$$

We are interested in the joint density between $\mathbf{x}$ and $\mathbf{y}$, $p(\mathbf{x}, \mathbf{y})$. Vectors $\mathbf{x}$ and $\mathbf{y}$ are jointly Gaussian if

$$p(\mathbf{x}, \mathbf{y}) = \frac{1}{\sqrt{(2\pi)^{n+m}|\mathbf{P_z}|}} \exp\left\{-\frac{1}{2}(\mathbf{z} - \mathbf{m_z})'\mathbf{P_z}^{-1}(\mathbf{z} - \mathbf{m_z})\right\} \tag{12-7}$$

where

$$\mathbf{z} = \text{col}\,(\mathbf{x}, \mathbf{y}) \tag{12-8}$$

$$\mathbf{m_z} = \text{col}\,(\mathbf{m_x}, \mathbf{m_y}) \tag{12-9}$$

and

$$\mathbf{P_z} = \begin{pmatrix} \mathbf{P_x} & \mathbf{P_{xy}} \\ \mathbf{P_{yx}} & \mathbf{P_y} \end{pmatrix} \tag{12-10}$$

Note that if **x** and **y** are jointly Gaussian then they are marginally (i.e., individually) Gaussian. The converse is true if **x** and **y** are independent, but it is not necessarily true if they are not independent (Papoulis, 1965, pg. 184).

In order to evaluate $p(\mathbf{x}, \mathbf{y})$ in (12-7), we need $|\mathbf{P}_z|$ and $\mathbf{P}_z^{-1}$. It is straightforward to compute $|\mathbf{P}_z|$ once the given values for $\mathbf{P}_x$, $\mathbf{P}_y$, $\mathbf{P}_{xy}$, and $\mathbf{P}_{yx}$ are substituted into (12-10). It is often useful to be able to express the components of $\mathbf{P}_z^{-1}$ directly in terms of the components of $\mathbf{P}_z$. It is a straightforward exercise in algebra (just form $\mathbf{P}_z\mathbf{P}_z^{-1} = \mathbf{I}$ and equate elements on both sides) to show that

$$\mathbf{P}_z^{-1} = \begin{pmatrix} \mathbf{A} & \mathbf{B} \\ \mathbf{B}' & \mathbf{C} \end{pmatrix} \tag{12-11}$$

where

$$\mathbf{A} = (\mathbf{P}_x - \mathbf{P}_{xy}\mathbf{P}_y^{-1}\mathbf{P}_{yx})^{-1} = \mathbf{P}_x^{-1} + \mathbf{P}_x^{-1}\mathbf{P}_{xy}\mathbf{C}\,\mathbf{P}_{yx}\mathbf{P}_x^{-1} \tag{12-12}$$

$$\mathbf{B} = -\mathbf{A}\mathbf{P}_{xy}\mathbf{P}_y^{-1} = -\mathbf{P}_x^{-1}\mathbf{P}_{xy}\mathbf{C} \tag{12-13}$$

and

$$\mathbf{C} = (\mathbf{P}_y - \mathbf{P}_{yx}\mathbf{P}_x^{-1}\mathbf{P}_{xy})^{-1} = \mathbf{P}_y^{-1} + \mathbf{P}_y^{-1}\mathbf{P}_{yx}\mathbf{A}\,\mathbf{P}_{xy}\mathbf{P}_y^{-1} \tag{12-14}$$

THE CONDITIONAL DENSITY FUNCTION

One of the most important density functions we will be interested in is the conditional density function $p(\mathbf{x}|\mathbf{y})$. Recall, from probability theory (e.g., Papoulis, 1965), that

$$p(\mathbf{x}|\mathbf{y}) = \frac{p(\mathbf{x}, \mathbf{y})}{p(\mathbf{y})} \tag{12-15}$$

Theorem 12-1. *Let* **x** *and* **y** *be* n- *and* m-*dimensional vectors that are jointly Gaussian. Then*

$$p(\mathbf{x}|\mathbf{y}) = \frac{1}{\sqrt{(2\pi)^n|\mathcal{Q}|}} \exp\left\{-\frac{1}{2}(\mathbf{x} - \mathbf{m})'\mathcal{Q}^{-1}(\mathbf{x} - \mathbf{m})\right\} \tag{12-16}$$

where

$$\mathbf{m} = \mathrm{E}\{\mathbf{x}|\mathbf{y}\} = \mathbf{m}_x + \mathbf{P}_{xy}\mathbf{P}_y^{-1}(\mathbf{y} - \mathbf{m}_y) \tag{12-17}$$

and

$$\mathcal{Q} = \mathbf{A}^{-1} = \mathbf{P}_x - \mathbf{P}_{xy}\mathbf{P}_y^{-1}\mathbf{P}_{yx} \tag{12-18}$$

This means that $p(\mathbf{x}|\mathbf{y})$ *is also multivariate Gaussian with (conditional) mean* **m** *and covariance* $\mathcal{Q}$.

Proof. From (12-15), (12-7), and (12-2), we find that

$$p(\mathbf{x}|\mathbf{y}) = \frac{1}{\sqrt{(2\pi)^n \dfrac{|\mathbf{P}_z|}{|\mathbf{P}_y|}}} \exp\left[-\frac{1}{2}(\mathbf{z} - \mathbf{m}_z)'\begin{pmatrix} \mathbf{A} & \mathbf{B} \\ \mathbf{B}' & \mathbf{C} - \mathbf{P}_y^{-1} \end{pmatrix}(\mathbf{z} - \mathbf{m}_z)\right] \tag{12-19}$$

Taking a closer look at the quadratic exponent, which we denote $E(\mathbf{x},\mathbf{y})$, we find that

$$\begin{aligned}E(\mathbf{x},\mathbf{y}) &= (\mathbf{x} - \mathbf{m}_x)'\mathbf{A}(\mathbf{x} - \mathbf{m}_x) + 2(\mathbf{x} - \mathbf{m}_x)'\mathbf{B}(\mathbf{y} - \mathbf{m}_y) \\ &\quad + (\mathbf{y} - \mathbf{m}_y)'(\mathbf{C} - \mathbf{P}_y^{-1})(\mathbf{y} - \mathbf{m}_y) \\ &= (\mathbf{x} - \mathbf{m}_x)'\mathbf{A}(\mathbf{x} - \mathbf{m}_x) - 2(\mathbf{x} - \mathbf{m}_x)'\mathbf{A}\,\mathbf{P}_{xy}\mathbf{P}_y^{-1}(\mathbf{y} - \mathbf{m}_y) \\ &\quad + (\mathbf{y} - \mathbf{m}_y)'\mathbf{P}_y^{-1}\mathbf{P}_{yx}\mathbf{A}\,\mathbf{P}_{xy}\mathbf{P}_y^{-1}(\mathbf{y} - \mathbf{m}_y)\end{aligned} \tag{12-20}$$

In obtaining (12-20) we have used (12-13) and (12-14). We now recognize that (12-20) looks like a quadratic expression in $\mathbf{x} - \mathbf{m}_x$ and $\mathbf{P}_{xy}\mathbf{P}_y^{-1}(\mathbf{y} - \mathbf{m}_y)$, and express it in factored form, as

$$\begin{aligned}E(\mathbf{x}, \mathbf{y}) = [(&\mathbf{x} - \mathbf{m}_x) \\ &- \mathbf{P}_{xy}\mathbf{P}_y^{-1}(\mathbf{y} - \mathbf{m}_y)]'\mathbf{A}[(\mathbf{x} - \mathbf{m}_x) - \mathbf{P}_{xy}\mathbf{P}_y^{-1}(\mathbf{y} - \mathbf{m}_y)]\end{aligned} \tag{12-21}$$

Defining $\mathbf{m}$ and $\mathcal{Q}$ as in (12-18) we see that

$$E(\mathbf{x},\mathbf{y}) = (\mathbf{x} - \mathbf{m})'\mathcal{Q}^{-1}(\mathbf{x} - \mathbf{m}) \tag{12-22}$$

If we can show that $|\mathbf{P}_z|/|\mathbf{P}_y| = |\mathcal{Q}|$, then we will have shown that $p(\mathbf{x}|\mathbf{y})$ is given by (12-16), which will mean that $p(\mathbf{x}|\mathbf{y})$ is multivariate Gaussian with mean $\mathbf{m}$ and covariance $\mathcal{Q}$.

It is not at all obvious that $|\mathcal{Q}| = |\mathbf{P}_z|/|\mathbf{P}_y|$. We shall reexpress matrix $\mathbf{P}_z$ so that $|\mathbf{P}_z|$ can be determined by appealing to the following theorems (Graybill, 1961, pg. 6):

i. If $\mathbf{V}$ and $\mathbf{G}$ are $n \times n$ matrices, then $|\mathbf{VG}| = |\mathbf{V}|\,|\mathbf{G}|$.

ii. If $\mathbf{M}$ is a square matrix such that

$$\mathbf{M} = \begin{pmatrix}\mathbf{M}_{11} & \mathbf{M}_{12} \\ \mathbf{M}_{21} & \mathbf{M}_{22}\end{pmatrix}$$

where $\mathbf{M}_{11}$ and $\mathbf{M}_{22}$ are square matrices, and if $\mathbf{M}_{12} = \mathbf{0}$ or $\mathbf{M}_{21} = \mathbf{0}$, then $|\mathbf{M}| = |\mathbf{M}_{11}|\,|\mathbf{M}_{22}|$.

We now show that two matrices, $\mathbf{L}$ and $\mathbf{N}$, can be found so that

$$\mathbf{P}_z = \begin{pmatrix}\mathbf{P}_x & \mathbf{P}_{xy} \\ \mathbf{P}_{yx} & \mathbf{P}_y\end{pmatrix} = \begin{pmatrix}\mathbf{L} & \mathbf{P}_{xy} \\ \mathbf{0} & \mathbf{P}_y\end{pmatrix}\begin{pmatrix}\mathbf{I}_n & \mathbf{0} \\ \mathbf{N} & \mathbf{I}_m\end{pmatrix} \tag{12-23}$$

Multiplying the two matrices on the right-hand side of (12-23), and equating the 1-1 and 2-1 components from both sides of the resulting equation,we find that

$$\mathbf{P}_x = \mathbf{L} + \mathbf{P}_{xy}\mathbf{N} \tag{12-24}$$

and

$$\mathbf{P}_{yx} = \mathbf{P}_y\mathbf{N} \tag{12-25}$$

from which it follows that

$$\mathbf{N} = \mathbf{P}_y^{-1}\mathbf{P}_{yx} \tag{12-26}$$

and

$$\mathbf{L} = \mathbf{P}_x - \mathbf{P}_{xy}\mathbf{P}_y^{-1}\mathbf{P}_{yx} \tag{12-27}$$

From (12-23), we see that

$$|\mathbf{P}_z| = |\mathbf{L}|\,|\mathbf{P}_y|$$

or,

$$|\mathbf{L}| = |\mathbf{P}_z|/|\mathbf{P}_y| \tag{12-28}$$

Comparing the equations for $\mathbf{L}$ and $\mathfrak{Q}$, we find they are the same; thus, we have proven that

$$|\mathfrak{Q}| = |\mathbf{P}_z|/|\mathbf{P}_y| \tag{12-29}$$

which completes the proof of this theorem. □

PROPERTIES OF MULTIVARIATE GAUSSIAN RANDOM VARIABLES

From the preceding formulas for $p(\mathbf{y})$, $p(\mathbf{x},\mathbf{y})$ and $p(\mathbf{x}|\mathbf{y})$, we see that *multivariate Gaussian probability density functions are completely characterized by their first two moments*, i.e., their mean vector and covariance matrix. All other moments can be expressed in terms of their first two moments (see, e.g., Papoulis, 1965).

From probability theory, we also recall the following two important facts about Gaussian random variables:

1. Statistically independent Gaussian random variables are uncorrelated and vice-versa; thus, $\mathbf{P}_{xy}$ and $\mathbf{P}_{yx}$ are both zero matrices.
2. Linear (or affine) transformations on and linear (or affine) combinations of Gaussian random variables are themselves Gaussian random variables; thus, if $\mathbf{x}$ and $\mathbf{y}$ are jointly Gaussian, then $\mathbf{z} = \mathbf{A}\mathbf{x} + \mathbf{B}\mathbf{y} + \mathbf{c}$ is also Gaussian. We refer to this property as the *linearity property*.

PROPERTIES OF CONDITIONAL MEAN

We learned, in Theorem 12-1, that

$$\mathbf{E}\{\mathbf{x}|\mathbf{y}\} = \mathbf{m}_x + \mathbf{P}_{xy}\mathbf{P}_y^{-1}(\mathbf{y} - \mathbf{m}_y) \tag{12-30}$$

Because $\mathbf{E}\{\mathbf{x}|\mathbf{y}\}$ depends on $\mathbf{y}$, which is random, it is also random.

Theorem 12-2. *When* $\mathbf{x}$ *and* $\mathbf{y}$ *are jointly Gaussian,* $\mathbf{E}\{\mathbf{x}|\mathbf{y}\}$ *is multivariate Gaussian, and is an affine combination of the elements of* $\mathbf{y}$.

Proof. That $\mathbf{E}\{\mathbf{x}|\mathbf{y}\}$ is Gaussian follows from the linearity property applied to (12-30). An affine transformation of $\mathbf{y}$ has the structure $\mathbf{Ty} + \mathbf{f}$. $\mathbf{E}\{\mathbf{x}|\mathbf{y}\}$ has this structure; thus, it is an affine transformation. Note that if $\mathbf{m}_x = \mathbf{0}$ and $\mathbf{m}_y = \mathbf{0}$, then $\mathbf{E}\{\mathbf{x}|\mathbf{y}\}$ is a linear transformation of $\mathbf{y}$. □

Theorem 12-3. *Let* $\mathbf{x}$, $\mathbf{y}$, *and* $\mathbf{z}$ *be* $n \times 1$, $m \times 1$ *and* $r \times 1$ *jointly Gaussian random vectors. If* $\mathbf{y}$ and $\mathbf{z}$ *are statistically independent, then*

$$\mathbf{E}\{\mathbf{x}|\mathbf{y}, \mathbf{z}\} = \mathbf{E}\{\mathbf{x}|\mathbf{y}\} + \mathbf{E}\{\mathbf{x}|\mathbf{z}\} - \mathbf{m}_x \tag{12-31}$$

Proof. Let $\boldsymbol{\xi} = \text{col}\,(\mathbf{y},\mathbf{z})$; then

$$\mathbf{E}\{\mathbf{x}|\boldsymbol{\xi}\} = \mathbf{m}_x + \mathbf{P}_{x\xi}\mathbf{P}_{\xi}^{-1}(\boldsymbol{\xi} - \mathbf{m}_{\xi}) \tag{12-32}$$

We leave it to the reader to show that

$$\mathbf{P}_{x\xi} = (\mathbf{P}_{xy}|\mathbf{P}_{xz}) \tag{12-33}$$

and

$$\mathbf{P}_{\xi} = \left(\begin{array}{c|c} \mathbf{P}_y & \mathbf{0} \\ \hline \mathbf{0} & \mathbf{P}_z \end{array}\right) \tag{12-34}$$

(the off-diagonal elements in $\mathbf{P}_{\xi}$ are zero if $\mathbf{y}$ and $\mathbf{z}$ are statistically independent; this is also true if $\mathbf{y}$ and $\mathbf{z}$ are uncorrelated, because $\mathbf{y}$ and $\mathbf{z}$ are jointly Gaussian); thus,

$$\begin{aligned}\mathbf{E}\{\mathbf{x}|\boldsymbol{\xi}\} &= \mathbf{m}_x + \mathbf{P}_{xy}\mathbf{P}_y^{-1}(\mathbf{y} - \mathbf{m}_y) + \mathbf{P}_{xz}\mathbf{P}_z^{-1}(\mathbf{z} - \mathbf{m}_z) \\ &= \mathbf{E}\{\mathbf{x}|\mathbf{y}\} + \mathbf{E}\{\mathbf{x}|\mathbf{z}\} - \mathbf{m}_x \quad \square\end{aligned}$$

In our developments of recursive estimators, $\mathbf{y}$ and $\mathbf{z}$ (which will be associated with a partitioning of measurement vectors $\mathcal{Z}$) are not necessarily independent. The following important generalization of Theorem 12-3 will be needed.

Theorem 12-4. *Let* $\mathbf{x}$, $\mathbf{y}$, *and* $\mathbf{z}$ *be* $n \times 1$, $m \times 1$ *and* $r \times 1$ *jointly Gaussian random vectors. If* $\mathbf{y}$ *and* $\mathbf{z}$ *are not necessarily statistically independent, then*

$$\mathbf{E}\{\mathbf{x}|\mathbf{y}, \mathbf{z}\} = \mathbf{E}\{\mathbf{x}|\mathbf{y}, \tilde{\mathbf{z}}\} \tag{12-35}$$

where

$$\tilde{\mathbf{z}} = \mathbf{z} - \mathbf{E}\{\mathbf{z}|\mathbf{y}\} \tag{12-36}$$

so that

$$\mathbf{E}\{\mathbf{x}|\mathbf{y},\mathbf{z}\} = \mathbf{E}\{\mathbf{x}|\mathbf{y}\} + \mathbf{E}\{\mathbf{x}|\tilde{\mathbf{z}}\} - \mathbf{m}_x \tag{12-37}$$

Proof (Mendel, 1983b, pg. 53). The proof proceeds in two stages: (a) assume (12-35) is true and demonstrate the truth of (12-37), and (b) demonstrate the truth of (12-35).

a. If we can show that $\mathbf{y}$ and $\tilde{\mathbf{z}}$ are statistically independent, then (12-37) follows from Theorem 12-3. For Gaussian random vectors, however, uncorrelatedness implies independence.

To begin, we assert that $\mathbf{y}$ and $\tilde{\mathbf{z}}$ are jointly Gaussian, because $\tilde{\mathbf{z}} = \mathbf{z} - \mathbf{E}\{\mathbf{z}|\mathbf{y}\} = \mathbf{z} - \mathbf{m}_z - \mathbf{P}_{zy}\mathbf{P}_y^{-1}(\mathbf{y} - \mathbf{m}_y)$ depends on $\mathbf{y}$ and $\mathbf{z}$, which are jointly Gaussian.

Next, we show that $\tilde{\mathbf{z}}$ is zero mean. This follows from the calculation

$$\mathbf{m}_{\tilde{z}} = \mathbf{E}\{\mathbf{z} - \mathbf{E}\{\mathbf{z}|\mathbf{y}\}\} = \mathbf{E}\{\mathbf{z}\} - \mathbf{E}\{\mathbf{E}\{\mathbf{z}|\mathbf{y}\}\} \tag{12-38}$$

where the outer expectation in the second term on the right-hand side of (12-38) is with respect to $\mathbf{y}$. From probability theory (Papoulis, 1965, pg. 208), $\mathbf{E}\{\mathbf{z}\}$ can be expressed as

$$\mathbf{E}\{\mathbf{z}\} = \mathbf{E}\{\mathbf{E}\{\mathbf{z}|\mathbf{y}\}\} \tag{12-39}$$

From (12-38) and (12-39), we see that $\mathbf{m}_{\tilde{z}} = \mathbf{0}$.

Finally, we show that $\mathbf{y}$ and $\tilde{\mathbf{z}}$ are uncorrelated. This follows from the calculation

$$\begin{aligned}\mathbf{E}\{(\mathbf{y} - \mathbf{m}_y)(\tilde{\mathbf{z}} - \mathbf{m}_{\tilde{z}})'\} &= \mathbf{E}\{(\mathbf{y} - \mathbf{m}_y)\tilde{\mathbf{z}}'\} = \mathbf{E}\{\mathbf{y}\tilde{\mathbf{z}}'\} \\ &= \mathbf{E}\{\mathbf{y}\mathbf{z}'\} - \mathbf{E}\{\mathbf{y}\mathbf{E}\{\mathbf{z}'|\mathbf{y}\}\} \\ &= \mathbf{E}\{\mathbf{y}\mathbf{z}'\} - \mathbf{E}\{\mathbf{y}\mathbf{z}'\} = 0\end{aligned} \tag{12-40}$$

b. A detailed proof is given by Meditch (1969, pp. 101–102). The idea is to (1) compute $\mathbf{E}\{\mathbf{x}|\mathbf{y},\mathbf{z}\}$ in expanded form, (2) compute $\mathbf{E}\{\mathbf{x}|\mathbf{y},\tilde{\mathbf{z}}\}$ in expanded form, using $\tilde{\mathbf{z}}$ given in (12-36), and (3) compare the results from (1) and (2) to prove the truth of (12-35). □

Equation (12-35) is very important. It states that, when $\mathbf{z}$ and $\mathbf{y}$ are dependent, conditioning on $\mathbf{z}$ can always be replaced by conditioning on another Gaussian random vector $\tilde{\mathbf{z}}$, where $\tilde{\mathbf{z}}$ and $\mathbf{y}$ are statistically independent.

The results in Theorems 12-3 and 12-4 depend on all random vectors being jointly Gaussian. Very similar results, which are distribution free, are described in Problem 13-4; however, these results are restricted in yet another way.

PROBLEMS

12-1. Fill in all of the details required to prove part b of Theorem 12-4.

12-2. Let $\mathbf{x}$, $\mathbf{y}$ and $\mathbf{z}$ be jointly distributed random vectors; c and h fixed constants; and $g(\cdot)$ a scalar-valued function. Assume $\mathbf{E}\{\mathbf{x}\}$, $\mathbf{E}\{\mathbf{z}\}$, and $\mathbf{E}\{g(\mathbf{y})\mathbf{x}\}$ exist. Prove the following useful properties of conditional expectation:

(a) $\mathbf{E}\{\mathbf{x}|\mathbf{y}\} = \mathbf{E}\{\mathbf{x}\}$ if $\mathbf{x}$ and $\mathbf{y}$ are independent
(b) $\mathbf{E}\{g(\mathbf{y})\mathbf{x}|\mathbf{y}\} = g(\mathbf{y})\mathbf{E}\{\mathbf{x}|\mathbf{y}\}$
(c) $\mathbf{E}\{c|\mathbf{y}\} = c$
(d) $\mathbf{E}\{g(\mathbf{y})|\mathbf{y}\} = g(\mathbf{y})$
(e) $\mathbf{E}\{c\mathbf{x} + h\mathbf{z}|\mathbf{y}\} = c\mathbf{E}\{\mathbf{x}|\mathbf{y}\} + h\mathbf{E}\{\mathbf{z}|\mathbf{y}\}$
(f) $\mathbf{E}\{\mathbf{x}\} = \mathbf{E}\{\mathbf{E}\{\mathbf{x}|\mathbf{y}\}\}$ where the outer expectation is with respect to $\mathbf{y}$
(g) $\mathbf{E}\{g(\mathbf{y})\mathbf{x}\} = \mathbf{E}\{g(\mathbf{y})\mathbf{E}\{\mathbf{x}|\mathbf{y}\}\}$ where the outer expectation is with respect to $\mathbf{y}$.

12-3. Prove that the cross-covariance matrix of two uncorrelated random vectors is zero.

Lesson 13

Estimation of Random Parameters: General Results

INTRODUCTION

In Lesson 2 we showed that state estimation and deconvolution can be viewed as problems in which we are interested in estimating a vector of random parameters. For us, state estimation and deconvolution serve as the primary motivation for studying methods for estimating random parameters; however, the statistics literature is filled with other applications for these methods.

We now view $\boldsymbol{\theta}$ as an $n \times 1$ vector of random unknown parameters. The information available to us are measurements $\mathbf{z}(1), \mathbf{z}(2), \ldots, \mathbf{z}(k)$, which are assumed to depend upon $\boldsymbol{\theta}$. In this lesson, we do not begin by assuming a specific structural dependency between $\mathbf{z}(i)$ and $\boldsymbol{\theta}$. This is quite different than what we did in WLSE and BLUE. Those methods were studied for the linear model $\mathbf{z}(i) = \mathbf{H}(i)\boldsymbol{\theta} + \mathbf{v}(i)$, and, closed-form solutions for $\hat{\boldsymbol{\theta}}_{WLS}(k)$ and $\hat{\boldsymbol{\theta}}_{BLU}(k)$ could not have been obtained had we not begun by assuming the linear model. We shall study the estimation of random $\boldsymbol{\theta}$ for the linear model in Lesson 14.

In this lesson we examine two methods for estimating a vector of random parameters. The first method is based upon minimizing the mean-squared error between $\boldsymbol{\theta}$ and $\hat{\boldsymbol{\theta}}(k)$. The resulting estimator is called a *mean-squared estimator*, and is denoted $\hat{\boldsymbol{\theta}}_{MS}(k)$. The second method is based upon maximizing an unconditional likelihood function, one that not only requires knowledge of $p(\mathfrak{Z}|\boldsymbol{\theta})$ but also of $p(\boldsymbol{\theta})$. The resulting estimator is called a *maximum a posteriori estimator*, and is denoted $\hat{\boldsymbol{\theta}}_{MAP}(k)$.

MEAN-SQUARED ESTIMATION

Many different measures of parameter estimation error can be minimized in order to obtain an estimate of $\boldsymbol{\theta}$ [see Jazwinski (1970), Meditch (1969), van Trees (1968), and Sorenson (1980), for example]; but, by far the most widely studied measure is the mean-squared error.

Objective Function and Problem Statement

Given measurements $\mathbf{z}(1), \ldots, \mathbf{z}(k)$, we shall determine an estimator of $\boldsymbol{\theta}$, namely

$$\hat{\boldsymbol{\theta}}_{MS}(k) = \boldsymbol{\phi}[\mathbf{z}(i), i = 1, 2, \ldots, k] \tag{13-1}$$

such that the mean-squared error

$$J[\tilde{\boldsymbol{\theta}}_{MS}(k)] = \mathbf{E}\{\tilde{\boldsymbol{\theta}}'_{MS}(k)\tilde{\boldsymbol{\theta}}_{MS}(k)\} \tag{13-2}$$

is minimized. In (13-2), $\tilde{\boldsymbol{\theta}}_{MS}(k) = \boldsymbol{\theta} - \hat{\boldsymbol{\theta}}_{MS}(k)$.

The right-hand side of (13-1) means that we have some arbitrary and as yet unknown function of all the measurements. The n components of $\hat{\boldsymbol{\theta}}_{MS}(k)$ may each depend differently on the measurements. The function $\boldsymbol{\phi}[\mathbf{z}(i), i = 1, 2, \ldots, k]$ may be nonlinear or linear. Its exact structure will be determined by minimizing $J[\tilde{\boldsymbol{\theta}}_{MS}(k)]$. If perchance $\hat{\boldsymbol{\theta}}_{MS}(k)$ is a linear estimator, then

$$\hat{\boldsymbol{\theta}}_{MS}(k) = \sum_{i=1}^{k} \mathbf{A}(i)\mathbf{z}(i) \tag{13-3}$$

We now show that the notion of conditional expectation is central to the calculation of $\hat{\boldsymbol{\theta}}_{MS}(k)$. As usual, we let

$$\mathfrak{Z}(k) = \text{col}[\mathbf{z}(k), \mathbf{z}(k-1), \ldots, \mathbf{z}(1)] \tag{13-4}$$

The underlying random quantities in our estimation problem are $\boldsymbol{\theta}$ and $\mathfrak{Z}(k)$. We assume that their joint density function $p[\boldsymbol{\theta}, \mathfrak{Z}(k)]$ exists, so that

$$\mathbf{E}\{\tilde{\boldsymbol{\theta}}'_{MS}(k)\tilde{\boldsymbol{\theta}}_{MS}(k)\} = \int_{-\infty}^{\infty} \cdots \int_{-\infty}^{\infty} \tilde{\boldsymbol{\theta}}'_{MS}(k)\tilde{\boldsymbol{\theta}}_{MS}(k) p[\boldsymbol{\theta}, \mathfrak{Z}(k)] d\boldsymbol{\theta} d\mathfrak{Z}(k) \tag{13-5}$$

where $d\boldsymbol{\theta} = d\theta_1 d\theta_2 \ldots d\theta_n$, $d\mathfrak{Z}(k) = dz_1(1) \ldots dz_1(k) dz_2(1) \ldots dz_2(k) \ldots dz_m(1) \ldots dz_m(k)$, and there are $n + km$ integrals. Using the fact that

$$p[\boldsymbol{\theta}, \mathfrak{Z}(k)] = p[\boldsymbol{\theta}|\mathfrak{Z}(k)] p[\mathfrak{Z}(k)] \tag{13-6}$$

we rewrite (13-5) as

$$\begin{aligned}
\mathbf{E}\{\tilde{\boldsymbol{\theta}}'_{MS}(k)\tilde{\boldsymbol{\theta}}_{MS}(k)\} &= \int_{-\infty}^{\infty} \cdots \int_{-\infty}^{\infty} \left\{ \int_{-\infty}^{\infty} \cdots \int_{-\infty}^{\infty} \tilde{\boldsymbol{\theta}}'_{MS}(k)\tilde{\boldsymbol{\theta}}_{MS}(k) p[\boldsymbol{\theta}|\mathfrak{Z}(k)] d\boldsymbol{\theta} \right\} \\
&\quad p[\mathfrak{Z}(k)] d\mathfrak{Z}(k) \\
&= \int_{-\infty}^{\infty} \cdots \int_{-\infty}^{\infty} \mathbf{E}\{\tilde{\boldsymbol{\theta}}'_{MS}(k)\tilde{\boldsymbol{\theta}}_{MS}(k)|\mathfrak{Z}(k)\} p[\mathfrak{Z}(k)] d\mathfrak{Z}(k)
\end{aligned} \tag{13-7}$$

From this equation we see that minimizing the conditional expectation $\mathbf{E}\{\tilde{\boldsymbol{\theta}}'_{MS}(k)\tilde{\boldsymbol{\theta}}_{MS}(k)|\mathfrak{Z}(k)\}$ with respect to $\hat{\boldsymbol{\theta}}_{MS}(k)$ is equivalent to our original objective of minimizing the total expectation $\mathbf{E}\{\tilde{\boldsymbol{\theta}}'_{MS}(k)\tilde{\boldsymbol{\theta}}_{MS}(k)\}$. Note that the integrals on the right-hand side of (13-7) remove the dependency of the integrand on the data $\mathfrak{Z}(k)$.

In summary, we have the following *mean-squared estimation problem*: *Given the measurements* $\mathbf{z}(1), \mathbf{z}(2), \ldots, \mathbf{z}(k)$, *determine an estimator of* $\boldsymbol{\theta}$, *namely,*

$$\hat{\boldsymbol{\theta}}_{MS}(k) = \boldsymbol{\phi}[\mathbf{z}(i), i = 1, 2, \ldots, k]$$

such that the conditional mean-squared error

$$J_1[\tilde{\boldsymbol{\theta}}_{MS}(k)] = \mathbf{E}\{\tilde{\boldsymbol{\theta}}'_{MS}(k)\tilde{\boldsymbol{\theta}}_{MS}(k)|\mathbf{z}(1), \ldots, \mathbf{z}(k)\} \tag{13-8}$$

is minimized.

Derivation of Estimator

The solution to the mean-squared estimation problem is given in Theorem 13-1, which is known as the *Fundamental Theorem of Estimation Theory.*

Theorem 13-1. *The estimator that minimizes the mean-squared error is*

$$\hat{\boldsymbol{\theta}}_{MS}(k) = \mathbf{E}\{\boldsymbol{\theta}|\mathfrak{Z}(k)\} \tag{13-9}$$

Proof (Mendel, 1983b). In this proof we omit all functional dependences on k, for notational simplicity. Our approach is to substitute $\tilde{\boldsymbol{\theta}}_{MS}(k) = \boldsymbol{\theta} - \hat{\boldsymbol{\theta}}_{MS}(k)$ into (13-8) and to complete the square, as follows:

$$\begin{aligned} J_1[\tilde{\boldsymbol{\theta}}_{MS}(k)] &= \mathbf{E}\{(\boldsymbol{\theta} - \hat{\boldsymbol{\theta}}_{MS})'(\boldsymbol{\theta} - \hat{\boldsymbol{\theta}}_{MS})|\mathfrak{Z}\} \\ &= \mathbf{E}\{\boldsymbol{\theta}'\boldsymbol{\theta} - \boldsymbol{\theta}'\hat{\boldsymbol{\theta}}_{MS} - \hat{\boldsymbol{\theta}}'_{MS}\boldsymbol{\theta} + \hat{\boldsymbol{\theta}}'_{MS}\hat{\boldsymbol{\theta}}_{MS}|\mathfrak{Z}\} \\ &= \mathbf{E}\{\boldsymbol{\theta}'\boldsymbol{\theta}|\mathfrak{Z}\} - \mathbf{E}\{\boldsymbol{\theta}'|\mathfrak{Z}\}\hat{\boldsymbol{\theta}}_{MS} - \hat{\boldsymbol{\theta}}'_{MS}\mathbf{E}\{\boldsymbol{\theta}|\mathfrak{Z}\} + \hat{\boldsymbol{\theta}}'_{MS}\hat{\boldsymbol{\theta}}_{MS} \\ &= \mathbf{E}\{\boldsymbol{\theta}'\boldsymbol{\theta}|\mathfrak{Z}\} + [\hat{\boldsymbol{\theta}}_{MS} - \mathbf{E}\{\boldsymbol{\theta}|\mathfrak{Z}\}]'[\hat{\boldsymbol{\theta}}_{MS} - \mathbf{E}\{\boldsymbol{\theta}|\mathfrak{Z}\}] \\ &\quad - \mathbf{E}\{\boldsymbol{\theta}'|\mathfrak{Z}\}\mathbf{E}\{\boldsymbol{\theta}|\mathfrak{Z}\} \end{aligned} \tag{13-10}$$

To obtain the third line we used the fact that $\hat{\boldsymbol{\theta}}_{MS}$, by definition, is a function of $\mathfrak{Z}$; hence, $\mathbf{E}\{\hat{\boldsymbol{\theta}}_{MS}|\mathfrak{Z}\} = \hat{\boldsymbol{\theta}}_{MS}$. The first and last terms in (13-10) do not depend on $\hat{\boldsymbol{\theta}}_{MS}$; hence, the smallest value of $J_1[\tilde{\boldsymbol{\theta}}_{MS}(k)]$ is obviously attained by setting the bracketed terms equal to zero. This means that $\hat{\boldsymbol{\theta}}_{MS}$ must be chosen as in (13-9). □

Let $J_1^*[\tilde{\boldsymbol{\theta}}_{MS}(k)]$ denote the minimum value of $J_1[\tilde{\boldsymbol{\theta}}_{MS}(k)]$. We see, from (13-10) and (13-9), that

$$J_1^*[\tilde{\boldsymbol{\theta}}_{MS}(k)] = \mathbf{E}\{\boldsymbol{\theta}'\boldsymbol{\theta}|\mathfrak{Z}\} - \hat{\boldsymbol{\theta}}'_{MS}(k)\hat{\boldsymbol{\theta}}_{MS}(k) \tag{13-11}$$

As it stands, (13-9) is not terribly useful for computing $\hat{\boldsymbol{\theta}}_{MS}(k)$. In general, we must first compute $p[\boldsymbol{\theta}|\mathfrak{Z}(k)]$ and then perform the requisite number of integrations of $\boldsymbol{\theta}p[\boldsymbol{\theta}|\mathfrak{Z}(k)]$ to obtain $\hat{\boldsymbol{\theta}}_{MS}(k)$. In the special but important case when $\boldsymbol{\theta}$ and $\mathcal{Z}(k)$ are jointly Gaussian we have a very important and practical corollary to Theorem 13-1.

Corollary 13-1. *When* $\boldsymbol{\theta}$ *and* $\mathcal{Z}(k)$ *are jointly Gaussian, the estimator that minimizes the mean-squared error is*

$$\hat{\boldsymbol{\theta}}_{MS}(k) = \mathbf{m}_{\theta} + \mathbf{P}_{\theta\mathcal{Z}}(k)\mathbf{P}_{\mathcal{Z}}^{-1}(k)[\mathcal{Z}(k) - \mathbf{m}_{\mathcal{Z}}(k)] \tag{13-12}$$

Proof. When $\boldsymbol{\theta}$ and $\mathcal{Z}(k)$ are jointly Gaussian then $\mathbf{E}\{\boldsymbol{\theta}|\mathcal{Z}(k)\}$ can be evaluated using (12-17) of Lesson 12. Doing this we obtain (13-12). □

Corollary 13-1 gives us an explicit structure for $\hat{\boldsymbol{\theta}}_{MS}(k)$. We see that $\hat{\boldsymbol{\theta}}_{MS}(k)$ is an affine transformation of $\mathcal{Z}(k)$. If $\mathbf{m}_{\theta} = \mathbf{0}$ and $\mathbf{m}_{\mathcal{Z}}(k) = \mathbf{0}$, then $\hat{\boldsymbol{\theta}}_{MS}(k)$ is a linear transformation of $\mathcal{Z}(k)$.

In order to compute $\hat{\boldsymbol{\theta}}_{MS}(k)$ using (13-12), we must know $\mathbf{m}_{\theta}$ and $\mathbf{m}_{\mathcal{Z}}(k)$ and we must first compute $\mathbf{P}_{\theta\mathcal{Z}}(k)$ and $\mathbf{P}_{\mathcal{Z}}^{-1}(k)$. We perform these computations in Lesson 14 for the linear model, $\mathcal{Z}(k) = \mathcal{H}(k)\boldsymbol{\theta} + \mathcal{V}(k)$.

Corollary 13-2. *Suppose* $\boldsymbol{\theta}$ *and* $\mathcal{Z}(k)$ *are not necessarily jointly Gaussian*, and *that we know* $\mathbf{m}_{\theta}$, $\mathbf{m}_{\mathcal{Z}}(k)$, $\mathbf{P}_{\mathcal{Z}}(k)$ *and* $\mathbf{P}_{\theta\mathcal{Z}}(k)$. *In this case, the estimator that is constrained to be an affine transformation of* $\mathcal{Z}(k)$, *and that minimizes the mean-squared error is also given by (13-12).*

Proof. This corollary can be proved in a number of ways. A direct proof begins by assuming that $\hat{\boldsymbol{\theta}}_{MS}(k) = \mathbf{A}(k)\mathcal{Z}(k) + \mathbf{b}(k)$ and choosing $\mathbf{A}(k)$ and $\mathbf{b}(k)$ so that $\hat{\boldsymbol{\theta}}_{MS}(k)$ is an unbiased estimator of $\boldsymbol{\theta}$ and $\mathbf{E}\{\tilde{\boldsymbol{\theta}}'_{MS}(k)\tilde{\boldsymbol{\theta}}_{MS}(k)\} = \text{trace } \mathbf{E}\{\tilde{\boldsymbol{\theta}}_{MS}(k)\tilde{\boldsymbol{\theta}}'_{MS}(k)\}$ is minimized. We leave the details of this direct proof to the reader.

A less direct proof is based upon the following Gedanken experiment. Using known first and second moments of $\boldsymbol{\theta}$ and $\mathcal{Z}(k)$, we can conceptualize unique Gaussian random vectors that have these same first and second moments. For these statistically-equivalent (through second-order moments) Gaussian vectors, we know, from Corollary 13-1, that the mean-squared estimator is given by the affine transformation of $\mathcal{Z}(k)$ in (13-12). □

Corollaries 13-1 and 13-2, as well as Theorem 13-1, provide us with the answer to the following important question: *When is the linear (affine) mean-squared estimator the same as the mean-squared estimator*? The answer is, when $\boldsymbol{\theta}$ and $\mathcal{Z}(k)$ are jointly Gaussian. If $\boldsymbol{\theta}$ and $\mathcal{Z}(k)$ are not jointly Gaussian, then $\hat{\boldsymbol{\theta}}_{MS}(k) = \mathbf{E}\{\boldsymbol{\theta}|\mathcal{Z}(k)\}$, which, in general, is a nonlinear function of measurements $\mathcal{Z}(k)$, i.e., it is a nonlinear estimator.

Corollary 13-3 (Orthogonality Principle). *Suppose* $\mathbf{f}[\mathcal{Z}(k)]$ *is any function of the data* $\mathcal{Z}(k)$. *Then the error in the mean-squared estimator is orthogonal to* $\mathbf{f}[\mathcal{Z}(k)]$ *in the sense that*

$$\mathbf{E}\{[\boldsymbol{\theta} - \hat{\boldsymbol{\theta}}_{MS}(k)]\mathbf{f}'[\mathcal{Z}(k)]\} = \mathbf{0} \tag{13-13}$$

Proof (Mendel, 1983b, pp. 46–47). We use the following result from probability theory (Papoulis, 1965; see, also, Problem 12-2(g)). Let $\boldsymbol{\alpha}$ and $\boldsymbol{\beta}$ be jointly distributed random vectors and $g(\boldsymbol{\beta})$ be a scalar-valued function; then

$$\mathbf{E}\{\boldsymbol{\alpha}\, g(\boldsymbol{\beta})\} = \mathbf{E}\{\mathbf{E}\{\boldsymbol{\alpha}|\boldsymbol{\beta}\}\, g(\boldsymbol{\beta})\} \tag{13-14}$$

where the outer expectation on the right-hand side is with respect to $\boldsymbol{\beta}$. We proceed as follows (again, omitting the argument k):

$$\begin{aligned}\mathbf{E}\{(\boldsymbol{\theta} - \hat{\boldsymbol{\theta}}_{MS})\mathbf{f}'(\mathcal{Z})\} &= \mathbf{E}\{\mathbf{E}\{(\boldsymbol{\theta} - \hat{\boldsymbol{\theta}}_{MS})|\mathcal{Z}\}\mathbf{f}'(\mathcal{Z})\} \\ &= \mathbf{E}\{(\hat{\boldsymbol{\theta}}_{MS} - \hat{\boldsymbol{\theta}}_{MS})\mathbf{f}'(\mathcal{Z})\} = \mathbf{0}\end{aligned}$$

where we have used the facts that $\hat{\boldsymbol{\theta}}_{MS}$ is no longer random when $\mathcal{Z}$ is specified and $\mathbf{E}\{\boldsymbol{\theta}|\mathcal{Z}\} = \hat{\boldsymbol{\theta}}_{MS}$. □

A frequently encountered special case of (13-13) occurs when $\mathbf{f}[\mathcal{Z}(k)] = \hat{\boldsymbol{\theta}}_{MS}(k)$; then Corollary 13-13 can be written as

$$\mathbf{E}\{\tilde{\boldsymbol{\theta}}_{MS}(k)\hat{\boldsymbol{\theta}}'_{MS}(k)\} = \mathbf{0} \tag{13-15}$$

Properties of Mean-Squared Estimates When $\boldsymbol{\theta}$ and $\mathcal{Z}(k)$ are Gaussian

In this section we present a collection of important and useful properties associated with $\hat{\boldsymbol{\theta}}_{MS}(k)$ for the case when $\boldsymbol{\theta}$ and $\mathcal{Z}(k)$ are jointly Gaussian. In this case $\hat{\boldsymbol{\theta}}_{MS}(k)$ is given by (13-12).

Property 1 (*Unbiasedness*). The mean-squared estimator, $\hat{\boldsymbol{\theta}}_{MS}(k)$ in (13-12), is unbiased.

Proof. Taking the expected value of (13-12), we see that $\mathbf{E}\{\hat{\boldsymbol{\theta}}_{MS}(k)\} = \mathbf{m}_{\boldsymbol{\theta}}$; thus, $\hat{\boldsymbol{\theta}}_{MS}(k)$ is an unbiased estimator of $\boldsymbol{\theta}$. □

Property 2 (*Minimum Variance*). Dispersion about the mean value of $\hat{\theta}_{i,MS}(k)$ is measured by the error variance $\sigma^2_{\tilde{\theta}_{i,MS}}(k)$, where $i = 1, 2, \ldots, n$. An estimator that has the smallest error variance is a minimum-variance estimator (an MVE). The mean-squared estimator in (13-12) is an MVE.

Proof. From Property 1 and the definition of error variance, we see that

$$\sigma^2_{\tilde{\theta}_{i,MS}}(k) = \mathbf{E}\{\tilde{\theta}^2_{i,MS}(k)\} \qquad i = 1, 2, \ldots, n \tag{13-16}$$

Our mean-squared estimator was obtained by minimizing $J[\tilde{\boldsymbol{\theta}}_{MS}(k)]$ in (13-2), which can now be expressed as

$$J[\tilde{\boldsymbol{\theta}}_{MS}(k)] = \sum_{i=1}^{n} \sigma^2_{\tilde{\theta}_{i,MS}}(k) \tag{13-17}$$

Because variances are always positive the minimum value of $J[\tilde{\theta}_{MS}(k)]$ must be achieved when each of the n variances is minimized; hence, our mean-squared estimator is equivalent to an MVE. □

Property 3 (*Linearity*). $\hat{\theta}_{MS}(k)$ in (13-12) is a "linear" (i.e., affine) estimator.

Proof. This is obvious from the form of (13-12). □

Linearity of $\hat{\theta}_{MS}(k)$ permits us to infer the following very important property about both $\hat{\theta}_{MS}(k)$ and $\tilde{\theta}_{MS}(k)$.

Property 4 (*Gaussian*). Both $\hat{\theta}_{MS}(k)$ and $\tilde{\theta}_{MS}(k)$ are multivariate Gaussian.

Proof. We use the linearity property of jointly Gaussian random vectors stated in Lesson 12. Estimator $\hat{\theta}_{MS}(k)$ in (13-12) is an affine transformation of Gaussian random vector $\mathfrak{Z}(k)$; hence, $\hat{\theta}_{MS}(k)$ is multivariate Gaussian. Estimation error $\tilde{\theta}_{MS}(k) = \theta - \hat{\theta}_{MS}(k)$ is an affine transformation of jointly Gaussian vectors θ and $\mathfrak{Z}(k)$; hence, $\tilde{\theta}_{MS}(k)$ is also Gaussian. □

Estimate $\hat{\theta}_{MS}(k)$ in (13-12) is itself random, because measurements $\mathfrak{Z}(k)$ are random. To characterize it completely in a statistical sense, we must specify its probability density function. Generally, this is very difficult to do, and often requires that the probability density function of $\hat{\theta}_{MS}(k)$ be approximated using many moments (in theory an infinite number are required). In the Gaussian case, we have just learned that the structure of the probability density function for $\hat{\theta}_{MS}(k)$ [and $\tilde{\theta}_{MS}(k)$] is known. Additionally, we know that a Gaussian density function is completely specified by exactly two moments, its mean and covariance; thus, tremendous simplifications occur when θ and $\mathfrak{Z}(k)$ are jointly Gaussian.

Property 5 (*Uniqueness*). Mean-squared estimator $\hat{\theta}_{MS}(k)$, in (13-12), is unique.

The proof of this property is not central to our developments; hence, it is omitted.

Generalizations

Many of the results presented in this section are applicable to objective functions other than the mean-squared objective function in (13-2). See Meditch (1969) for discussions on a wide number of objective functions that lead to $\mathbf{E}\{\theta|\mathfrak{Z}(k)\}$ as the optimal estimator of θ.

MAXIMUM A POSTERIORI ESTIMATION

Recall Bayes's rule (Papoulis, 1965, pg. 39):

$$p(\boldsymbol{\theta}|\mathcal{Z}(k)) = p(\mathcal{Z}(k)|\boldsymbol{\theta})p(\boldsymbol{\theta})/p(\mathcal{Z}(k)) \tag{13-18}$$

in which density function $p(\boldsymbol{\theta}|\mathcal{Z}(k))$ is known as the a posteriori (or posterior) conditional density function, and $p(\boldsymbol{\theta})$ is the prior probability density function for $\boldsymbol{\theta}$. Observe that $p(\boldsymbol{\theta}|\mathcal{Z}(k))$ is related to likelihood function $l\{\boldsymbol{\theta}|\mathcal{Z}(k)\}$, because $l\{\boldsymbol{\theta}|\mathcal{Z}(k)\} \propto p(\mathcal{Z}(k)|\boldsymbol{\theta})$. Additionally, because $p(\mathcal{Z}(k))$ does not depend on $\boldsymbol{\theta}$,

$$p(\boldsymbol{\theta}|\mathcal{Z}(k)) \propto p(\mathcal{Z}(k)|\boldsymbol{\theta})p(\boldsymbol{\theta}) \tag{13-19}$$

In maximum a posteriori (MAP) estimation, values of $\boldsymbol{\theta}$ are found that maximize $p(\boldsymbol{\theta}|\mathcal{Z}(k))$ in (13-19); such estimates are known as MAP estimates, and will be denoted as $\hat{\boldsymbol{\theta}}_{\text{MAP}}(k)$.

If $\theta_1, \theta_2, \ldots, \theta_n$ are uniformly distributed, then $p(\boldsymbol{\theta}|\mathcal{Z}(k)) \propto p(\mathcal{Z}(k)|\boldsymbol{\theta})$, and the MAP estimator of $\boldsymbol{\theta}$ equals the ML estimator of $\boldsymbol{\theta}$. Generally, MAP estimates are quite different from ML estimates. For example, the invariance property of MLE's usually does not carry over to MAP estimates. One reason for this can be seen from (13-19). Suppose, for example, that $\boldsymbol{\phi} = \mathbf{g}(\boldsymbol{\theta})$ and we want to determine $\hat{\boldsymbol{\phi}}_{\text{MAP}}$ by first computing $\hat{\boldsymbol{\theta}}_{\text{MAP}}$. Because $p(\boldsymbol{\theta})$ depends on the Jacobian matrix of $\mathbf{g}^{-1}(\boldsymbol{\phi})$, $\hat{\boldsymbol{\phi}}_{\text{MAP}} \neq \mathbf{g}(\hat{\boldsymbol{\theta}}_{\text{MAP}})$. Kashyap and Rao (1976, pg. 137) note, "the two estimates are usually asymptotically identical to one another since in the large sample case the knowledge of the observations *swamps that of the prior distribution*." For additional discussions on the asymptotic properties of MAP estimators, see Zacks (1971).

Quantity $p(\boldsymbol{\theta}|\mathcal{Z}(k))$ in (13-19) is sometimes called an *unconditional likelihood function*, because the random nature of $\boldsymbol{\theta}$ has been accounted for by $p(\boldsymbol{\theta})$. Density $p(\mathcal{Z}(k)|\boldsymbol{\theta})$ is then called a *conditional likelihood function* (Nahi, 1969).

Obtaining a MAP estimate involves specifying both $p(\mathcal{Z}(k)|\boldsymbol{\theta})$ and $p(\boldsymbol{\theta})$ and finding those values of $\boldsymbol{\theta}$ that maximize $p(\boldsymbol{\theta}|\mathcal{Z}(k))$, or $\ln p(\boldsymbol{\theta}|\mathcal{Z}(k))$. Generally speaking, mathematical programming must be used to compute $\hat{\boldsymbol{\theta}}_{\text{MAP}}(k)$. When $\mathcal{Z}(k)$ is related to $\boldsymbol{\theta}$ by our linear model, $\mathcal{Z}(k) = \mathcal{H}(k)\boldsymbol{\theta} + \mathcal{V}(k)$, then it may be possible to obtain $\hat{\boldsymbol{\theta}}_{\text{MAP}}(k)$ in closed form. We examine this situation in some detail in Lesson 14.

Example 13-1

This is a continuation of Example 11-1. We observe a random sample $\{z(1), z(2), \ldots, z(N)\}$ at the output of a Gaussian random number generator, i.e., $z(i) \sim N(z(i); \mu, \sigma_z^2)$. Now, however, μ is a random variable with prior distribution $N(\mu; 0, \sigma_\mu^2)$. Both σ_z^2 and σ_μ^2 are assumed known, and we wish to determine the MAP estimator of μ.

We can view this random number generator as a cascade of two random number generators. The first is characterized by $N(\mu; 0, \sigma_\mu^2)$ and provides at its output a single realization for μ, say μ_{R}. The second is characterized by $N(z(i); \mu_{\text{R}}, \sigma_z^2)$. Observe that

μ_R, which is unknown to us in this example, is transferred from the first random number generator to the second one before we can obtain the given random sample $\{z(1), z(2), \ldots, z(N)\}$.

Using the facts that

$$p(z(i)|\mu) = (2\pi\sigma_z^2)^{-1/2} \exp\left\{-\frac{1}{2}[z(i) - \mu]^2/\sigma_z^2\right\} \tag{13-20}$$

$$p(\mu) = (2\pi\sigma_\mu^2)^{-1/2} \exp\left\{-\frac{1}{2}\mu^2/\sigma_\mu^2\right\} \tag{13-21}$$

and

$$p(\mathfrak{Z}(N)|\mu) = \prod_{i=1}^{N} p(z(i)|\mu) \tag{13-22}$$

we find

$$p(\mu|\mathfrak{Z}(N)) \propto (2\pi\sigma_z^2)^{-N/2} \exp\left\{-\frac{1}{2}\sum_{i=1}^{N}[z(i) - \mu]^2/\sigma_z^2\right\} \cdot (2\pi\sigma_\mu^2)^{-1/2} \exp\left\{-\frac{1}{2}\mu^2/\sigma_\mu^2\right\} \tag{13-23}$$

Taking the logarithm of (13-23) and neglecting the terms which do not depend upon μ, we obtain

$$L_{MAP}(\mu|\mathfrak{Z}(N)) = -\frac{1}{2}\sum_{i=1}^{N}\{[z(i) - \mu]^2/\sigma_z^2\} - \frac{1}{2}\mu^2/\sigma_\mu^2 \tag{13-24}$$

Setting $\partial L_{MAP}/\partial\mu = 0$, and solving for $\hat{\mu}_{MAP}(N)$, we find that

$$\hat{\mu}_{MAP}(N) = \frac{\sigma_\mu^2}{\sigma_z^2 + N\sigma_\mu^2}\sum_{i=1}^{N} z(i) \tag{13-25}$$

Next, we compare $\hat{\mu}_{MAP}(N)$ and $\hat{\mu}_{ML}(N)$, where [see (19) from Lesson 11]

$$\hat{\mu}_{ML}(N) = \frac{1}{N}\sum_{i=1}^{N} z(i) \tag{13-26}$$

In general $\hat{\mu}_{MAP}(N) \neq \hat{\mu}_{ML}(N)$. If, however, no a priori information about μ is available, then we let $\sigma_\mu^2 \to \infty$, in which case $\hat{\mu}_{MAP}(N) = \hat{\mu}_{ML}(N)$. Observe, also that, as $N \to \infty$ $\hat{\mu}_{MAP}(N) = \hat{\mu}_{ML}(N)$, which implies (Sorenson, 1980) that the influence of the prior knowledge about μ [i.e., $\mu \sim N(\mu; 0, \sigma_\mu^2)$] diminishes as the number of measurements increase. □

Theorem 13-2. *If* $\mathfrak{Z}(k)$ *and* $\boldsymbol{\theta}$ *are jointly Gaussian, then* $\hat{\boldsymbol{\theta}}_{MAP}(k) = \hat{\boldsymbol{\theta}}_{MS}(k)$.

Proof. If $\mathfrak{Z}(k)$ and $\boldsymbol{\theta}$ are jointly Gaussian, then (see Theorem 12-1)

$$p(\boldsymbol{\theta}|\mathfrak{Z}(k)) = \frac{1}{\sqrt{(2\pi)^n|\mathfrak{X}(k)|}} \exp\left\{-\frac{1}{2}[\boldsymbol{\theta} - \mathbf{m}(k)]'\mathfrak{X}^{-1}(k)[\boldsymbol{\theta} - \mathbf{m}(k)]\right\} \tag{13-27}$$

where

$$\mathbf{m}(k) = \mathbf{E}\{\boldsymbol{\theta}|\mathcal{Z}(k)\} \tag{13-28}$$

$\hat{\boldsymbol{\theta}}_{\text{MAP}}(k)$ is found by maximizing $p(\boldsymbol{\theta}|\mathcal{Z}(k))$, or equivalently by minimizing the argument of the exponential in (13-27). The minimum value of $[\boldsymbol{\theta} - \mathbf{m}(k)]'\mathcal{Z}^{-1}(k)[\boldsymbol{\theta} - \mathbf{m}(k)]$ is zero, and this occurs when

$$[\boldsymbol{\theta} - \mathbf{m}(k)]\Big|_{\boldsymbol{\theta} = \hat{\boldsymbol{\theta}}_{\text{MAP}}(k)} = \mathbf{0} \tag{13-29}$$

i.e.,

$$\hat{\boldsymbol{\theta}}_{\text{MAP}}(k) = \mathbf{E}\{\boldsymbol{\theta}|\mathcal{Z}(k)\} \tag{13-30}$$

Comparing (13-30) and (13-9) we conclude that $\hat{\boldsymbol{\theta}}_{\text{MAP}}(k) = \hat{\boldsymbol{\theta}}_{\text{MS}}(k)$. □

The result in Theorems 13-2 is true regardless of the nature of the model relating $\boldsymbol{\theta}$ to $\mathcal{Z}(k)$. Of course, in order to use it, we must first establish that $\mathcal{Z}(k)$ and $\boldsymbol{\theta}$ are jointly Gaussian. Except for the linear model, which we examine in Lesson 14, this is very difficult to do.

PROBLEMS

13-1. Prove that $\hat{\boldsymbol{\theta}}_{\text{MS}}(k)$, given in (13-12), is unique.

13-2. Prove Corollary 13-2 by means of a direct proof.

13-3. Let $\boldsymbol{\theta}$ and $\mathcal{Z}(N)$ be zero mean $n \times 1$ and $N \times 1$ random vectors, respectively, with known second-order statistics, $\mathbf{P}_{\boldsymbol{\theta}}$, $\mathbf{P}_{\mathcal{Z}}$, $\mathbf{P}_{\boldsymbol{\theta}\mathcal{Z}}$, and $\mathbf{P}_{\mathcal{Z}\boldsymbol{\theta}}$. View $\mathcal{Z}(N)$ as a vector of measurements. It is desired to determine a linear estimator of $\boldsymbol{\theta}$,

$$\hat{\boldsymbol{\theta}}(N) = \mathbf{K}_L(N)\mathcal{Z}(N)$$

where $\mathbf{K}_L(N)$ is an $n \times N$ matrix that is chosen to minimize the mean-squared error $\mathbf{E}\{[\boldsymbol{\theta} - \mathbf{K}_L(N)\mathcal{Z}(N)]'[\boldsymbol{\theta} - \mathbf{K}_L(N)\mathcal{Z}(N)]\}$.

(a) Show that the gain matrix, which minimizes the mean-squared error, is $\mathbf{K}_L(N) = \mathbf{E}\{\boldsymbol{\theta}\mathcal{Z}'(N)\}[\mathbf{E}\{\mathcal{Z}(N)\mathcal{Z}'(N)\}]^{-1} = \mathbf{P}_{\boldsymbol{\theta}\mathcal{Z}}\,\mathbf{P}_{\mathcal{Z}}^{-1}$.

(b) Show that the covariance matrix, $\mathbf{P}(N)$, of the estimation error, $\tilde{\boldsymbol{\theta}}(N) = \boldsymbol{\theta} - \hat{\boldsymbol{\theta}}(N)$, is

$$\mathbf{P}(N) = \mathbf{P}_{\boldsymbol{\theta}} - \mathbf{P}_{\boldsymbol{\theta}\mathcal{Z}}\mathbf{P}_{\mathcal{Z}}^{-1}\mathbf{P}_{\mathcal{Z}\boldsymbol{\theta}}$$

(c) Relate the results obtained in this problem to those in Corollary 13-2.

13-4. For random vectors $\boldsymbol{\theta}$ and $\mathcal{Z}(k)$, the *linear projection*, $\boldsymbol{\theta}^*(k)$, of $\boldsymbol{\theta}$ on a Hilbert space spanned by $\mathcal{Z}(k)$ is defined as $\boldsymbol{\theta}^*(k) = \mathbf{a} + \mathbf{B}\mathcal{Z}(k)$, where $\mathbf{E}\{\boldsymbol{\theta}^*(k)\} = \mathbf{E}\{\boldsymbol{\theta}\}$ and $\mathbf{E}\{[\boldsymbol{\theta} - \boldsymbol{\theta}^*(k)]\mathcal{Z}'(k)\} = \mathbf{0}$. We denote the linear projection, $\boldsymbol{\theta}^*(k)$, as $\hat{\mathbf{E}}\{\boldsymbol{\theta}|\mathcal{Z}(k)\}$.

(a) Prove that the linear (i.e., affine), unbiased mean-squared estimator of $\boldsymbol{\theta}$, $\hat{\boldsymbol{\theta}}(k)$, is the linear projection of $\boldsymbol{\theta}$ on $\mathcal{Z}(k)$.

(b) Prove that the linear projection, $\boldsymbol{\theta}^*(k)$, of $\boldsymbol{\theta}$ on the Hilbert space spanned by

$\mathfrak{X}(k)$ is uniquely equal to the linear (i.e., affine), unbiased mean-squared estimator of $\boldsymbol{\theta}$, $\hat{\boldsymbol{\theta}}(k)$.

(c) For random vectors $\mathbf{x}$, $\mathbf{y}$, $\mathbf{z}$, where $\mathbf{y}$ and $\mathbf{z}$ are uncorrelated, prove that

$$\hat{\mathbf{E}}\{\mathbf{x}|\mathbf{y},\mathbf{z}\} = \hat{\mathbf{E}}\{\mathbf{x}|\mathbf{y}\} + \hat{\mathbf{E}}\{\mathbf{x}|\mathbf{z}\} - \mathbf{m}_\mathbf{x}$$

(d) For random vectors $\mathbf{x}$, $\mathbf{y}$, $\mathbf{z}$, where $\mathbf{y}$ and $\mathbf{z}$ are correlated, prove that

$$\hat{\mathbf{E}}\{\mathbf{x}|\mathbf{y},\mathbf{z}\} = \hat{\mathbf{E}}\{\mathbf{x}|\mathbf{y},\tilde{\mathbf{z}}\}$$

where

$$\tilde{\mathbf{z}} = \mathbf{z} - \hat{\mathbf{E}}\{\mathbf{z}|\mathbf{y}\}$$

so that

$$\hat{\mathbf{E}}\{\mathbf{x}|\mathbf{y},\mathbf{z}\} = \hat{\mathbf{E}}\{\mathbf{x}|\mathbf{y}\} + \hat{\mathbf{E}}\{\mathbf{x}|\tilde{\mathbf{z}}\} - \mathbf{m}_\mathbf{x}$$

Parts (c) and (d) show that the results given in Theorems 12-3 and 12-4 are "distribution free" within the class of linear (i.e., affine), unbiased, mean-squared estimators.

13-5. Consider the linear model $z(k) = 2\theta + n(k)$, where

$$p[n(k)] = \begin{cases} 1 - |n(k)|, & |n(k)| \le 1 \\ 0\quad, & \text{otherwise} \end{cases}$$

and

$$p(\theta) = \begin{cases} \frac{1}{2}, & |\theta| \le 1 \\ 0, & \text{otherwise} \end{cases}$$

A random sample of N measurements is available. Explain how to find the ML and MAP estimators of θ, and be sure to list all of the assumptions needed to obtain a solution (they are not all given).

Lesson 14

Estimation of Random Parameters: The Linear and Gaussian Model

INTRODUCTION

In this lesson we begin with the linear model

$$\mathcal{Z}(k) = \mathcal{H}(k)\boldsymbol{\theta} + \mathcal{V}(k) \tag{14-1}$$

where $\boldsymbol{\theta}$ is an $n \times 1$ vector of random unknown parameters, $\mathcal{H}(k)$ is deterministic, and $\mathcal{V}(k)$ is white Gaussian noise with known covariance matrix $\mathcal{R}(k)$. We also assume that $\boldsymbol{\theta}$ is multivariate Gaussian with known mean, $\mathbf{m}_\theta$, and covariance, $\mathbf{P}_\theta$, i.e.,

$$\boldsymbol{\theta} \sim N(\boldsymbol{\theta}; \mathbf{m}_\theta, \mathbf{P}_\theta) \tag{14-2}$$

and, that $\boldsymbol{\theta}$ and $\mathcal{V}(k)$ are Gaussian and mutually uncorrelated. Our main objectives are to compute $\hat{\boldsymbol{\theta}}_{MS}(k)$ and $\hat{\boldsymbol{\theta}}_{MAP}(k)$ for this linear Gaussian model, and to see how these estimators are related. We shall also illustrate many of our results for the deconvolution and state estimation examples which were described in Lesson 2.

MEAN-SQUARED ESTIMATOR

Because $\boldsymbol{\theta}$ and $\mathcal{V}(k)$ are Gaussian and mutually uncorrelated, they are jointly Gaussian. Consequently, $\mathcal{Z}(k)$ and $\boldsymbol{\theta}$ are also jointly Gaussian; thus (Corollary 13-1),

$$\hat{\boldsymbol{\theta}}_{MS}(k) = \mathbf{m}_\theta + \mathbf{P}_{\theta\mathcal{Z}}(k)\mathbf{P}_{\mathcal{Z}}^{-1}(k)[\mathcal{Z}(k) - \mathbf{m}_{\mathcal{Z}}(k)] \tag{14-3}$$

It is straightforward, using (14-1) and (14-2), to show that

$$\mathbf{m}_{\mathcal{Z}}(k) = \mathcal{H}(k)\mathbf{m}_\theta \tag{14-4}$$

$$\mathbf{P}_{\mathcal{Z}}(k) = \mathcal{H}(k)\mathbf{P}_\theta\mathcal{H}'(k) + \mathcal{R}(k) \tag{14-5}$$

and

$$\mathbf{P}_{\theta\mathcal{Z}}(k) = \mathbf{P}_\theta\mathcal{H}'(k) \tag{14-6}$$

consequently,

$$\boxed{\hat{\boldsymbol{\theta}}_{MS}(k) = \mathbf{m}_\theta + \mathbf{P}_\theta\mathcal{H}'(k)[\mathcal{H}(k)\mathbf{P}_\theta\mathcal{H}'(k) + \mathcal{R}(k)]^{-1}[\mathcal{Z}(k) - \mathcal{H}(k)\mathbf{m}_\theta]} \tag{14-7}$$

Observe that $\hat{\boldsymbol{\theta}}_{MS}(k)$ depends on all the given information, namely, $\mathcal{Z}(k)$, $\mathcal{H}(k)$, $\mathbf{m}_\theta$ and $\mathbf{P}_\theta$.

Next, we compute the error-covariance matrix, $\mathbf{P}_{MS}(k)$, that is associated with $\hat{\boldsymbol{\theta}}_{MS}(k)$ in (14-7). Because $\hat{\boldsymbol{\theta}}_{MS}(k)$ is unbiased, we know that $\mathbf{E}\{\tilde{\boldsymbol{\theta}}_{MS}(k)\} = \mathbf{0}$ for all k; thus,

$$\mathbf{P}_{MS}(k) = \mathbf{E}\{\tilde{\boldsymbol{\theta}}_{MS}(k)\tilde{\boldsymbol{\theta}}'_{MS}(k)\} \tag{14-8}$$

From (14-7), and the fact that $\tilde{\boldsymbol{\theta}}_{MS}(k) = \boldsymbol{\theta} - \hat{\boldsymbol{\theta}}_{MS}(k)$, we see that

$$\tilde{\boldsymbol{\theta}}_{MS}(k) = (\boldsymbol{\theta} - \mathbf{m}_\theta) - \mathbf{P}_\theta\mathcal{H}'(\mathcal{H}\mathbf{P}_\theta\mathcal{H}' + \mathcal{R})^{-1}(\mathcal{Z} - \mathcal{H}\mathbf{m}_\theta) \tag{14-9}$$

From (14-9), (14-8) and (14-1) it is a straightforward exercise to show that

$$\boxed{\mathbf{P}_{MS}(k) = \mathbf{P}_\theta - \mathbf{P}_\theta\mathcal{H}'(k)[\mathcal{H}(k)\mathbf{P}_\theta\mathcal{H}'(k) + \mathcal{R}(k)]^{-1}\mathcal{H}(k)\mathbf{P}_\theta} \tag{14-10}$$

Applying Matrix Inversion Lemma 4-1 to (14-10), we obtain the following alternate formula for $\mathbf{P}_{MS}(k)$,

$$\boxed{\mathbf{P}_{MS}(k) = [\mathbf{P}_\theta^{-1} + \mathcal{H}'(k)\mathcal{R}^{-1}(k)\mathcal{H}(k)]^{-1}} \tag{14-11}$$

Next, we express $\hat{\boldsymbol{\theta}}_{MS}(k)$ as an explicit function of $\mathbf{P}_{MS}(k)$. To do this, we note that

$$\begin{aligned}
\mathbf{P}_\theta\mathcal{H}'(\mathcal{H}\mathbf{P}_\theta\mathcal{H}' + \mathcal{R})^{-1} &= \mathbf{P}_\theta\mathcal{H}'(\mathcal{H}\mathbf{P}_\theta\mathcal{H}' + \mathcal{R})^{-1} \\
&\quad (\mathcal{H}\mathbf{P}_\theta\mathcal{H}' + \mathcal{R} - \mathcal{H}\mathbf{P}_\theta\mathcal{H}')\mathcal{R}^{-1} \\
&= \mathbf{P}_\theta\mathcal{H}'[\mathbf{I} - (\mathcal{H}\mathbf{P}_\theta\mathcal{H}' + \mathcal{R})^{-1}\mathcal{H}\mathbf{P}_\theta\mathcal{H}']\mathcal{R}^{-1} \\
&= [\mathbf{P}_\theta - \mathbf{P}_\theta\mathcal{H}'(\mathcal{H}\mathbf{P}_\theta\mathcal{H}' + \mathcal{R})^{-1}\mathcal{H}\mathbf{P}_\theta]\mathcal{H}'\mathcal{R}^{-1} \\
&= \mathbf{P}_{MS}\mathcal{H}'\mathcal{R}^{-1}
\end{aligned} \tag{14-12}$$

hence,

$$\boxed{\hat{\boldsymbol{\theta}}_{MS}(k) = \mathbf{m}_\theta + \mathbf{P}_{MS}(k)\mathcal{H}'(k)\mathcal{R}^{-1}(k)[\mathcal{Z}(k) - \mathcal{H}(k)\mathbf{m}_\theta]} \tag{14-13}$$

Theorem 14-1. *If* $\mathbf{P}_\theta^{-1} = \mathbf{0}$, *and,* $\mathcal{H}(k)$ *is deterministic, then*

$$\hat{\boldsymbol{\theta}}_{MS}(k) = \hat{\boldsymbol{\theta}}_{BLU}(k) \quad (14\text{-}14)$$

Proof. Set $\mathbf{P}_\theta^{-1} = \mathbf{0}$ in (14-11), to see that

$$\mathbf{P}_{MS}(k) = [\mathcal{H}'(k)\mathcal{R}^{-1}(k)\mathcal{H}(k)]^{-1} \quad (14\text{-}15)$$

and, therefore,

$$\hat{\boldsymbol{\theta}}_{MS}(k) = [\mathcal{H}'(k)\mathcal{R}^{-1}(k)\mathcal{H}(k)]^{-1}\mathcal{H}'(k)\mathcal{R}^{-1}(k)\mathcal{Z}(k) \quad (14\text{-}16)$$

Compare (14-16) and (9-22), to conclude that $\hat{\boldsymbol{\theta}}_{MS}(k) = \hat{\boldsymbol{\theta}}_{BLU}(k)$. □

One of the most startling aspects of Theorem 14-1 is that it shows us that *BLU estimation applies to random parameters as well as to deterministic parameters.* We return to a reexamination of BLUE below.

What does the condition $\mathbf{P}_\theta^{-1} = \mathbf{0}$, given in Theorem 14-1, mean? Suppose, for example, that the elements of $\boldsymbol{\theta}$ are uncorrelated; then, $\mathbf{P}_\theta$ is a diagonal matrix, with diagonal elements $\sigma_{i,\theta}^2$. When all of these variances are very large, then $\mathbf{P}_\theta^{-1} = \mathbf{0}$. A large variance for θ_i means we have no idea where θ_i is located about its mean value.

Example 14-1 (Minimum-Variance Deconvolution)

In Example 2-6 we showed that, for the application of deconvolution, our linear model is

$$\mathcal{Z}(N) = \mathcal{H}(N-1)\boldsymbol{\mu} + \mathcal{V}(N) \quad (14\text{-}17)$$

We shall assume that $\boldsymbol{\mu}$ and $\mathcal{V}(N)$ are jointly Gaussian, and, that $\mathbf{m}_\mu = \mathbf{0}$ and $\mathbf{m}_\mathcal{V} = \mathbf{0}$; hence, $\mathbf{m}_\mathcal{Z} = \mathbf{0}$. Additionally, we assume that $\text{cov}[\mathcal{V}(N)] = \rho\mathbf{I}$. From (14-7), we determine the following formula for $\hat{\boldsymbol{\mu}}_{MS}(N)$,

$$\hat{\boldsymbol{\mu}}_{MS}(N) = \mathbf{P}_\mu\mathcal{H}'(N-1)[\mathcal{H}(N-1)\mathbf{P}_\mu\mathcal{H}'(N-1) + \rho\mathbf{I}]^{-1}\mathcal{Z}(N) \quad (14\text{-}18)$$

Recall, from Example 2-6, that when $\mu(k)$ is described by the product model $\mu(k) = q(k)r(k)$, then

$$\boldsymbol{\mu} = \mathbf{Q}_q\mathbf{r} \quad (14\text{-}19)$$

where

$$\mathbf{Q}_q = \text{diag}[q(1), q(2), \ldots, q(N)] \quad (14\text{-}20)$$

and

$$\mathbf{r} = \text{col}(r(1), r(2), \ldots, r(N)) \quad (14\text{-}21)$$

In the product model, $r(k)$ is white Gaussian noise with variance σ_r^2, and $q(k)$ is a Bernoulli sequence. Obviously, if we know $\mathbf{Q}_q$ then $\boldsymbol{\mu}$ is Gaussian, in which case

$$\mathbf{P}_\mu = \mathbf{Q}_q^2\sigma_r^2 = \mathbf{Q}_q\sigma_r^2 \quad (14\text{-}22)$$

where we have used the fact that $\mathbf{Q}_q^2 = \mathbf{Q}_q$, because $q(k) = 0$ or 1. When $\mathbf{Q}_q$ is known, (14-18) becomes

$$\hat{\boldsymbol{\mu}}_{MS}(N|\mathbf{Q}_q) = \sigma_r^2\mathbf{Q}_q\mathcal{H}'(N-1)[\sigma_r^2\mathcal{H}(N-1)\mathbf{Q}_q\mathcal{H}'(N-1) + \rho\mathbf{I}]^{-1}\mathcal{Z}(N) \quad (14\text{-}23)$$

Although $\hat{\boldsymbol{\mu}}_{MS}$ is a mean-squared estimator, so that it enjoys all of the properties of such an estimator (e.g., unbiased, minimum-variance, etc.), $\hat{\boldsymbol{\mu}}_{MS}$ is not a consistent estimator of $\boldsymbol{\mu}$. Consistency is a large sample property of an estimator; however, as N increases, the dimension of $\boldsymbol{\mu}$ increases, because $\boldsymbol{\mu}$ is $N \times 1$. Consequently, we cannot prove consistency of $\hat{\boldsymbol{\mu}}_{MS}$ (recall that, in all other problems, $\boldsymbol{\theta}$ is $n \times 1$, where n is data independent; in these problems we can study consistency of $\hat{\boldsymbol{\theta}}$).

Equations (14-18) and (14-23) are not very practical for actually computing $\hat{\boldsymbol{\mu}}_{MS}$, because both require the inversion of an $N \times N$ matrix, and N can become quite large (it equals the number of measurements). We return to a more practical way for computing $\hat{\boldsymbol{\mu}}_{MS}$ in Lesson 22. □

BEST LINEAR UNBIASED ESTIMATION, REVISITED

In Lesson 9 we derived the BLUE of $\boldsymbol{\theta}$ for the linear model (14-1), under the following assumptions about this model:

1. $\boldsymbol{\theta}$ is a deterministic but unknown vector of parameters,
2. $\mathcal{H}(k)$ is deterministic, and
3. $\mathcal{V}(k)$ is zero-mean noise with covariance matrix $\mathcal{R}(k)$.

We assumed that $\hat{\boldsymbol{\theta}}_{BLU}(k) = \mathbf{F}(k)\mathcal{Z}(k)$ and chose $\mathbf{F}_{BLU}(k)$ so that $\hat{\boldsymbol{\theta}}_{BLU}(k)$ is an unbiased estimator of $\boldsymbol{\theta}$, and the error variance for each one of the n elements of $\boldsymbol{\theta}$ is minimized. The reader should return to the derivation of $\hat{\boldsymbol{\theta}}_{BLU}(k)$ to see that the assumption "$\boldsymbol{\theta}$ is deterministic" is never needed, either in the derivation of the unbiasedness constraint [see the proof of Theorem 6-1, in which Equation (6-9) becomes $[\mathbf{I} - \mathbf{F}(k)\mathcal{H}(k)]\mathbf{E}\{\boldsymbol{\theta}\} = \mathbf{0}$, if $\boldsymbol{\theta}$ is random], or in the derivation of $J_i(\mathbf{f}_i, \boldsymbol{\lambda}_i)$ in Equation (9-14) (due to some remarkable cancellations); thus, $\hat{\boldsymbol{\theta}}_{BLU}(k)$, *given in (9-22), is applicable to random as well as deterministic parameters in our linear model (14-1); and, because the BLUE of* $\boldsymbol{\theta}$ *is the special case of the WLSE when* $\mathbf{W}(k) = \mathcal{R}^{-1}(k)$, $\hat{\boldsymbol{\theta}}_{WLS}(k)$, *given in (3-10), is also applicable to random as well as deterministic parameters in our linear model.*

Theorem 14-1 relates $\hat{\boldsymbol{\theta}}_{MS}(k)$ and $\hat{\boldsymbol{\theta}}_{BLU}(k)$ under some very stringent conditions that are needed in order to remove the dependence of $\hat{\boldsymbol{\theta}}_{MS}$ on the a priori statistical information about $\boldsymbol{\theta}$ (i.e., $\mathbf{m}_\theta$ and $\mathbf{P}_\theta$), because this information was never used in the derivation of $\hat{\boldsymbol{\theta}}_{BLU}(k)$.

Next, we derive a different BLUE of $\boldsymbol{\theta}$, one that incorporates the a priori statistical information about $\boldsymbol{\theta}$. To do this (Sorenson, 1980, pg. 210), we treat $\mathbf{m}_\theta$ as an additional measurement which will be augmented to $\mathcal{Z}(k)$. Our additional measurement equation is obtained by adding and subtracting $\boldsymbol{\theta}$ in the identity $\mathbf{m}_\theta = \mathbf{m}_\theta$, i.e.,

$$\mathbf{m}_\theta = \boldsymbol{\theta} + (\mathbf{m}_\theta - \boldsymbol{\theta}) \tag{14-24}$$

Quantity $\mathbf{m}_\theta - \boldsymbol{\theta}$ is now treated as zero-mean noise with covariance matrix $\mathbf{P}_\theta$. Our augmented linear model is

$$\underbrace{\left(\frac{\mathcal{Z}(k)}{\mathbf{m}_\theta}\right)}_{\mathcal{Z}_a} = \underbrace{\left(\frac{\mathcal{H}(k)}{\mathbf{I}}\right)}_{\mathcal{H}_a}\boldsymbol{\theta} + \underbrace{\left(\frac{\mathcal{V}(k)}{\mathbf{m}_\theta - \boldsymbol{\theta}}\right)}_{\mathcal{V}_a} \tag{14-25}$$

which can be written, as

$$\mathcal{Z}_a(k) = \mathcal{H}_a(k)\boldsymbol{\theta} + \mathcal{V}_a(k) \tag{14-26}$$

where $\mathcal{Z}_a(k)$, $\mathcal{H}_a(k)$, and $\mathcal{V}_a(k)$ are defined in (14-25). Additionally,

$$\mathbf{E}\{\mathcal{V}_a(k)\mathcal{V}_a'(k)\} \triangleq \mathcal{R}_a(k) = \left(\begin{array}{c|c} \mathcal{R}(\mathbf{k}) & \mathbf{0} \\ \hline \mathbf{0} & \mathbf{P}_\theta \end{array}\right) \tag{14-27}$$

We now treat (14-26) as the starting point for derivation of a BLUE of $\boldsymbol{\theta}$, which we denote $\hat{\boldsymbol{\theta}}^a_{\text{BLU}}(k)$. Obviously,

$$\boxed{\hat{\boldsymbol{\theta}}^a_{\text{BLU}}(k) = [\mathcal{H}_a'(k)\mathcal{R}_a^{-1}(k)\mathcal{H}_a(k)]^{-1}\mathcal{H}_a'(k)\mathcal{R}_a^{-1}(k)\mathcal{Z}_a(k)} \tag{14-28}$$

Theorem 14-2. *For the linear Gaussian model, when $\mathcal{H}(\mathrm{k})$ is deterministic it is always true that*

$$\hat{\boldsymbol{\theta}}_{\text{MS}}(k) = \hat{\boldsymbol{\theta}}^a_{\text{BLU}}(k) \tag{14-29}$$

Proof. To begin, substitute the definitions of $\mathcal{H}_a$, $\mathcal{R}_a$, and $\mathcal{Z}_a$ into (14-28), and multiply out all the partitioned matrices, to show that

$$\hat{\boldsymbol{\theta}}^a_{\text{BLU}}(k) = (\mathbf{P}_\theta^{-1} + \mathcal{H}'\mathcal{R}^{-1}\mathcal{H})^{-1}(\mathbf{P}_\theta^{-1}\mathbf{m}_\theta + \mathcal{H}'\mathcal{R}^{-1}\mathcal{Z}) \tag{14-30}$$

From (14-11), we recognize $(\mathbf{P}_\theta^{-1} + \mathcal{H}'\mathcal{R}^{-1}\mathcal{H})^{-1}$ as $\mathbf{P}_{\text{MS}}$; hence,

$$\hat{\boldsymbol{\theta}}^a_{\text{BLU}}(k) = \mathbf{P}_{\text{MS}}(k)[\mathbf{P}_\theta^{-1}\mathbf{m}_\theta + \mathcal{H}'(k)\mathcal{R}^{-1}(k)\mathcal{Z}(k)] \tag{14-31}$$

Next, we rewrite $\hat{\boldsymbol{\theta}}_{\text{MS}}(k)$ in (14-13), as

$$\hat{\boldsymbol{\theta}}_{\text{MS}}(k) = [\mathbf{I} - \mathbf{P}_{\text{MS}}\mathcal{H}'\mathcal{R}^{-1}\mathcal{H}]\mathbf{m}_\theta + \mathbf{P}_{\text{MS}}\mathcal{H}'\mathcal{R}^{-1}\mathcal{Z} \tag{14-32}$$

however,

$$\mathbf{I} - \mathbf{P}_{\text{MS}}\mathcal{H}'\mathcal{R}^{-1}\mathcal{H} = \mathbf{P}_{\text{MS}}(\mathbf{P}_{\text{MS}}^{-1} - \mathcal{H}'\mathcal{R}^{-1}\mathcal{H}) = \mathbf{P}_{\text{MS}}\mathbf{P}_\theta^{-1} \tag{14-33}$$

where we have again used (14-11). Substitute (14-33) into (14-32) to see that

$$\hat{\boldsymbol{\theta}}_{\text{MS}}(k) = \mathbf{P}_{\text{MS}}(k)[\mathbf{P}_\theta^{-1}\mathbf{m}_\theta + \mathcal{H}'(k)\mathcal{R}^{-1}(k)\mathcal{Z}(k)] \tag{14-34}$$

hence, (14-29) follows when we compare (14-31) and (14-34). □

To conclude this section, we note that the weighted-least squares objective function that is associated with $\hat{\boldsymbol{\theta}}^a_{\text{BLU}}(k)$ is

$$\begin{aligned} J_a[\hat{\boldsymbol{\theta}}^a(k)] &= \tilde{\mathcal{Z}}_a'(k)\mathcal{R}_a^{-1}(k)\tilde{\mathcal{Z}}_a(k) \\ &= (\mathbf{m}_\theta - \boldsymbol{\theta})'\mathbf{P}_\theta^{-1}(\mathbf{m}_\theta - \boldsymbol{\theta}) + \tilde{\mathcal{Z}}'(k)\mathcal{R}^{-1}(k)\tilde{\mathcal{Z}}(k) \end{aligned} \tag{14-35}$$

The first term in (14-35) contains all the a priori information about $\boldsymbol{\theta}$. Quantity $\mathbf{m}_\theta - \boldsymbol{\theta}$ is treated as the difference between "measurement" $\mathbf{m}_\theta$ and its noise-free model, $\boldsymbol{\theta}$.

MAXIMUM A POSTERIORI ESTIMATOR

In order to determine $\hat{\boldsymbol{\theta}}_{\text{MAP}}(k)$ for the linear model in (14-1) we first need to determine $p(\boldsymbol{\theta}|\mathcal{Z}(k))$. Using the facts that $\boldsymbol{\theta} \sim N(\boldsymbol{\theta}; \mathbf{m}_\theta, \mathbf{P}_\theta)$ and $\mathcal{V}(k) \sim N(\mathcal{V}(k); \mathbf{0}, \mathcal{R}(k))$, it follows that

$$p(\boldsymbol{\theta}) = \frac{1}{\sqrt{(2\pi)^n|\mathbf{P}_\theta|}} \exp\left\{-\frac{1}{2}(\boldsymbol{\theta} - \mathbf{m}_\theta)'\mathbf{P}_\theta^{-1}(\boldsymbol{\theta} - \mathbf{m}_\theta)\right\} \tag{14-36}$$

and

$$p(\mathcal{Z}(k)|\boldsymbol{\theta}) = \frac{1}{\sqrt{(2\pi)^N|\mathcal{R}(k)|}} \exp\left\{-\frac{1}{2}[\mathcal{Z}(k) - \mathcal{H}(k)\boldsymbol{\theta}]'\mathcal{R}^{-1}(k)[\mathcal{Z}(k) - \mathcal{H}(k)\boldsymbol{\theta}]\right\} \tag{14-37}$$

hence,

$$\ln p(\boldsymbol{\theta}|\mathcal{Z}) \propto -\frac{1}{2}(\boldsymbol{\theta} - \mathbf{m}_\theta)'\mathbf{P}_\theta^{-1}(\boldsymbol{\theta} - \mathbf{m}_\theta) - \frac{1}{2}(\mathcal{Z} - \mathcal{H}\boldsymbol{\theta})'\mathcal{R}^{-1}(\mathcal{Z} - \mathcal{H}\boldsymbol{\theta}) \tag{14-38}$$

To find $\hat{\boldsymbol{\theta}}_{\text{MAP}}(k)$ we must maximize $\ln p(\boldsymbol{\theta}|\mathcal{Z})$ in (14-38). Note that to find $\hat{\boldsymbol{\theta}}^a_{\text{BLU}}(k)$ we had to maximize $J_a[\hat{\boldsymbol{\theta}}^a(k)]$ in (14-35); but,

$$\ln p(\boldsymbol{\theta}|\mathcal{Z}) \propto -J_a[\hat{\boldsymbol{\theta}}^a(k)] \tag{14-39}$$

This means that maximizing $\ln p(\boldsymbol{\theta}|\mathcal{Z})$ leads to the same value of $\hat{\boldsymbol{\theta}}$ as does minimizing $J_a[\hat{\boldsymbol{\theta}}^a(k)]$. We have, therefore, shown that $\hat{\boldsymbol{\theta}}_{\text{MAP}}(k) = \hat{\boldsymbol{\theta}}^a_{\text{BLU}}(k)$.

Theorem 14-3. *For the linear Gaussian model, when* $\mathcal{H}(\text{k})$ *is deterministic it is always true that*

$$\hat{\boldsymbol{\theta}}_{\text{MAP}}(k) = \hat{\boldsymbol{\theta}}^a_{\text{BLU}}(k) \quad \square \tag{14-40}$$

Combining the results in Theorems 14-2 and 14-3, we have the very important result, that *for the linear Gaussian model*

$$\hat{\boldsymbol{\theta}}_{\text{MS}}(k) = \hat{\boldsymbol{\theta}}^a_{\text{BLU}}(k) = \hat{\boldsymbol{\theta}}_{\text{MAP}}(k) \tag{14-41}$$

Put another way, for the linear Gaussian model, all roads lead to the same estimator.

Of course, the fact that $\hat{\boldsymbol{\theta}}_{MS}(k) = \hat{\boldsymbol{\theta}}_{MAP}(k)$ should not come as any surprise, because we already established it (in a model-free environment) in Theorem 13-2.

Example 14-2 (Maximum-Likelihood Deconvolution)

As in Example 14-1, we begin with the deconvolution linear model

$$\mathcal{Z}(N) = \mathcal{H}(N-1)\boldsymbol{\mu} + \mathcal{V}(N) \tag{14-42}$$

Now, however, we use the product model for $\boldsymbol{\mu}$, given in (14-19), to express $\mathcal{Z}(N)$ as

$$\mathcal{Z}(N) = \mathcal{H}(N-1)\mathbf{Q_q r} + \mathcal{V}(N) \tag{14-43}$$

For notational convenience, let

$$\mathbf{q} = \text{col}\,(q(1), q(2), \ldots, q(N)) \tag{14-44}$$

Our objectives in this example are to obtain MAP estimators for both $\mathbf{q}$ and $\mathbf{r}$. In the literature on maximum-likelihood deconvolution (e.g., Mendel, 1983) these estimators are referred to as unconditional ML estimators, and are denoted $\hat{\mathbf{r}}$ and $\hat{\mathbf{q}}$. We denote these estimators as $\hat{\mathbf{r}}_{MAP}$ and $\hat{\mathbf{q}}_{MAP}$, in order to be consistent with this book's notation.

The starting point for determining $\hat{\mathbf{q}}_{MAP}$ and $\hat{\mathbf{r}}_{MAP}$ is the joint density function $p(\mathbf{r}, \mathbf{q}|\mathcal{Z}(N))$, where

$$p(\mathbf{r},\mathbf{q}|\mathcal{Z}(N)) \propto p(\mathcal{Z}(N)|\mathbf{r},\mathbf{q})p(\mathbf{r},\mathbf{q}) \tag{14-45}$$

but

$$p(\mathbf{r},\mathbf{q}) = p(\mathbf{r}|\mathbf{q})\Pr(\mathbf{q}) \tag{14-46}$$

Equation (14-46) uses a probability function for $\mathbf{q}$ rather than a probability density function, because $q(k)$ takes on only two discrete values, 0 and 1. Substituting (14-46) into (14-45), we find that

$$p(\mathbf{r},\mathbf{q}|\mathcal{Z}(N)) \propto p(\mathcal{Z}(N)|\mathbf{r},\mathbf{q})p(\mathbf{r}|\mathbf{q})\Pr(\mathbf{q}) \tag{14-47}$$

Note, also, that

$$\begin{aligned} p(\mathbf{r},\mathcal{Z}(N)|\mathbf{q}) &= p(\mathcal{Z}(N)|\mathbf{r},\mathbf{q})p(\mathbf{r}|\mathbf{q}) \\ &= p(\mathcal{Z}(N)|\mathbf{r},\mathbf{q})p(\mathbf{r}) \end{aligned} \tag{14-48}$$

where we have used the fact that $\mathbf{r}$ and $\mathbf{q}$ are statistically independent. Substituting (14-48) for $p(\mathcal{Z}(N)|\mathbf{r},\mathbf{q})$ into (14-47), we see that

$$p(\mathbf{r},\mathbf{q}|\mathcal{Z}(N)) \propto p(\mathbf{r},\mathcal{Z}(N)|\mathbf{q})\Pr(\mathbf{q}) \tag{14-49}$$

Observe that the $\mathbf{r}$ dependence of the MAP-likelihood function is completely contained in $p(\mathbf{r},\mathcal{Z}(N)|\mathbf{q})$. Additionally, when $\mathbf{q}$ is given, the only remaining random quantities are the zero-mean Gaussian quantities $\mathbf{r}$ and $\mathcal{Z}(N)$; hence, $p(\mathbf{r},\mathcal{Z}(N)|\mathbf{q})$ is multivariate Gaussian. Letting

$$\mathbf{x} = \text{col}\,(\mathbf{r},\mathcal{Z}(N)) \tag{14-50}$$

and

$$\mathbf{P}_{\mathbf{x}} = \mathbf{E}\{\mathbf{x}\mathbf{x}'|\mathbf{q}\} \tag{14-51}$$

then

$$p(\mathbf{r},\mathcal{Z}(N)|\mathbf{q}) = (2\pi)^{-N}|\mathbf{P}_{\mathbf{x}}|^{-1/2}\exp\left(-\frac{1}{2}\mathbf{x}'\mathbf{P}_{\mathbf{x}}^{-1}\mathbf{x}\right) \tag{14-52}$$

We leave, as an exercise for the reader, the maximization of $p(\mathbf{r},\mathbf{q}|\mathcal{Z}(N))$, from which it follows (Mendel, 1983b, pp. 112–114) that

$$\mathbf{r}^*(N|\mathbf{q}) = \sigma_r^2\mathbf{Q}_{\mathbf{q}}\mathcal{H}'(N-1)[\sigma_r^2\mathcal{H}(N-1)\mathbf{Q}_{\mathbf{q}}\mathcal{H}'(N-1) + \rho\mathbf{I}]^{-1}\mathcal{Z}(N) \tag{14-53}$$

$\hat{\mathbf{q}}_{\text{MAP}}$ can be found by maximizing

$$p(\mathbf{r}^*,\mathbf{q}|\mathcal{Z}(N)) \propto p(\mathbf{r}^*,\mathcal{Z}(N)|\mathbf{q})\Pr(\mathbf{q}) = (2\pi)^{-N}(\sigma_r^2\rho)^{-N/2}$$
$$\exp\left\{-\frac{1}{2}\mathcal{Z}'(\mathbf{N})[\sigma_r^2\mathcal{H}(N-1)\mathbf{Q}_{\mathbf{q}}\mathcal{H}'(N-1) + \rho\mathbf{I}]^{-1}\mathcal{Z}(N)\right\}\Pr(\mathbf{q}) \tag{14-54}$$

where

$$\Pr(\mathbf{q}) = \prod_{k=1}^{N}\Pr[q(k)] = \lambda^{m_q}(1-\lambda)^{N-m_q} \tag{14-55}$$

and

$$m_q = \sum_{k=1}^{N} q(k) \tag{14-56}$$

and finally,

$$\hat{\mathbf{r}}_{\text{MAP}} = \mathbf{r}^*(N|\hat{\mathbf{q}}_{\text{MAP}}) \tag{14-57}$$

Equations (14-54) and (14-57) are quite interesting results. Observe that we are permitted first to direct our attention to finding $\hat{\mathbf{q}}_{\text{MAP}}$ and then to finding $\hat{\mathbf{r}}_{\text{MAP}}$. Observe, also, that $\hat{\mathbf{r}}_{\text{MAP}} = \hat{\boldsymbol{\mu}}_{\text{MS}}(N|\mathbf{Q}_{\mathbf{q}})$ [compare (14-53) and (14-23)].

There is no simple solution for determining $\hat{\mathbf{q}}_{\text{MAP}}$. Because the elements of $\mathbf{q}$ in $p(\mathbf{r}^*,\mathbf{q}|\mathcal{Z}(N))$ have nonlinear interactions, and because the elements of $\mathbf{q}$ are constrained to take on binary values, it is necessary to evaluate $p(r^*,\mathbf{q}|\mathcal{Z}(N))$ for every possible $\mathbf{q}$ sequence to find the $\mathbf{q}$ for which $p(r^*,\mathbf{q}|\mathcal{Z}(\mathbf{N}))$ is a global maximum. Because $q(k)$ can take on one of two possible values, there are 2^N possible sequences where N is the number of elements in $\mathbf{q}$. For reasonable values of N (such as $N = 400$), finding the global maximum of $p(r^*,\mathbf{q}|\mathcal{Z}(N))$ would require several centuries of computer time.

We can always design a method for *detecting* significant values of $\mu(k)$ so that the resulting $\hat{\mathbf{q}}$ will be nearly as likely as the unconditional maximum-likelihood estimate, $\hat{\mathbf{q}}_{\text{MAP}}$. Two MAP detectors for accomplishing this are described in Mendel (1983b, pp. 127–137). [Note: The reader who is interested in more details about these MAP detectors should first review Chapter 5 in Mendel, 1983b, because they are designed using a slightly different likelihood function than the one in (14-54).] □

Example 14-3 (State-Estimation)

In Example 2-4 we showed that, for the application of state estimation, our linear model is

$$\mathscr{Z}(N) = \mathscr{H}(N, k_1)\mathbf{x}(k_1) + \mathscr{V}(N, k_1) \tag{14-58}$$

From (2-17), we see that

$$\mathbf{x}(k_1) = \mathbf{\Phi}^{k_1}\mathbf{x}(0) + \sum_{i=1}^{k_1} \mathbf{\Phi}^{k_1 - i}\gamma u(i-1) = \mathbf{\Phi}^{k_1}\mathbf{x}(0) + \mathbf{L}\mathbf{u} \tag{14-59}$$

where

$$\mathbf{u} = \text{col}\,(u(0), u(1), \ldots, u(N)) \tag{14-60}$$

and $\mathbf{L}$ is an $n \times (N+1)$ matrix, the exact structure of which is not important for this example. Additionally, from (2-21), we see that

$$\mathscr{V}(N, k_1) = \mathbf{M}(N, k_1)\mathbf{u} + \mathbf{v} \tag{14-61}$$

where

$$\mathbf{v} = \text{col}\,(v(1), v(2), \ldots, v(N)) \tag{14-62}$$

Observe that both $\mathbf{x}(k_1)$ and $\mathscr{V}(N, k_1)$ can be viewed as linear functions of $\mathbf{x}(0)$, $\mathbf{u}$ and $\mathbf{v}$.

We now assume that $\mathbf{x}(0)$, $\mathbf{u}$ and $\mathbf{v}$ are jointly Gaussian. Reasons for doing this are discussed in Lesson 15. Then, because $\mathbf{x}(k_1)$ and $\mathscr{V}(N, k_1)$ are linear functions of $\mathbf{x}(0)$, $\mathbf{u}$ and $\mathbf{v}$, $\mathbf{x}(k_1)$ and $\mathscr{V}(N, k_1)$ are jointly Gaussian (Papoulis, 1965, pg. 253); hence,

$$\begin{aligned}\hat{\mathbf{x}}_{\text{MS}}(k_1|N) = \mathbf{m}_{\mathbf{x}(\mathbf{k}_1)} + \mathbf{P}_{\mathbf{x}(\mathbf{k}_1)}\mathscr{H}'(N, k_1)[\mathscr{H}(N, k_1)\mathbf{P}_{\mathbf{x}(\mathbf{k}_1)}\mathscr{H}'(N, k_1) \\ + \mathscr{R}(N, k_1)]^{-1}[\mathscr{Z}(N) - \mathscr{H}(N, k_1)\mathbf{m}_{\mathbf{x}(k_1)}]\end{aligned} \tag{14-63}$$

and

$$\boxed{\hat{\mathbf{x}}_{\text{MAP}}(k_1|N) = \hat{\mathbf{x}}_{\text{MS}}(k_1|N)} \tag{14-64}$$

In order to evaluate $\hat{\mathbf{x}}_{\text{MS}}(k_1|N)$ we must first compute $\mathbf{m}_{\mathbf{x}(k_1)}$ and $\mathbf{P}_{\mathbf{x}(k_1)}$. We show how to do this in Lesson 15.

Formula (14-63) is very cumbersome. It appears that its right-hand side changes as a function of k_1 (and N). *We conjecture, however, that it ought to be possible to express $\hat{\mathbf{x}}_{\text{MS}}(k_1|N)$ as an affine transformation of $\hat{\mathbf{x}}_{\text{MS}}(k_1 - 1|N)$*, because $\mathbf{x}(k_1)$ is an affine transformation of $\mathbf{x}(k_1 - 1)$, i.e., $\mathbf{x}(k_1) = \mathbf{\Phi}\mathbf{x}(k_1 - 1) + \gamma u(k_1 - 1)$.

Because of the importance of state estimation in many different fields (e.g., control theory, communication theory, signal processing, etc.) we shall examine it in great detail in many of our succeeding lessons. □

PROBLEMS

14-1. Derive Eq. (14-10) for $\mathbf{P}_{\text{MS}}(k)$.

14-2. Show that $\mathbf{r}^*(N|\mathbf{q})$ is given by Eq. (14-53), and that $\hat{\mathbf{q}}_{\text{MAP}}$ can be found by maximizing (14-54).

14-3. x and v are independent Gaussian random variables with zero means and variances σ_x^2 and σ_v^2, respectively. We observe the single measurement $z = x + v = 1$.

(a) Find $\hat{x}_{\text{ML}}$.

(b) Find $\hat{x}_{\text{MAP}}$.

14-4. For the linear Gaussian model in which $\mathcal{H}(k)$ is deterministic, prove that $\hat{\boldsymbol{\theta}}_{\text{MAP}}(k)$ is a most efficient estimator of $\boldsymbol{\theta}$. Do this in two different ways. Is $\hat{\boldsymbol{\theta}}_{\text{MS}}(k)$ a most efficient estimator of $\boldsymbol{\theta}$?

Lesson 15

Elements of Discrete-Time Gauss-Markov Random Processes

INTRODUCTION

Lessons 13 and 14 have demonstrated the importance of Gaussian random variables in estimation theory. In this lesson we extend some of the basic concepts that were introduced in Lesson 12, for Gaussian random variables, to indexed random variables, namely random processes. These extensions are needed in order to develop state estimators.

DEFINITIONS AND PROPERTIES OF DISCRETE-TIME GAUSS-MARKOV RANDOM PROCESSES

Recall that a random process is a collection of random variables in which the notion of time plays a role.

Definition 15-1 (Meditch, 1969, pg. 106). *A vector random process is a family of random vectors* $\{\mathbf{s}(t), t \in \mathcal{I}\}$ *indexed by a parameter* t *all of whose values lie in some appropriate index set* $\mathcal{I}$. *When* $\mathcal{I} = \{k: k = 0, 1, \ldots\}$ *we have a discrete-time random process.* □

Definition 15-2 (Meditch, 1969, pg. 117). *A vector random process* $\{\mathbf{s}(t), t \in \mathcal{I}\}$ *is defined to be multivariate Gaussian if, for* **any** ℓ *time points* $t_1, t_2, \ldots, t_\ell$ *in* $\mathcal{I}$ *where* ℓ *is an integer, the set of* ℓ *random* n *vectors* $\mathbf{s}(t_1), \mathbf{s}(t_2), \ldots, \mathbf{s}(t_\ell)$ *is jointly Gaussian distributed.* □

Let $\mathcal{S}(l)$ be defined as

$$\mathcal{S}(l) = \text{col}\,(\mathbf{s}(t_1), \mathbf{s}(t_2), \ldots, \mathbf{s}(t_l)) \tag{15-1}$$

then, Definition 15-2 means that

$$p[\mathcal{S}(l)] = (2\pi)^{-nl/2}|\mathbf{P}_{\mathcal{S}}|^{-1/2} \exp\left\{-\frac{1}{2}[\mathcal{S}(l) - \mathbf{m}_{\mathcal{S}}(l)]'\mathbf{P}_{\mathcal{S}}^{-1}(l)[\mathcal{S}(l) - \mathbf{m}_{\mathcal{S}}(l)]\right\} \tag{15-2}$$

in which

$$\mathbf{m}_{\mathcal{S}}(l) = \mathbf{E}\{\mathcal{S}(l)\} \tag{15-3}$$

and $\mathbf{P}_{\mathcal{S}}(l)$ is the $nl \times nl$ matrix $\mathbf{E}\{[\mathcal{S}(l) - \mathbf{m}_{\mathcal{S}}(l)][\mathcal{S}(l) - \mathbf{m}_{\mathcal{S}}(l)]'\}$ with elements $\mathbf{P}_{\mathcal{S}}(i, j)$, where

$$\mathbf{P}_{\mathcal{S}}(i, j) = \mathbf{E}\{[\mathbf{s}(t_i) - \mathbf{m}_s(t_i)][\mathbf{s}(t_j) - \mathbf{m}_s(t_j)]'\} \tag{15-4}$$

$i, j = 1, 2, \ldots, l$.

Definition 15-3 (Meditch, 1969, pg. 118). *A vector random process* $\{\mathrm{s(t)}, \mathrm{t}\epsilon\mathcal{J}\}$ *is a Markov process, if, for* **any** m *time points* $\mathrm{t}_1 < \mathrm{t}_2 < \ldots < \mathrm{t}_\mathrm{m}$ *in* $\mathcal{J}$, *where* m *is any integer, it is true that*

$$\Pr[\mathbf{s}(t_m) \le \mathbf{S}(t_m)|\mathbf{s}(t_{m-1}) = \mathbf{S}(t_{m-1}), \ldots, \mathbf{s}(t_1) = \mathbf{S}(t_1)] = \Pr[\mathbf{s}(t_m) \le \mathbf{S}(t_m)|\mathbf{s}(t_{m-1}) = \mathbf{S}(t_{m-1})] \tag{15-5}$$

For continuous random variables, this means that

$$p[\mathbf{s}(t_m)|\mathbf{s}(t_{m-1}), \ldots, \mathbf{s}(t_1)] = p[\mathbf{s}(t_m)|\mathbf{s}(t_{m-1})] \quad \square \tag{15-6}$$

Note that, in (15-5), $\mathbf{s}(t_m) \le \mathbf{S}(t_m)$ means $s_i(t_m) \le S_i(t_m)$ for $i = 1, 2, \ldots, n$. If we view time point t_m as the present time and time points $t_{m-1}, \ldots, t_1$ as the past, then a Markov process is one whose probability law (e.g., probability density function) depends only on the immediate past value, t_{m-1}. This is often referred to as the *Markov property* for a vector random process. Because the probability law depends only on the immediate past value we often refer to such a process as a *first-order Markov process* (if it depended on the immediate two past values it would be a second-order Markov process).

Theorem 15-1. *Let* $\{\mathrm{s(t)}, \mathrm{t}\epsilon\mathcal{J}\}$ *be a first-order Markov process, and* $\mathrm{t}_1 < \mathrm{t}_2 < \ldots < \mathrm{t}_\mathrm{m}$ *be any time points in* $\mathcal{J}$, *where* m *is an integer. Then*

$$p[\mathbf{s}(t_m), \mathbf{s}(t_{m-1}), \ldots, \mathbf{s}(t_1)] = p[\mathbf{s}(t_m)|\mathbf{s}(t_{m-1})]p[\mathbf{s}(t_{m-1})|\mathbf{s}(t_{m-2})] \ldots p[\mathbf{s}(t_2)|\mathbf{s}(t_1)]p[\mathbf{s}(t_1)] \tag{15-7}$$

Proof. From probability theory (e.g., Papoulis, 1965) and the Markov property of $\mathbf{s}(t)$, we know that

$$\begin{aligned} p[\mathbf{s}(t_m), \mathbf{s}(t_{m-1}), \ldots, \mathbf{s}(t_1)] \\ &= p[\mathbf{s}(t_m)|\mathbf{s}(t_{m-1}), \ldots, \mathbf{s}(t_1)]p[\mathbf{s}(t_{m-1}), \ldots, \mathbf{s}(t_1)] \\ &= p[\mathbf{s}(t_m)|\mathbf{s}(t_{m-1})]p[\mathbf{s}(t_{m-1}), \ldots, \mathbf{s}(t_1)] \end{aligned} \tag{15-8}$$

In a similar manner, we find that

$$\left.\begin{aligned} p[\mathbf{s}(t_{m-1}), \ldots, \mathbf{s}(t_1)] &= p[\mathbf{s}(t_{m-1})|\mathbf{s}(t_{m-2})]p[\mathbf{s}(t_{m-2}), \ldots, \mathbf{s}(t_1)] \\ p[\mathbf{s}(t_{m-2}), \ldots, \mathbf{s}(t_1)] &= p[\mathbf{s}(t_{m-2})|\mathbf{s}(t_{m-3})]p[\mathbf{s}(t_{m-3}), \ldots, \mathbf{s}(t_1)] \\ &\ldots \\ p[\mathbf{s}(t_2), \mathbf{s}(t_1)] &= p[\mathbf{s}(t_2)|\mathbf{s}(t_1)]p[\mathbf{s}(t_1)] \end{aligned}\right\} \tag{15-9}$$

Equation (15-7) is obtained by successively substituting each one of the equations in (15-9) into (15-8). □

Theorem 15-1 demonstrates that a first-order Markov process is completely characterized by two probability density functions, namely, the *transition probability density function*, $p[\mathbf{s}(t_i)|\mathbf{s}(t_{i-1})]$, and the initial (prior) *probability density function* $p[\mathbf{s}(t_1)]$. Note that generally the transition probability density functions can all be different, in which case they should be subscripted [e.g., $p_m[\mathbf{s}(t_m)|\mathbf{s}(t_{m-1})]$ and $p_{m-1}[\mathbf{s}(t_{m-1})|\mathbf{s}(t_{m-2})]$].

Theorem 15-2. *For a first-order Markov process,*

$$\mathbf{E}\{\mathbf{s}(t_m)|\mathbf{s}(t_{m-1}), \ldots, \mathbf{s}(t_1)\} = E\{\mathbf{s}(t_m)|\mathbf{s}(t_{m-1})\} \quad \square \tag{15-10}$$

We leave the proof of this useful result as an exercise.

A vector random process that is both Gaussian and a first-order Markov process will be referred to in the sequel as a Gauss-Markov process.

Definition 15-4. *A vector random process* $\{\mathbf{s}(t), t \in \mathcal{J}\}$ *is said to be a Gaussian white process if*, for **any** m *time points* $t_1, t_2, \ldots, t_m$ *in* $\mathcal{J}$, *where* m *is any integer, the* m *random vectors* $\mathbf{s}(t_1), \mathbf{s}(t_2), \ldots, \mathbf{s}(t_m)$ *are uncorrelated Gaussian random vectors.* □

White noise is zero mean, or else it cannot have a flat spectrum. For white noise

$$\mathbf{E}\{\mathbf{s}(t_i)\mathbf{s}'(t_j)\} = \mathbf{0} \qquad \text{for all } i \neq j \tag{15-11}$$

Additionally, for Gaussian white noise

$$p[\mathcal{S}(l)] = p[\mathbf{s}(t_1)]p[\mathbf{s}(t_2)]\ldots p[\mathbf{s}(t_l)] \tag{15-12}$$

[because uncorrelatedness implies statistical independence (see Lesson 12)] where $p[\mathbf{s}(t_i)]$ is a multivariate Gaussian probability density function.

Theorem 15-3. *A vector Gaussian white process* $\mathbf{s}(t)$ *can be veiwed as a first-order Gauss-Markov process for which*

$$p[\mathbf{s}(t)|\mathbf{s}(\tau)] = p[\mathbf{s}(t)] \tag{15-13}$$

for all t, $\tau \epsilon \mathcal{I}$ *and* t $\neq \tau$.

Proof. For a Gaussian white process, we know, from (15-12), that

$$p[\mathbf{s}(t), \mathbf{s}(\tau)] = p[\mathbf{s}(t)]p[\mathbf{s}(\tau)] \tag{15-14}$$

but, we also know that

$$p[\mathbf{s}(t), \mathbf{s}(\tau)] = p[\mathbf{s}(t)|\mathbf{s}(\tau)]p[\mathbf{s}(\tau)] \tag{15-15}$$

Equating (15-14) and (15-15), we obtain (15-13). □

Theorem 15-3 means that past and future values of $\mathbf{s}(t)$ in no way help determine present values of $\mathbf{s}(t)$. For Gaussian white processes, the transition probability density function equals the marginal density function, $p[\mathbf{s}(t)]$, which is multivariate Gaussian. Additionally,

$$\mathbf{E}\{\mathbf{s}(t_m)|\mathbf{s}(t_{m-1}), \ldots, \mathbf{s}(t_1)\} = \mathbf{E}\{\mathbf{s}(t_m)\} \tag{15-16}$$

A BASIC STATE-VARIABLE MODEL

In succeeding lessons we shall develop a variety of state estimators for the following basic linear, (possibly) time-varying, discrete-time dynamical system (our *basic state-variable model*), which is characterized by $n \times 1$ state vector $\mathbf{x}(k)$ and $m \times 1$ measurement vector $\mathbf{z}(k)$:

$$\boxed{\begin{aligned}\mathbf{x}(k+1) = \mathbf{\Phi}(k+1,k)\mathbf{x}(k) \\ + \mathbf{\Gamma}(k+1,k)\mathbf{w}(k) + \mathbf{\Psi}(k+1,k)\mathbf{u}(k)\end{aligned}} \tag{15-17}$$

and

$$\boxed{\mathbf{z}(k+1) = \mathbf{H}(k+1)\mathbf{x}(k+1) + \mathbf{v}(k+1)} \tag{15-18}$$

where $k = 0, 1, \ldots$. In this model $\mathbf{w}(k)$ and $\mathbf{v}(k)$ are $p \times 1$ and $m \times 1$ mutually uncorrelated (possibly nonstationary) jointly Gaussian white noise sequences; i.e.,

$$\boxed{\mathbf{E}\{\mathbf{w}(i)\mathbf{w}'(j)\} = \mathbf{Q}(i)\delta_{ij}} \tag{15-19}$$

$$\boxed{\mathbf{E}\{\mathbf{v}(i)\mathbf{v}'(j)\} = \mathbf{R}(i)\delta_{ij}} \tag{15-20}$$

and

$$\boxed{E\{\mathbf{w}(i)\mathbf{v}'(j)\} = \mathbf{S} = \mathbf{0} \qquad \text{for all } i \text{ and } j} \tag{15-21}$$

Covariance matrix $\mathbf{Q}(i)$ is positive semidefinite and $\mathbf{R}(i)$ is positive definite [so that $\mathbf{R}^{-1}(i)$ exists]. Additionally, $\mathbf{u}(k)$ is an $l \times 1$ vector of known system inputs, and initial state vector $\mathbf{x}(0)$ is multivariate Gaussian, with mean $\mathbf{m}_\mathbf{x}(0)$ and covariance $\mathbf{P}_\mathbf{x}(0)$, i.e.,

$$\boxed{\mathbf{x}(0) \sim N(\mathbf{x}(0); \mathbf{m}_\mathbf{x}(0), \mathbf{P}_\mathbf{x}(0))} \tag{15-22}$$

and, $\mathbf{x}(0)$ is not correlated with $\mathbf{w}(k)$ and $\mathbf{v}(k)$. The dimensions of matrices $\mathbf{\Phi}$, $\mathbf{\Gamma}$, $\mathbf{\Psi}$, $\mathbf{H}$, $\mathbf{Q}$ and $\mathbf{R}$ are $n \times n$, $n \times p$, $n \times l$, $m \times n$, $p \times p$, and $m \times m$, respectively.

Disturbance $\mathbf{w}(k)$ is often used to model the following types of uncertainty:

1. disturbance forces acting on the system (e.g., wind that buffets an airplane);
2. errors in modeling the system (e.g., neglected effects); and
3. errors, due to actuators, in the translation of the known input, $\mathbf{u}(k)$, into physical signals.

Vector $\mathbf{v}(k)$ is often used to model the following types of uncertainty:

1. errors in measurements made by sensing instruments;
2. unavoidable disturbances that act directly on the sensors; and
3. errors in the realization of feedback compensators using physical components [this is valid only when the measurement equation contains a direct throughput of the input $\mathbf{u}(k)$, i.e., when $\mathbf{z}(k+1) = \mathbf{H}(k+1)\mathbf{x}(k+1) + \mathbf{G}(k+1)\mathbf{u}(k+1) + \mathbf{v}(k+1)$; we shall examine this situation in Lesson 23].

Of course, not all dynamical systems are described by this basic model. In general, $\mathbf{w}(k)$ and $\mathbf{v}(k)$ may be correlated, some measurements may be made so accurate that, for all practical purposes, they are "perfect" (i.e., there is no measurement noise associated with them), and either $\mathbf{u}(k)$ or $\mathbf{v}(k)$, or both, may be colored noise processes. We shall consider the modification of our basic state-variable model for each of these important situations in Lesson 23.

PROPERTIES OF THE BASIC STATE-VARIABLE MODEL

In this section we state and prove a number of important statistical properties for our basic state-variable model.

Theorem 15-4. *When* $\mathbf{x}(0)$ *and* $\mathbf{w}(\mathrm{k})$ *are jointly Gaussian then* $\{\mathbf{x}(\mathrm{k}), \mathrm{k} = 0, 1, \ldots\}$ *is a Gauss-Markov sequence.*

Note that if $\mathbf{x}(0)$ and $\mathbf{w}(k)$ are individually Gaussian and statistically independent (or uncorrelated), then they will be jointly Gaussian (Papoulis, 1965).

Proof

a. Gaussian Property [assuming $\mathbf{u}(k)$ nonrandom]. Because $\mathbf{u}(k)$ is nonrandom, it has no effect on determining whether $\mathbf{x}(k)$ is Gaussian; hence, for this part of the proof we assume $\mathbf{u}(k) = \mathbf{0}$. The solution to (15-17) is

$$\mathbf{x}(k) = \mathbf{\Phi}(k,0)\mathbf{x}(0) + \sum_{i=1}^{k} \mathbf{\Phi}(k,i)\mathbf{\Gamma}(i,i-1)\mathbf{w}(i-1) \tag{15-23}$$

where

$$\mathbf{\Phi}(k,i) = \mathbf{\Phi}(k,k-1)\mathbf{\Phi}(k-1,k-2)\ldots\mathbf{\Phi}(i+1,i) \tag{15-24}$$

Observe that $\mathbf{x}(k)$ is a linear transformation of jointly Gaussian random vectors $\mathbf{x}(0)$, $\mathbf{w}(0)$, $\mathbf{w}(1), \ldots, \mathbf{w}(k-1)$; hence, $\mathbf{x}(k)$ is Gaussian.

b. Markov Property. This property does not require $\mathbf{x}(k)$ or $\mathbf{w}(k)$ to be Gaussian. Because $\mathbf{x}$ satisfies state equation (15-17), we see that $\mathbf{x}(k)$ depends only on its immediate past value; hence, $\mathbf{x}(k)$ is Markov. □

We have been able to show that our dynamical system is Markov because we specified a model for it. Without such a specification, it would be quite difficult (or impossible) to test for the Markov nature of a random process.

By stacking up $\mathbf{x}(1), \mathbf{x}(2), \ldots$ into a supervector it is easily seen that this supervector is just a linear transformation of jointly Gaussian quantities $\mathbf{x}(0)$, $\mathbf{w}(0), \mathbf{w}(1), \ldots$; hence, $\mathbf{x}(1), \mathbf{x}(2), \ldots$ are themselves *jointly Gaussian*.

A Gauss-Markov sequence can be completely characterized in two ways:

1. specify the marginal density of the initial state vector, $p[\mathbf{x}(0)]$, and the transition density $p[\mathbf{x}(k+1)|\mathbf{x}(k)]$, or
2. specify the mean and covariance of the state vector sequence. The second characterization is a complete one because Gaussian random vectors are completely characterized by their means and covariances (Lesson 12). We shall find the second characterization more useful than the first.

The Gaussian density function for state vector $\mathbf{x}(k)$ is

$$p[\mathbf{x}(k)] = [(2\pi)^n|\mathbf{P}_\mathbf{x}(k)|]^{-1/2} \exp\left\{-\frac{1}{2}[\mathbf{x}(k) - \mathbf{m}_\mathbf{x}(k)]'\mathbf{P}_\mathbf{x}^{-1}(k)[\mathbf{x}(k) - \mathbf{m}_\mathbf{x}(k)]\right\} \tag{15-25}$$

where

$$\mathbf{m}_\mathbf{x}(k) = \mathbf{E}\{\mathbf{x}(k)\} \tag{15-26}$$

and

$$\mathbf{P}_\mathbf{x}(k) = \mathbf{E}\{[\mathbf{x}(k) - \mathbf{m}_\mathbf{x}(k)][\mathbf{x}(k) - \mathbf{m}_\mathbf{x}(k)]'\} \tag{15-27}$$

We now demonstrate that $\mathbf{m}_\mathbf{x}(k)$ and $\mathbf{P}_\mathbf{x}(k)$ can be computed by means of recursive equations.

Theorem 15-5. *For our basic state-variable model,*

a. $\mathbf{m}_\mathbf{x}(\mathrm{k})$ *can be computed from the vector recursive equation*

$$\mathbf{m}_\mathbf{x}(k+1) = \mathbf{\Phi}(k+1,k)\mathbf{m}_\mathbf{x}(k) + \mathbf{\Psi}(k+1,k)\mathbf{u}(k) \tag{15-28}$$

where $\mathrm{k} = 0, 1, \ldots,$ *and* $\mathbf{m}_\mathbf{x}(0)$ *initializes (15-28),*

b. $\mathbf{P}_\mathbf{x}(\mathrm{k})$ *can be computed from the matrix recursive equation*

$$\mathbf{P}_\mathbf{x}(k+1) = \mathbf{\Phi}(k+1,k)\mathbf{P}_\mathbf{x}(k)\mathbf{\Phi}'(k+1,k) + \mathbf{\Gamma}(k+1,k)\mathbf{Q}(k)\mathbf{\Gamma}'(k+1,k) \tag{15-29}$$

where $\mathrm{k} = 0, 1, \ldots,$ *and* $\mathbf{P}_\mathbf{x}(0)$ *initializes (15-29), and*

c. $\mathbf{E}\{[\mathbf{x}(\mathrm{i}) - \mathbf{m}_\mathbf{x}(\mathrm{i})][\mathbf{x}(\mathrm{j}) - \mathbf{m}_\mathbf{x}(\mathrm{j})]'\} \triangleq \mathbf{P}_\mathbf{x}(\mathrm{i}, \mathrm{j})$ *can be computed from*

$$\mathbf{P}_\mathbf{x}(i,j) = \begin{cases} \mathbf{\Phi}(i,j)\mathbf{P}_\mathbf{x}(j) & \text{when } i > j \\ \mathbf{P}_\mathbf{x}(i)\mathbf{\Phi}'(j,i) & \text{when } i < j \end{cases} \tag{15-30}$$

Proof

a. Take the expected value of both sides of (15-17), using the facts that expectation is a linear operation (Papoulis, 1965) and $\mathbf{w}(k)$ is zero mean, to obtain (15-28).
b. For notational simplicity, we omit the temporal arguments of $\mathbf{\Phi}$ and $\mathbf{\Gamma}$ in this part of the proof. Using (15-17) and (15-28), we obtain

$$\begin{aligned}\mathbf{P}_\mathbf{x}(k+1) &= \mathbf{E}\{[\mathbf{x}(k+1) - \mathbf{m}_\mathbf{x}(k+1)][\mathbf{x}(k+1) - \mathbf{m}_\mathbf{x}(k+1)]'\} \\ &= \mathbf{E}\{[\mathbf{\Phi}[\mathbf{x}(k) - \mathbf{m}_\mathbf{x}(k)] + \mathbf{\Gamma}\mathbf{w}(k)] \\ &\quad [\mathbf{\Phi}[\mathbf{x}(k) - \mathbf{m}_\mathbf{x}(k)] + \mathbf{\Gamma}\mathbf{w}(k)]'\} \\ &= \mathbf{\Phi}\mathbf{P}_\mathbf{x}(k)\mathbf{\Phi}' + \mathbf{\Gamma}\mathbf{Q}(k)\mathbf{\Gamma}' + \mathbf{\Phi}\mathbf{E}\{[\mathbf{x}(k) - \mathbf{m}_\mathbf{x}(k)]\mathbf{w}'(k)\}\mathbf{\Gamma}' \\ &\quad + \mathbf{\Gamma}\mathbf{E}\{\mathbf{w}(k)[\mathbf{x}(k) - \mathbf{m}_\mathbf{x}(k)]'\}\mathbf{\Phi}'\end{aligned} \tag{15-31}$$

Because $\mathbf{m}_\mathbf{x}(k)$ is not random and $\mathbf{w}(k)$ is zero mean, $\mathbf{E}\{\mathbf{m}_\mathbf{x}(k)\mathbf{w}'(k)\} =$

$\mathbf{m_x}(k)\mathbf{E}\{\mathbf{w}'(k)\} = \mathbf{0}$, and $\mathbf{E}\{\mathbf{w}(k)\mathbf{m}'_\mathbf{x}(k)\} = \mathbf{0}$. State vector $\mathbf{x}(k)$ depends at most on random input $\mathbf{w}(k-1)$ [see (15-17)]; hence,

$$\mathbf{E}\{\mathbf{x}(k)\mathbf{w}'(k)\} = \mathbf{E}\{\mathbf{x}(k)\}\mathbf{E}\{\mathbf{w}'(k)\} = \mathbf{0} \tag{15-32}$$

and $\mathbf{E}\{\mathbf{w}(k)\mathbf{x}'(k)\} = \mathbf{0}$ as well. The last two terms in (15-31) are therefore equal to zero, and the equation reduces to (15-29).

c. We leave the proof of (15-30) as an exercise. Observe that once we know covariance matrix $\mathbf{P_x}(k)$ it is an easy matter to determine any cross-covariance matrix between state $\mathbf{x}(k)$ and $\mathbf{x}(i)(i \neq k)$. The Markov nature of our basic state-variable model is responsible for this. □

Observe that mean vector $\mathbf{m_x}(k)$ satisfies a deterministic vector state equation, (15-28), covariance matrix $\mathbf{P_x}(k)$ satisfies a deterministic matrix state equation, (15-29), and (15-28) and (15-29) are easily programmed for digital computation.

Next we direct our attention to the statistics of measurement vector $\mathbf{z}(k)$.

Theorem 15-6. *For our basic state-variable model, when* $\mathbf{x}(0)$, $\mathbf{w}(\mathrm{k})$ *and* $\mathbf{v}(\mathrm{k})$ *are jointly Gaussian, then* $\{\mathbf{z}(\mathrm{k}), \mathrm{k} = 1, 2, \ldots\}$ *is Gaussian, and*

$$\mathbf{m_z}(k+1) = \mathbf{H}(k+1)\mathbf{m_x}(k+1) \tag{15-33}$$

and

$$\mathbf{P_z}(k+1) = \mathbf{H}(k+1)\mathbf{P_x}(k+1)\mathbf{H}'(k+1) + \mathbf{R}(k+1) \tag{15-34}$$

where $\mathbf{m_x}(\mathrm{k}+1)$ *and* $\mathbf{P_x}(\mathrm{k}+1)$ *are computed from (15-28) and (15-29), respectively.* □

We leave the proof as an exercise for the reader. Note that if $\mathbf{x}(0)$, $\mathbf{w}(k)$ and $\mathbf{v}(k)$ are statistically independent and Gaussian they will be jointly Gaussian.

Example 15-1

Consider the simple single-input single-output first-order system

$$x(k+1) = \frac{1}{2}x(k) + w(k) \tag{15-35}$$

$$z(k+1) = x(k+1) + v(k+1) \tag{15-36}$$

where $w(k)$ and $v(k)$ are wide-sense stationary white noise processes, for which $q = 20$ and $r = 5$. Additionally, $m_x(0) = 4$ and $p_x(0) = 10$.

The mean of $x(k)$ is computed from the following homogeneous equation

$$m_x(k+1) = \frac{1}{2}m_x(k) \qquad m_x(0) = 4 \tag{15-37}$$

and, the variance of $x(k)$ is computed from Equation (15-29), which in this case simplifies to

$$p_x(k+1) = \frac{1}{4}p_x(k) + 20 \qquad p_x(0) = 10 \tag{15-38}$$

Additionally, the mean and variance of $z(k)$ are computed from

$$m_z(k+1) = m_x(k+1) \tag{15-39}$$

and

$$p_z(k+1) = p_x(k+1) + 5 \tag{15-40}$$

Figure 15-1 depicts $m_x(k)$ and $p_x^{1/2}(k)$. Observe that $m_x(k)$ decays to zero very rapidly and that $p_x^{1/2}(k)$ approaches a steady-state value. $\bar{p}_x^{1/2} = 5.163$. This steady-state value can be computed from equation (15-38) by setting $p_x(k) = p_x(k+1) = \bar{p}_x$. The existence of $\bar{p}_x$ is guaranteed by our first-order system being stable.

Although $m_x(k) \rightarrow 0$ there is a lot of uncertainty about $x(k)$, as evidenced by the large value of $\bar{p}_x$.There will be an even larger uncertainty about $z(k)$, because $\bar{p}_z \rightarrow 31.66$. These large values for $\bar{p}_z$ and $\bar{p}_x$ are due to the large values of q and r. In many practical applications, both q and r will be much less than unity in which case $\bar{p}_z$ and $\bar{p}_x$ will be quite small. □

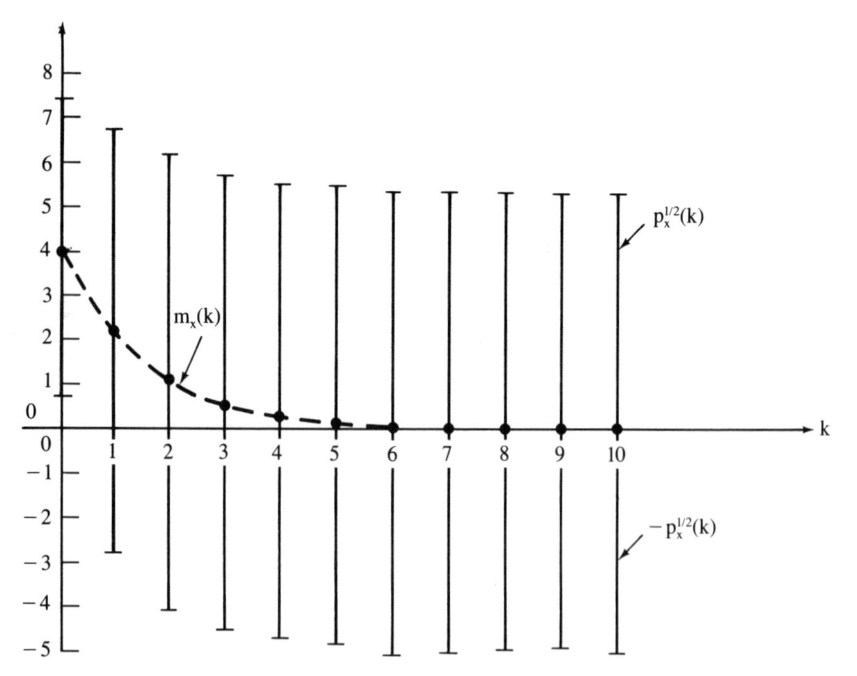

Figure 15-1 Mean (dashed) and standard deviation (bars) for first-order system (15-35) and (15-36).

If our basic state-variable model is time-invariant and stationary, and, if $\mathbf{\Phi}$ is associated with an asymptotically stable system (i.e., one whose poles all lie within the unit circle), then (Anderson and Moore, 1979) matrix $\mathbf{P_x}(k)$ reaches a limiting (steady-state) solution $\overline{\mathbf{P}}_\mathbf{x}$, i.e.,

$$\lim_{k\to\infty} \mathbf{P_x}(k) = \overline{\mathbf{P}}_\mathbf{x} \tag{15-41}$$

Matrix $\overline{\mathbf{P}}_\mathbf{x}$ is the solution of the following steady-state version of (15-29),

$$\overline{\mathbf{P}}_\mathbf{x} = \mathbf{\Phi}\overline{\mathbf{P}}_\mathbf{x}\mathbf{\Phi}' + \mathbf{\Gamma Q \Gamma}' \tag{15-42}$$

This equation is called a *discrete-time Lyapunov equation*. See Laub (1979) for an excellent numerical method that can be used to solve (15-42) for $\overline{\mathbf{P}}_\mathbf{x}$.

SIGNAL-TO-NOISE RATIO

In this section we simplify our basic state-variable model (15-17) and (15-18) to a time-invariant, stationary, single-input single-output model:

$$\mathbf{x}(k+1) = \mathbf{\Phi}\mathbf{x}(k) + \boldsymbol{\gamma} w(k) + \boldsymbol{\psi} u(k) \tag{15-43}$$

$$z(k+1) = \mathbf{h}'\mathbf{x}(k+1) + v(k+1) \tag{15-44}$$

Measurement $z(k)$ is of the classical form of signal plus noise, where "signal" $s(k) = \mathbf{h}'\mathbf{x}(k)$.

The *signal-to-noise ratio* is an often-used measure of quality of measurement $z(k)$. Here we define that ratio, denoted by SNR(k), as

$$\text{SNR}(k) = \frac{\sigma_s^2(k)}{r} \tag{15-45}$$

From preceding analyses, we see that

$$\text{SNR}(k) = \frac{\mathbf{h}'\mathbf{P_x}(k)\mathbf{h}}{r} \tag{15-46}$$

Because $\mathbf{P_x}(k)$ is in general a function of time, SNR(k) is also a function of time. If, however, $\mathbf{\Phi}$ is associated with an asymptotically stable system then (15-41) is true. In this case we can use $\overline{\mathbf{P}}_\mathbf{x}$ in (15-46) to provide us with a single number, $\overline{\text{SNR}}$, for the signal-to-noise ratio, i.e.,

$$\overline{\text{SNR}} = \frac{\mathbf{h}'\overline{\mathbf{P}}_\mathbf{x}\mathbf{h}}{r} \tag{15-47}$$

Finally, we demonstrate that SNR(k) (or $\overline{\text{SNR}}$) can be computed without knowing q and r explicitly; all that is needed is the ratio q/r. Multiplying and dividing the right-hand side of (15-46) by q, we find that

$$\text{SNR}(k) = \left[\mathbf{h}'\frac{\mathbf{P_x}(k)}{q}\mathbf{h}\right](q/r) \tag{15-48}$$

Scaled covariance matrix $\mathbf{P}_x(k)/q$ is computed from the following version of (15-29)

$$\frac{\mathbf{P}_x(k+1)}{q} = \mathbf{\Phi}\frac{\mathbf{P}_x(k)}{q}\mathbf{\Phi}' + \boldsymbol{\gamma}\boldsymbol{\gamma}' \tag{15-49}$$

One of the most useful ways for using (15-48) is to compute q/r for a given signal-to-noise ratio $\overline{\text{SNR}}$, i.e.,

$$\frac{q}{r} = \frac{\overline{\text{SNR}}}{\mathbf{h}'\dfrac{\overline{\mathbf{P}}_x}{q}\mathbf{h}} \tag{15-50}$$

In Lesson 18 we show that q/r can be viewed as an estimator tuning parameter; hence, signal-to-noise ratio, $\overline{\text{SNR}}$, can also be treated as such a parameter.

Example 15-2 (Mendel, 1981)

Consider the first-order system

$$x(k+1) = \phi x(k) + \gamma w(k) \tag{15-51}$$

$$z(k+1) = hx(k+1) + v(k) \tag{15-52}$$

In this case, it is easy to solve (15-49), to show that

$$\overline{P}_x/q = \gamma^2/(1-\phi^2) \tag{15-53}$$

hence,

$$\overline{\text{SNR}} = \left(\frac{h^2\gamma^2}{1-\phi^2}\right)\left(\frac{q}{r}\right) \tag{15-54}$$

Observe that, if $h^2\gamma^2 = 1-\phi^2$, then $\overline{\text{SNR}} = q/r$. The condition $h^2\gamma^2 = 1-\phi^2$ is satisfied if, for example, $\gamma = 1$, $\phi = 1/\sqrt{2}$ and $h = 1/\sqrt{2}$. □

PROBLEMS

15-1. Prove Theorem 15-2, and then show that for Gaussian white noise $\mathbf{E}\{\mathbf{s}(t_m)|\mathbf{s}(t_{m-1}), \ldots, \mathbf{s}(t_1)\} = \mathbf{E}\{\mathbf{s}(t_m)\}$.

15-2. Derive the formula for the cross-covariance of $\mathbf{x}(k)$, $\mathbf{P}_x(i,j)$, given in (15-30).

15-3. Derive the first- and second-order statistics of measurement vector $\mathbf{z}(k)$, that are summarized in Theorem 15-6.

15-4. Reconsider the basic state-variable model when $\mathbf{x}(0)$ is correlated with $\mathbf{w}(0)$, and $\mathbf{w}(k)$ and $\mathbf{v}(k)$ are correlated $[\mathbf{E}\{\mathbf{w}(k)\mathbf{v}'(k)\} = \mathbf{S}(k)]$.

(a) Show that the covariance equation for $\mathbf{z}(k)$ remains unchanged.

(b) Show that the covariance equation for $\mathbf{x}(k)$ is changed, but only at $k = 1$.

(c) Compute $\mathbf{E}\{\mathbf{z}(k+1)\mathbf{z}'(k)\}$.

15-5.

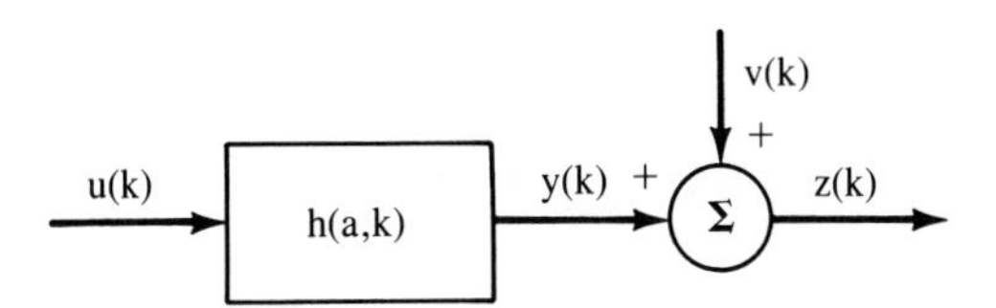

In this problem, assume that $u(k)$ and $v(k)$ are individually Gaussian and uncorrelated. Impulse response h depends on parameter a, where a is a Gaussian random variable that is statistically independent of $u(k)$ and $v(k)$.

(a) Evaluate $\mathbf{E}\{z(k)\}$.

(b) Explain whether or not $y(k)$ is Gaussian.

Lesson 16

State Estimation: Prediction

INTRODUCTION

We have mentioned, a number of times in this book, that in state estimation three situations are possible depending upon the relative relationship of total number of available measurements, N, and the time point, k, at which we estimate state vector $\mathbf{x}(k)$, namely: prediction ($N < k$), filtering ($N = k$), and smoothing ($N > k$). In this lesson we develop algorithms for mean-squared predicted estimates, $\hat{\mathbf{x}}_{MS}(k|j)$, of state $\mathbf{x}(k)$. In order to simplify our notation, we shall abbreviate $\hat{\mathbf{x}}_{MS}(k|j)$ as $\hat{\mathbf{x}}(k|j)$. (Just in case you have forgotten what the notation $\hat{\mathbf{x}}(k|j)$ stands for, see Lesson 2.) Note that, in prediction, $k > j$.

SINGLE-STAGE PREDICTOR

The most important predictor of $\mathbf{x}(k)$ for our future work on filtering and smoothing is the *single-stage predictor* $\hat{\mathbf{x}}(k|k-1)$. From the Fundamental Theorem of Estimation Theory (Theorem 13-1), we know that

$$\hat{\mathbf{x}}(k|k-1) = \mathbf{E}\{\mathbf{x}(k)|\mathfrak{Z}(k-1)\} \tag{16-1}$$

where

$$\mathfrak{Z}(k-1) = \mathrm{col}\,(\mathbf{z}(1), \mathbf{z}(2), \ldots, \mathbf{z}(k-1)) \tag{16-2}$$

It is very easy to derive a formula for $\hat{\mathbf{x}}(k|k-1)$ by operating on both sides of the state equation

$$\mathbf{x}(k) = \mathbf{\Phi}(k,k-1)\mathbf{x}(k-1) + \mathbf{\Gamma}(k,k-1)\mathbf{w}(k-1) + \mathbf{\Psi}(k,k-1)\mathbf{u}(k-1) \tag{16-3}$$

with the linear expectation operator $\mathbf{E}\{\cdot|\mathfrak{X}(k-1)\}$. Doing this, we find that

$$\hat{\mathbf{x}}(k|k-1) = \mathbf{\Phi}(k,k-1)\hat{\mathbf{x}}(k-1|k-1) + \mathbf{\Psi}(k,k-1)\mathbf{u}(k-1) \tag{16-4}$$

where $k = 1, 2, \ldots$. To obtain (16-4) we have used the facts that $\mathbf{E}\{\mathbf{w}(k-1)\} = \mathbf{0}$ and $\mathbf{u}(k-1)$ is deterministic.

Observe, from (16-4), that the single-stage predicted estimate, $\hat{\mathbf{x}}(k|k-1)$, depends on the filtered estimate, $\hat{\mathbf{x}}(k-1|k-1)$, of the preceding state vector $\mathbf{x}(k-1)$. At this point, (16-4) is an interesting theoretical result; but, there is nothing much we can do with it, because we do not as yet know how to compute filtered state estimates. In Lesson 17 we shall begin our study into filtered state estimates, and shall learn that such estimates of $\mathbf{x}(k)$ depend on predicted estimates of $\mathbf{x}(k)$, just as predicted estimates of $\mathbf{x}(k)$ depend on filtered estimates of $\mathbf{x}(k-1)$; thus, *filtered and predicted state estimates are very tightly coupled together.*

Let $\mathbf{P}(k|k-1)$ denote the error-covariance matrix that is associated with $\hat{\mathbf{x}}(k|k-1)$, i.e.,

$$\mathbf{P}(k|k-1) = \mathbf{E}\{[\tilde{\mathbf{x}}(k|k-1) - \mathbf{m}_{\tilde{\mathbf{x}}}(k|k-1)][\tilde{\mathbf{x}}(k|k-1) - \mathbf{m}_{\tilde{\mathbf{x}}}(k|k-1)]'\} \tag{16-5}$$

where

$$\tilde{\mathbf{x}}(k|k-1) = \mathbf{x}(k) - \hat{\mathbf{x}}(k|k-1) \tag{16-6}$$

Additionally, let $\mathbf{P}(k-1|k-1)$ denote the error-covariance matrix that is associated with $\hat{\mathbf{x}}(k-1|k-1)$, i.e.,

$$\mathbf{P}(k-1|k-1) = \mathbf{E}\{[\tilde{\mathbf{x}}(k-1|k-1) - \mathbf{m}_{\tilde{\mathbf{x}}}(k-1|k-1)][\tilde{\mathbf{x}}(k-1|k-1) - \mathbf{m}_{\tilde{\mathbf{x}}}(k-1|k-1)]'\} \tag{16-7}$$

For our basic state-variable model (see Property 1 of Lesson 13), $\mathbf{m}_{\tilde{\mathbf{x}}}(k|k-1) = \mathbf{0}$ and $\mathbf{m}_{\tilde{\mathbf{x}}}(k-1|k-1) = \mathbf{0}$, so that

$$\mathbf{P}(k|k-1) = \mathbf{E}\{\tilde{\mathbf{x}}(k|k-1)\tilde{\mathbf{x}}'(k|k-1)\} \tag{16-8}$$

and

$$\mathbf{P}(k-1|k-1) = \mathbf{E}\{\tilde{\mathbf{x}}(k-1|k-1)\tilde{\mathbf{x}}'(k-1|k-1)\} \tag{16-9}$$

Combining (16-3) and (16-4), we see that

$$\tilde{\mathbf{x}}(k|k-1) = \mathbf{\Phi}(k,k-1)\tilde{\mathbf{x}}(k-1|k-1) + \mathbf{\Gamma}(k,k-1)\mathbf{w}(k-1) \tag{16-10}$$

A straightforward calculation leads to the following formula for $\mathbf{P}(k|k-1)$,

$$\boxed{\begin{aligned}\mathbf{P}(k|k-1) &= \mathbf{\Phi}(k,k-1)\mathbf{P}(k-1|k-1)\mathbf{\Phi}'(k,k-1)\\ &\quad+\mathbf{\Gamma}(k,k-1)\mathbf{Q}(k-1)\mathbf{\Gamma}'(k,k-1)\end{aligned}} \tag{16-11}$$

where $k = 1, 2, \ldots$.

Observe, from (16-4) and (16-11), that $\hat{\mathbf{x}}(0|0)$ and $\mathbf{P}(0|0)$ initialize the single-stage predictor and its error-covariance. Additionally,

$$\hat{\mathbf{x}}(0|0) = \mathbf{E}\{\mathbf{x}(0)|\text{no measurements}\} = \mathbf{m}_{\mathbf{x}}(0) \tag{16-12}$$

and

$$\begin{aligned}\mathbf{P}(0|0) &= \mathbf{E}\{\tilde{\mathbf{x}}(0|0)\tilde{\mathbf{x}}'(0|0)\}\\ &= \mathbf{E}\{[\mathbf{x}(0) - \mathbf{m}_{\mathbf{x}}(0)][\mathbf{x}(0) - \mathbf{m}_{\mathbf{x}}(0)]'\} = \mathbf{P}_{\mathbf{x}}(0)\end{aligned} \tag{16-13}$$

Finally, recall (Property 4 of Lesson 13) that both $\hat{\mathbf{x}}(k|k-1)$ and $\tilde{\mathbf{x}}(k|k-1)$ are Gaussian.

A GENERAL STATE PREDICTOR

In this section we generalize the results of the preceding section so as to obtain predicted values of $\mathbf{x}(k)$ that look further into the future than just one step. We shall determine $\hat{\mathbf{x}}(k|j)$ where $k > j$ under the assumption that filtered state estimate $\hat{\mathbf{x}}(j|j)$ and its error-covariance matrix $\mathbf{E}\{\tilde{\mathbf{x}}(j|j)\tilde{\mathbf{x}}'(j|j)\} = \mathbf{P}(j|j)$ are known for some $j = 0, 1, \ldots$.

Theorem 16-1

a. *If input* $\mathbf{u}(\mathrm{k})$ *is deterministic, or does not depend on any measurements, then the mean-squared predicted estimator of* $\mathbf{x}(\mathrm{k})$, $\hat{\mathbf{x}}(\mathrm{k}|\mathrm{j})$, *is given by the expression*

$$\hat{\mathbf{x}}(k|j) = \mathbf{\Phi}(k,j)\hat{\mathbf{x}}(j|j) + \sum_{i=j+1}^{k} \mathbf{\Phi}(k,i)\mathbf{\Psi}(i,i-1)\mathbf{u}(i-1) \qquad k > j \tag{16-14}$$

b. *The vector random process* $\{\tilde{\mathbf{x}}(\mathrm{k}|\mathrm{j}), \mathrm{k} = \mathrm{j}+1, \mathrm{j}+2, \ldots\}$ *is:*
   i. *zero mean,*
   ii. *Gaussian, and*
   iii. *first-order Markov, and*
   iv. *its covariance matrix is governed by*

$$\begin{aligned}\mathbf{P}(k|j) &= \mathbf{\Phi}(k,k-1)\mathbf{P}(k-1|j)\mathbf{\Phi}'(k,k-1)\\ &\quad+\mathbf{\Gamma}(k,k-1)\mathbf{Q}(k-1)\mathbf{\Gamma}'(k,k-1)\end{aligned} \tag{16-15}$$

Before proving this theorem, let us observe that the prediction formula (16-14) is intuitively what one would expect. Why is this so? Suppose we have processed all of the measurements $\mathbf{z}(1), \mathbf{z}(2), \ldots, \mathbf{z}(j)$ to obtain $\hat{\mathbf{x}}(j|j)$ and are asked to predict the value of $\mathbf{x}(k)$, where $k > j$. No additional measurements can be used during prediction. All that we can therefore use is our dynamical state equation. When that equation is used for purposes of prediction we neglect the random disturbance term, because the disturbances are not measurable. We can only use measured quantities to assist our prediction efforts. The simplified state equation is

$$\mathbf{x}(k+1) = \mathbf{\Phi}(k+1,k)\mathbf{x}(k) + \mathbf{\Psi}(k+1,k)\mathbf{u}(k) \qquad (16\text{-}16)$$

a solution of which is

$$\mathbf{x}(k) = \mathbf{\Phi}(k,j)\mathbf{x}(j) + \sum_{i=j+1}^{k} \mathbf{\Phi}(k,i)\mathbf{\Psi}(i,i-1)\mathbf{u}(i-1) \qquad (16\text{-}17)$$

Substituting $\hat{\mathbf{x}}(j|j)$ for $\mathbf{x}(j)$, we obtain the predictor in (16-14). In our proof of Theorem 16-1 we establish (16-14) in a more rigorous manner.

Proof

a. The solution to state equation (16-3), for $\mathbf{x}(k)$, can be expressed in terms of $\mathbf{x}(j)$, where $j < k$, as

$$\mathbf{x}(k) = \mathbf{\Phi}(k,j)\mathbf{x}(j) + \sum_{i=j+1}^{k} \mathbf{\Phi}(k,i)[\mathbf{\Gamma}(i,i-1)\mathbf{w}(i-1) + \mathbf{\Psi}(i,i-1)\mathbf{u}(i-1)] \qquad (16\text{-}18)$$

We apply the Fundamental Theorem of Estimation Theory to (16-18) by taking the conditional expectation with respect to $\mathfrak{Z}(j)$ on both sides of it. Doing this, we find that

$$\hat{\mathbf{x}}(k|j) = \mathbf{\Phi}(k,j)\hat{\mathbf{x}}(j|j) + \sum_{i=j+1}^{k} \mathbf{\Phi}(k,i)[\mathbf{\Gamma}(i,i-1)\mathbf{E}\{\mathbf{w}(i-1)|\mathfrak{Z}(j)\} + \mathbf{\Psi}(i,i-1)\mathbf{E}\{\mathbf{u}(i-1)|\mathfrak{Z}(j)\}] \qquad (16\text{-}19)$$

Note that $\mathfrak{Z}(j)$ depends at most on $\mathbf{x}(j)$ which, in turn, depends at most on $\mathbf{w}(j-1)$. Consequently,

$$\mathbf{E}\{\mathbf{w}(i-1)|\mathfrak{Z}(j)\} = \mathbf{E}\{\mathbf{w}(i-1)|\mathbf{w}(0), \mathbf{w}(1), \ldots, \mathbf{w}(j-1)\} \qquad (16\text{-}20)$$

where $i = j+1, j+2, \ldots, k$. Because of this range of values on argument i, $\mathbf{w}(i-1)$ is never included in the conditioning set of values $\mathbf{w}(0)$, $\mathbf{w}(1), \ldots, \mathbf{w}(j-1)$; hence,

$$\mathbf{E}\{\mathbf{w}(i-1)|\mathfrak{Z}(j)\} = \mathbf{E}\{\mathbf{w}(i-1)\} = \mathbf{0} \qquad (16\text{-}21)$$

for all $i = j+1, j+2, \ldots, k$.

Note, also, that

$$\mathbf{E}\{\mathbf{u}(i-1)|\mathfrak{Z}(j)\} = \mathbf{E}\{\mathbf{u}(i-1)\} = \mathbf{u}(i-1) \qquad (16\text{-}22)$$

because we have assumed that $\mathbf{u}(i-1)$ does not depend on any of the measurements. Substituting (16-21) and (16-22) into (16-19), we obtain the prediction formula (16-14).

b-i. and b-ii. have already been proved in Properties 1 and 4 of Lesson 13.

b-iii. Starting with $\tilde{\mathbf{x}}(k|j) = \mathbf{x}(k) - \hat{\mathbf{x}}(k|j)$, and substituting (16-18) and (16-14) into this relation, we find that

$$\tilde{\mathbf{x}}(k|j) = \mathbf{\Phi}(k,j)\tilde{\mathbf{x}}(j|j) + \sum_{i=j+1}^{k} \mathbf{\Phi}(k,i)\mathbf{\Gamma}(i,i-1)\mathbf{w}(i-1) \tag{16-23}$$

This equation looks quite similar to the solution of state equation (16-3), when $\mathbf{u}(k) = \mathbf{0}$ (for all k), e.g., see (16-18). In fact, $\tilde{\mathbf{x}}(k|j)$ also satisfies the state equation

$$\tilde{\mathbf{x}}(k|j) = \mathbf{\Phi}(k,k-1)\tilde{\mathbf{x}}(k-1|j) + \mathbf{\Gamma}(k,k-1)\mathbf{w}(k-1) \tag{16-24}$$

Because $\tilde{\mathbf{x}}(k|j)$ depends only upon its previous value, $\tilde{\mathbf{x}}(k-1|j)$, it is first-order Markov.

b-iv. We derived a recursive covariance equation for $\mathbf{x}(k)$ in Theorem 15-5. That equation is (15-29). Because $\tilde{\mathbf{x}}(k|j)$ satisfies the state equation for $\mathbf{x}(k)$, its covariance $\mathbf{P}(k|j)$ is also given by (15-29). We have rewritten this equation as in (16-15). □

Observe that by setting $j = k-1$ in (16-14) we obtain our previously derived single-stage predictor $\hat{\mathbf{x}}(k|k-1)$.

Theorem 16-1 is quite limited because presently the only values of $\hat{\mathbf{x}}(j|j)$ and $\mathbf{P}(j|j)$ that we know are those at $j = 0$. For $j = 0$, (16-14) becomes

$$\hat{\mathbf{x}}(k|0) = \mathbf{\Phi}(k,0)\mathbf{m}_x(0) + \sum_{i=1}^{k} \mathbf{\Phi}(k,i)\mathbf{\Psi}(i,i-1)\mathbf{u}(i-1) \tag{16-25}$$

The reader might feel that this predictor of $\mathbf{x}(k)$ becomes poorer and poorer as k gets farther and farther away from zero. The following example demonstrates that this is not necessarily true.

Example 16-1

Let us examine prediction performance, as measured by $p(k|0)$, for the first-order system

$$x(k+1) = \frac{1}{\sqrt{2}}x(k) + w(k) \tag{16-26}$$

where $q = 25$ and $p(0)$ is variable. Quantity $p(k|0)$, which in the case of a scalar state vector is a variance, is easily computed from the recursive equation

$$p(k|0) = \frac{1}{2}p(k-1|0) + 25 \tag{16-27}$$

for $k = 1, 2, \ldots$. Two cases are summarized in Figure 16-1. When $p(0) = 6$ we have

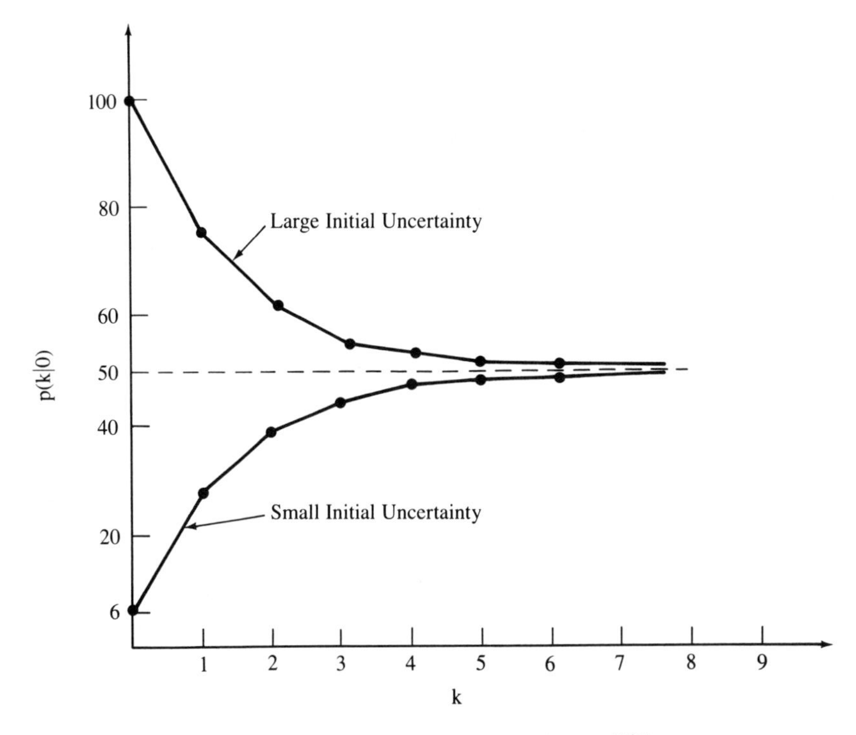

Figure 16-1 Prediction error variance $p(k|0)$.

relatively small uncertainty about $\hat{x}(0|0)$, and as we expected, our predictions of $x(k)$ for $k \geq 1$ do become worse, because $p(k|0) > 6$ for all $k \geq 1$. After a while $p(k|0)$ reaches a limiting value equal to 50. When this occurs we are estimating $\hat{x}(k|0)$ by a number that is very close to zero, because $\hat{x}(k|0) = \left(\frac{1}{\sqrt{2}}\right)^k \hat{x}(0|0)$, and $\left(\frac{1}{\sqrt{2}}\right)^k$ approaches zero for large values of k.

When $p(0) = 100$ we have large uncertainty about $\hat{x}(0|0)$, and, perhaps to our surprise, our predictions of $x(k)$ for $k \geq 1$ improve in performance, because $p(k|0) <$ 100 for all $k \geq 1$. In this case the predictor discounts the large initial uncertainty; however, as in the former case, $p(k|0)$ again reaches the limiting value of 50.

For suitably large values of k, the predictor is completely insensitive to $p(0)$. It reaches a *steady-state* level of performance equal to 50, which can be predetermined by setting $p(k|0)$ and $p(k-1|0)$ equal to $\overline{p}$, in (16-27), and solving the resulting equation for $\overline{p}$. □

Prediction is possible only because we have a known dynamical model, namely, our state-variable model. Without a model, prediction is dubious at best (e.g., try predicting tomorrow's price of a stock listed on any stock exchange using today's closing price).

THE INNOVATIONS PROCESS

Suppose we have just computed the single-stage, predicted estimate of $\mathbf{x}(k+1)$, $\hat{\mathbf{x}}(k+1|k)$. Then the single-stage predicted estimate of $\mathbf{z}(k+1)$, $\hat{\mathbf{z}}(k+1|k)$, is

$$\hat{\mathbf{z}}(k+1|k) = \mathbf{E}\{\mathbf{z}(k+1)|\mathfrak{Z}(k)\} = \mathbf{E}\{[\mathbf{H}(k+1)\mathbf{x}(k+1) + \mathbf{v}(k+1)]|\mathfrak{Z}(k)\}$$

or

$$\hat{\mathbf{z}}(k+1|k) = \mathbf{H}(k+1)\hat{\mathbf{x}}(k+1|k) \tag{16-28}$$

The error between $\mathbf{z}(k+1)$ and $\hat{\mathbf{z}}(k+1|k)$ is $\tilde{\mathbf{z}}(k+1|k)$, i.e.,

$$\tilde{\mathbf{z}}(k+1|k) = \mathbf{z}(k+1) - \hat{\mathbf{z}}(k+1|k) \tag{16-29}$$

Signal $\tilde{\mathbf{z}}(k+1|k)$ is often referred to either as the *innovations process, prediction error process*, or *measurement residual process*. We shall refer to it as the innovations process, because this is most commonly done in the estimation theory literature (e.g., Kailath, 1968). The innovations process plays a very important role in mean-squared filtering and smoothing. We summarize important facts about it in the following:

Theorem 16-2 (Innovations)

a. *The following representations of the innovations process* $\tilde{\mathbf{z}}(k+1|k)$ *are equivalent*:

$$\tilde{\mathbf{z}}(k+1|k) = \mathbf{z}(k+1) - \hat{\mathbf{z}}(k+1|k) \tag{16-30}$$

$$\tilde{\mathbf{z}}(k+1|k) = \mathbf{z}(k+1) - \mathbf{H}(k+1)\hat{\mathbf{x}}(k+1|k) \tag{16-31}$$

or

$$\tilde{\mathbf{z}}(k+1|k) = \mathbf{H}(k+1)\tilde{\mathbf{x}}(k+1|k) + \mathbf{v}(k+1) \tag{16-32}$$

b. *The innovations is a zero-mean Gaussian white noise sequence, with*

$$\begin{aligned}\mathbf{E}\{\tilde{\mathbf{z}}(k+1|k)\tilde{\mathbf{z}}'(k+1|k)\} &= \mathbf{P}_{\tilde{z}\tilde{z}}(k+1|k)\\ &= \mathbf{H}(k+1)\mathbf{P}(k+1|k)\mathbf{H}'(k+1) + \mathbf{R}(k+1)\end{aligned} \tag{16-33}$$

Proof (Mendel, 1983b)

a. Substitute (16-28) into (16-29) in order to obtain (16-31). Next, substitute the measurement equation $\mathbf{z}(k+1) = \mathbf{H}(k+1)\mathbf{x}(k+1) + \mathbf{v}(k+1)$ into (16-31), and use the fact that $\tilde{\mathbf{x}}(k+1|k) = \mathbf{x}(k+1) - \hat{\mathbf{x}}(k+1|k)$, to obtain (16-32).

b. Because $\tilde{\mathbf{x}}(k+1|k)$ and $\mathbf{v}(k+1)$ are both zero mean, $\mathbf{E}\{\tilde{\mathbf{z}}(k+1|k)\} = \mathbf{0}$. The innovations is Gaussian because $\mathbf{z}(k+1)$ and $\hat{\mathbf{x}}(k+1|k)$ are Gaussian, and, therefore, $\tilde{\mathbf{z}}(k+1|k)$ is a linear transformation of Gaussian random vectors. To prove that $\tilde{\mathbf{z}}(k+1|k)$ is white noise we must show that

$$\mathbf{E}\{\tilde{\mathbf{z}}(i+1|i)\tilde{\mathbf{z}}'(j+1|j)\} = \mathbf{P}_{\tilde{z}\tilde{z}}(i+1|i)\delta_{ij} \tag{16-34}$$

We shall consider the cases $i > j$ and $i = j$, leaving the case $i < j$ as an exercise for the reader. When $i > j$,

$$\begin{aligned}\mathbf{E}\{\tilde{\mathbf{z}}(i+1|i)\tilde{\mathbf{z}}'(j+1|j)\} &= \mathbf{E}\{[\mathbf{H}(i+1)\tilde{\mathbf{x}}(i+1|i) + \mathbf{v}(i+1)]\\ &\quad [\mathbf{H}(j+1)\tilde{\mathbf{x}}(j+1|j) + \mathbf{v}(j+1)]'\}\\ &= \mathbf{E}\{\mathbf{H}(i+1)\tilde{\mathbf{x}}(i+1|i)\\ &\quad [\mathbf{H}(j+1)\tilde{\mathbf{x}}(j+1|j) + \mathbf{v}(j+1)]'\}\end{aligned}$$

because $\mathbf{E}\{\mathbf{v}(i+1)\mathbf{v}'(j+1)\} = \mathbf{0}$ and $\mathbf{E}\{\mathbf{v}(i+1)\tilde{\mathbf{x}}'(j+1|j)\} = \mathbf{0}$. The latter is true because, for $i > j$, $\tilde{\mathbf{x}}(j+1|j)$ does not depend on measurement $\mathbf{z}(i+1)$; hence, for $i > j$, $\mathbf{v}(i+1)$ and $\tilde{\mathbf{x}}(j+1|j)$ are independent, so that $\mathbf{E}\{\mathbf{v}(i+1)\tilde{\mathbf{x}}'(j+1|j)\} = \mathbf{E}\{\mathbf{v}(i+1)\}\mathbf{E}\{\tilde{\mathbf{x}}'(j+1|j)\} = \mathbf{0}$. We continue, as follows:

$$\begin{aligned}&\mathbf{E}\{\tilde{\mathbf{z}}(i+1|i)\tilde{\mathbf{z}}'(j+1|j)\}\\ &\qquad = \mathbf{H}(i+1)\mathbf{E}\{\tilde{\mathbf{x}}(i+1|i)[\mathbf{z}(j+1) - \mathbf{H}(j+1)\hat{\mathbf{x}}(j+1|j)]'\} = \mathbf{0}\end{aligned}$$

by repeated application of the orthogonality principle (Corollary 13-3).

When $i = j$,

$$\begin{aligned}\mathbf{P}_{\tilde{z}\tilde{z}}(i+1|i) &= \mathbf{E}\{[\mathbf{H}(i+1)\tilde{\mathbf{x}}(i+1|i) + \mathbf{v}(i+1)][\mathbf{H}(i+1)\tilde{\mathbf{x}}(i+1|i)\\ &\quad + \mathbf{v}(i+1)]'\} = \mathbf{H}(i+1)\mathbf{P}(i+1|i)\mathbf{H}'(i+1) + \mathbf{R}(i+1)\end{aligned}$$

because, once again $\mathbf{E}\{\mathbf{v}(i+1)\tilde{\mathbf{x}}'(i+1|i)\} = \mathbf{0}$, and $\mathbf{P}(i+1) = \mathbf{E}\{\tilde{\mathbf{x}}(i+1|i)\tilde{\mathbf{x}}'(i+1|i)\}$. □

In Lesson 17 the inverse of $\mathbf{P}_{\tilde{z}\tilde{z}}(k+1|k)$ is needed; hence, we shall assume that $\mathbf{H}(k+1)\mathbf{P}(k+1|k)\mathbf{H}'(k+1) + \mathbf{R}(k+1)$ is nonsingular. This is usually true and will always be true if, as in our basic state-variable model, $\mathbf{R}(k+1)$ is positive definite.

PROBLEMS

16-1. Develop the counterpart to Theorem 16-1 for the case when input $\mathbf{u}(k)$ is random and independent of $\mathcal{Z}(j)$. What happens if $\mathbf{u}(k)$ is random and dependent upon $\mathcal{Z}(j)$?

16-2. For the innovations process $\tilde{\mathbf{z}}(k+1|k)$, prove that $\mathbf{E}\{\tilde{\mathbf{z}}(i+1|i)\tilde{\mathbf{z}}'(j+1|j)\} = \mathbf{0}$ when $i < j$.

16-3. In the proof of part (b) of Theorem 16-2 we make repeated use of the orthogonality principle, stated in Corollary 13-3. In the latter corollary $\mathbf{f}'[\mathcal{Z}(k)]$

appears to be a function of *all* of the measurements used in $\hat{\boldsymbol{\theta}}_{MS}(k)$. In the expression $\mathbf{E}\{\tilde{\mathbf{x}}(i+1|i)\mathbf{z}'(j+1)\}$, $i > j$, $\mathbf{z}'(j+1)$ certainly is not a function of all the measurements used in $\tilde{\mathbf{x}}(i+1|i)$. What is $\mathbf{f}[\cdot]$, when we apply the orthogonality principle to $\mathbf{E}\{\tilde{\mathbf{x}}(i+1|i)\mathbf{z}'(j+1)\}$, $i > j$, to conclude that this expectation is zero?

16-4. Refer to Problem 15-5. Assume that $u(k)$ can be measured [e.g., $u(k)$ might be the output of a random number generator], and that $a = a[z(1), z(2), \ldots, z(k-1)]$. What is $\hat{z}(k|k-1)$?

16-5. Consider the following autoregressive model, $z(k+n) = a_1 z(k+n-1) - a_2 z(k+n-2) - \cdots - a_n z(k) + w(k+n)$ in which $w(k)$ is white noise. Measurements $z(k), z(k+1), \ldots, z(k+n-1)$ are available.

(a) Compute $\hat{z}(k+n|k+n-1)$.

(b) Explain why the result in (a) is the overall mean-squared prediction of $z(k+n)$ even if $w(k+n)$ is non-Gaussian.

Lesson 17

State Estimation: Filtering (the Kalman Filter)

INTRODUCTION

In this lesson we shall develop the Kalman filter, which is a recursive mean-squared error filter for computing $\hat{\mathbf{x}}(k+1|k+1)$, $k = 0, 1, 2, \ldots$. As its name implies, this filter was developed by Kalman [circa 1959 (Kalman, 1960)].

From the Fundamental Theorem of Estimation Theory, Theorem 13-1, we know that

$$\hat{\mathbf{x}}(k+1|k+1) = \mathbf{E}\{\mathbf{x}(k+1)|\mathscr{Z}(k+1)\} \tag{17-1}$$

Our approach to developing the Kalman filter is to partition $\mathscr{Z}(k+1)$ into two sets of measurements, $\mathscr{Z}(k)$ and $\mathbf{z}(k+1)$, and to then expand the conditional expectation in terms of data sets $\mathscr{Z}(k)$ and $\mathbf{z}(k+1)$, i.e.,

$$\hat{\mathbf{x}}(k+1|k+1) = \mathbf{E}\{\mathbf{x}(k+1)|\mathscr{Z}(k), \mathbf{z}(k+1)\} \tag{17-2}$$

What complicates this expansion is the fact that $\mathscr{Z}(k)$ and $\mathbf{z}(k+1)$ are statistically dependent. Measurement vector $\mathscr{Z}(k)$ depends on state vectors $\mathbf{x}(1), \mathbf{x}(2), \ldots, \mathbf{x}(k)$, because $\mathbf{z}(j) = \mathbf{H}(j)\mathbf{x}(j) + \mathbf{v}(j)(j = 1, 2, \ldots, k)$. Measurement vector $\mathbf{z}(k+1)$ also depends on state vector $\mathbf{x}(k)$, because $\mathbf{z}(k+1) = \mathbf{H}(k+1)\mathbf{x}(k+1) + \mathbf{v}(k+1)$ and $\mathbf{x}(k+1) = \mathbf{\Phi}(k+1,k)\mathbf{x}(k) + \mathbf{\Gamma}(k+1,k)\mathbf{w}(k) + \mathbf{\Psi}(k+1,k)\mathbf{u}(k)$. Hence $\mathscr{Z}(k)$ and $\mathbf{z}(k+1)$ both depend on $\mathbf{x}(k)$ and are, therefore, dependent.

Recall that $\mathbf{x}(k+1)$, $\mathcal{Z}(k)$ and $\mathbf{z}(k+1)$ are jointly Gaussian random vectors; hence, we can use Theorem 12-4 to express (17-2) as

$$\hat{\mathbf{x}}(k+1|k+1) = \mathbf{E}\{\mathbf{x}(k+1)|\mathcal{Z}(k),\tilde{\mathbf{z}}\} \tag{17-3}$$

where

$$\tilde{\mathbf{z}} = \mathbf{z}(k+1) - \mathbf{E}\{\mathbf{z}(k+1)|\mathcal{Z}(k)\} \tag{17-4}$$

We immediately recognize $\tilde{\mathbf{z}}$ as the innovations process $\tilde{\mathbf{z}}(k+1|k)$ [see (16-29)]; thus, we rewrite (17-3) as

$$\hat{\mathbf{x}}(k+1|k+1) = \mathbf{E}\{\mathbf{x}(k+1)|\mathcal{Z}(k),\tilde{\mathbf{z}}(k+1|k)\} \tag{17-5}$$

Applying (12-37) to (17-5), we find that

$$\hat{\mathbf{x}}(k+1|k+1) = \mathbf{E}\{\mathbf{x}(k+1)|\mathcal{Z}(k)\} + \mathbf{E}\{\mathbf{x}(k+1)|\tilde{\mathbf{z}}(k+1|k)\} - \mathbf{m}_x(k+1) \tag{17-6}$$

We recognize the first term on the right-hand side of (17-6) as the single-stage predicted estimator of $\mathbf{x}(k+1)$, $\hat{\mathbf{x}}(k+1|k)$; hence,

$$\hat{\mathbf{x}}(k+1|k+1) = \hat{\mathbf{x}}(k+1|k) + \mathbf{E}\{\mathbf{x}(k+1)|\tilde{\mathbf{z}}(k+1|k)\} - \mathbf{m}_x(k+1) \tag{17-7}$$

This equation is the starting point for our derivation of the Kalman filter.

Before proceeding further, we observe, upon comparison of (17-2) and (17-5), that our original conditioning on $\mathbf{z}(k+1)$ has been replaced by conditioning on the innovations process $\tilde{\mathbf{z}}(k+1|k)$. One can show that $\tilde{\mathbf{z}}(k+1|k)$ is computable from $\mathbf{z}(k+1)$, and that $\mathbf{z}(k+1)$ is computable from $\tilde{\mathbf{z}}(k+1|k)$; hence, it is said that $\mathbf{z}(k+1)$ and $\tilde{\mathbf{z}}(k+1|k)$ are *causally invertible* (Anderson and Moore, 1979). We explain this statement more carefully at the end of this lesson.

A PRELIMINARY RESULT

In our derivation of the Kalman filter, we shall determine that

$$\hat{\mathbf{x}}(k+1|k+1) = \hat{\mathbf{x}}(k+1|k) + \mathbf{K}(k+1)\tilde{\mathbf{z}}(k+1|k) \tag{17-8}$$

where $\mathbf{K}(k+1)$ is an $n \times m$ (Kalman) gain matrix. We will calculate the optimal gain matrix in the next section.

Here let us view (17-8) as the structure of an abitrary recursive linear filter, which is written in so-called *predictor-corrector* format; i.e., the filtered estimate of $\mathbf{x}(k+1)$ is obtained by a predictor step, $\hat{\mathbf{x}}(k+1|k)$, and a corrector step, $\mathbf{K}(k+1)\tilde{\mathbf{z}}(k+1|k)$. The predictor step uses information from the state equation, because $\hat{\mathbf{x}}(k+1|k) = \mathbf{\Phi}(k+1,k)\hat{\mathbf{x}}(k|k) + \mathbf{\Psi}(k+1,k)\mathbf{u}(k)$. The corrector step uses the new measurement available at t_{k+1}. The correction is proportional to the difference between that measurement and its best predicted value, $\hat{\mathbf{z}}(k+1|k)$. The following result provides us with the means for evaluating $\hat{\mathbf{x}}(k+1|k+1)$ in terms of its error-covariance matrix $\mathbf{P}(k+1|k+1)$.

Preliminary Result. *Filtering error-covariance matrix* $\mathbf{P}(k+1|k+1)$ *for the arbitrary linear recursive filter (17-8) is computed from the following equation:*

$$\mathbf{P}(k+1|k+1) = [\mathbf{I} - \mathbf{K}(k+1)\mathbf{H}(k+1)]\mathbf{P}(k+1|k)[\mathbf{I} - \mathbf{K}(k+1)\mathbf{H}(k+1)]' + \mathbf{K}(k+1)\mathbf{R}(k+1)\mathbf{K}'(k+1) \quad (17\text{-}9)$$

Proof. Substitute (16-32) into (17-8) and then subtract the resulting equation from $\mathbf{x}(k+1)$ in order to obtain

$$\tilde{\mathbf{x}}(k+1|k+1) = [\mathbf{I} - \mathbf{K}(k+1)\mathbf{H}(k+1)]\tilde{\mathbf{x}}(k+1|k) - \mathbf{K}(k+1)\mathbf{v}(k+1) \quad (17\text{-}10)$$

Substitute this equation into $\mathbf{P}(k+1|k+1) = \mathbf{E}\{\tilde{\mathbf{x}}(k+1|k+1)\tilde{\mathbf{x}}'(k+1|k+1)\}$ to obtain equation (17-9). As in the proof of Theorem 16-2, we have used the fact that $\tilde{\mathbf{x}}(k+1|k)$ and $\mathbf{v}(k+1)$ are independent to show that $\mathbf{E}\{\tilde{\mathbf{x}}(k+1|k)\mathbf{v}'(k+1)\} = \mathbf{0}$. □

The state prediction-error covariance matrix $\mathbf{P}(k+1|k)$ is given by equation (16-11). Observe that (17-9) and (16-11) can be computed recursively, once gain matrix $\mathbf{K}(k+1)$ is specified, as follows: $\mathbf{P}(0|0) \rightarrow \mathbf{P}(1|0) \rightarrow \mathbf{P}(1|1) \rightarrow \mathbf{P}(2|1) \rightarrow \mathbf{P}(2|2) \rightarrow \cdots$ etc.

It is important to reiterate the fact that (17-9) is true for *any* gain matrix, including the optimal gain matrix given next in Theorem 17-1.

THE KALMAN FILTER

Theorem 17-1

a. *The mean-squared filtered estimator of* $\mathbf{x}(k+1)$, $\hat{\mathbf{x}}(k+1|k+1)$, *written in predictor-corrector format, is*

$$\hat{\mathbf{x}}(k+1|k+1) = \hat{\mathbf{x}}(k+1|k) + \mathbf{K}(k+1)\tilde{\mathbf{z}}(k+1|k) \quad (17\text{-}11)$$

for $k = 0, 1, \ldots,$ *where* $\hat{\mathbf{x}}(0|0) = \mathbf{m}_x(0)$, *and* $\tilde{\mathbf{z}}(k+1|k)$ *is the innovations process* $[\tilde{\mathbf{z}}(k+1|k) = \mathbf{z}(k+1) - \mathbf{H}(k+1)\hat{\mathbf{x}}(k+1|k)]$.

b. $\mathbf{K}(k+1)$ *is an* $n \times m$ *matrix (commonly referred to as the Kalman gain matrix, or weighting matrix) which is specified by the set of relations*

$$\mathbf{K}(k+1) = \mathbf{P}(k+1|k)\mathbf{H}'(k+1)[\mathbf{H}(k+1)\mathbf{P}(k+1|k)\mathbf{H}'(k+1) + \mathbf{R}(k+1)]^{-1} \quad (17\text{-}12)$$

$$\mathbf{P}(k+1|k) = \mathbf{\Phi}(k+1,k)\mathbf{P}(k|k)\mathbf{\Phi}'(k+1,k) + \mathbf{\Gamma}(k+1,k)\mathbf{Q}(k)\mathbf{\Gamma}'(k+1,k) \quad (17\text{-}13)$$

and

$$\mathbf{P}(k+1|k+1) = [\mathbf{I} - \mathbf{K}(k+1)\mathbf{H}(k+1)]\mathbf{P}(k+1|k) \qquad (17\text{-}14)$$

for k = 0, 1, . . . , *where* **I** *is the* n × n *identity matrix, and* $\mathbf{P}(0|0) = \mathbf{P}_\mathbf{x}(0)$.

c. *The stochastic process* $\{\tilde{\mathbf{x}}(k+1|k+1), k = 0, 1, \ldots\}$, *which is defined by the filtering error relation*

$$\tilde{\mathbf{x}}(k+1|k+1) = \mathbf{x}(k+1) - \hat{\mathbf{x}}(k+1|k+1) \qquad (17\text{-}15)$$

k = 0, 1, . . . , *is a zero-mean Gauss-Markov sequence whose covariance matrix is given by (17-14).*

Proof (Mendel, 1983b, pp. 56–57)

a. We begin with the formula for $\hat{\mathbf{x}}(k+1|k+1)$ in (17-7). Recall that $\mathbf{x}(k+1)$ and $\mathbf{z}(k+1)$ are jointly Gaussian. Because $\mathbf{z}(k+1)$ and $\tilde{\mathbf{z}}(k+1|k)$ are causally invertible, $\mathbf{x}(k+1)$ and $\tilde{\mathbf{z}}(k+1|k)$ are also jointly Gaussian. Additionally, $\mathbf{E}\{\tilde{\mathbf{z}}(k+1|k)\} = \mathbf{0}$; hence,

$$\mathbf{E}\{\mathbf{x}(k+1)|\tilde{\mathbf{z}}(k+1|k)\} = \mathbf{m}_\mathbf{x}(k+1) + \mathbf{P}_{\mathbf{x}\tilde{\mathbf{z}}}(k+1,k+1|k)\mathbf{P}_{\tilde{\mathbf{z}}\tilde{\mathbf{z}}}^{-1}(k+1|k)\tilde{\mathbf{z}}(k+1|k) \qquad (17\text{-}16)$$

We define gain matrix $\mathbf{K}(k+1)$ as

$$\mathbf{K}(k+1) = \mathbf{P}_{\mathbf{x}\tilde{\mathbf{z}}}(k+1,k+1|k)\mathbf{P}_{\tilde{\mathbf{z}}\tilde{\mathbf{z}}}^{-1}(k+1|k) \qquad (17\text{-}17)$$

Substituting (17-16) and (17-17) into (17-7) we obtain the Kalman filter equation (17-11). Because $\hat{\mathbf{x}}(k+1|k) = \mathbf{\Phi}(k+1,k)\hat{\mathbf{x}}(k|k) + \mathbf{\Psi}(k+1,k)\mathbf{u}(k)$, equation (17-11) must be initialized by $\hat{\mathbf{x}}(0|0)$, which we have shown must equal $\mathbf{m}_\mathbf{x}(0)$ [see Equation (16-12)].

b. In order to evaluate $\mathbf{K}(k+1)$ we must evaluate $\mathbf{P}_{\mathbf{x}\tilde{\mathbf{z}}}$ and $\mathbf{P}_{\tilde{\mathbf{z}}\tilde{\mathbf{z}}}^{-1}$. Matrix $\mathbf{P}_{\tilde{\mathbf{z}}\tilde{\mathbf{z}}}$ has been computed in (16-33). By definition of cross-covariance,

$$\mathbf{P}_{\mathbf{x}\tilde{\mathbf{z}}} = \mathbf{E}\{[\mathbf{x}(k+1) - \mathbf{m}_\mathbf{x}(k+1)]\tilde{\mathbf{z}}'(k+1|k)\} = \mathbf{E}\{\mathbf{x}(k+1)\tilde{\mathbf{z}}'(k+1|k)\} \qquad (17\text{-}18)$$

because $\tilde{\mathbf{z}}(k+1|k)$ is zero-mean. Substituting (16-32) into this expression, we find that

$$\mathbf{P}_{\mathbf{x}\tilde{\mathbf{z}}} = \mathbf{E}\{\mathbf{x}(k+1)\tilde{\mathbf{x}}'(k+1|k)\}\mathbf{H}'(k+1) \qquad (17\text{-}19)$$

because $\mathbf{E}\{\mathbf{x}(k+1)\mathbf{v}'(k+1)\} = \mathbf{0}$. Finally, expressing $\mathbf{x}(k+1)$ as $\tilde{\mathbf{x}}(k+1|k) + \hat{\mathbf{x}}(k+1|k)$ and applying the orthogonality principle (13-15), we find that

$$\mathbf{P}_{\mathbf{x}\tilde{\mathbf{z}}} = \mathbf{P}(k+1|k)\mathbf{H}'(k+1) \qquad (17\text{-}20)$$

Combining equations (17-20) and (16-33) into (17-17), we obtain equation (17-12) for the Kalman gain matrix.

State prediction-error covariance matrix $\mathbf{P}(k+1|k)$ was derived in Lesson 16.

State filtering-error covariance matrix $\mathbf{P}(k+1|k+1)$ is obtained by substituting (17-12) for $\mathbf{K}(k+1)$ into (17-9), as follows:

$$\begin{aligned}\mathbf{P}(k+1|k+1) &= (\mathbf{I}-\mathbf{KH})\mathbf{P}(\mathbf{I}-\mathbf{KH})' + \mathbf{KRK}' \\ &= (\mathbf{I}-\mathbf{KH})\mathbf{P} - \mathbf{PH}'\mathbf{K}' + \mathbf{KHPH}'\mathbf{K}' + \mathbf{KRK}' \\ &= (\mathbf{I}-\mathbf{KH})\mathbf{P} - \mathbf{PH}'\mathbf{K}' + \mathbf{K}(\mathbf{HPH}' + \mathbf{R})\mathbf{K}' \\ &= (\mathbf{I}-\mathbf{KH})\mathbf{P} - \mathbf{PH}'\mathbf{K}' + \mathbf{PH}'\mathbf{K}' \\ &= (\mathbf{I}-\mathbf{KH})\mathbf{P}\end{aligned} \tag{17-21}$$

c. The proof that $\tilde{\mathbf{x}}(k+1|k+1)$ is zero-mean, Gaussian and Markov is so similar to the proof of part b of Theorem 16-1 that we omit its details [see Meditch (1969, pp. 181–182)]. □

OBSERVATIONS ABOUT THE KALMAN FILTER

1. Figure 17-1 depicts the interconnection of our basic dynamical system [equations (15-17) and (15-18)] and Kalman filter system. The feedback nature of the Kalman filter is quite evident. Observe, also, that the Kalman filter contains within its structure a model of the plant.

   The feedback nature of the Kalman filter manifests itself in *two* different ways, namely in the calculation of $\hat{\mathbf{x}}(k+1|k+1)$ and also in the calculation of the matrix of gains, $\mathbf{K}(k+1)$, both of which we shall explore below.
2. The predictor-corrector form of the Kalman filter is illuminating from an information-usage viewpoint. Observe that the predictor equations, which compute $\hat{\mathbf{x}}(k+1|k)$ and $\mathbf{P}(k+1|k)$, use information only from the state equation, whereas the corrector equations, which compute $\mathbf{K}(k+1)$, $\hat{\mathbf{x}}(k+1|k+1)$ and $\mathbf{P}(k+1|k+1)$, use information only from the measurement equation.
3. Once the gain matrix is computed, then (17-11) represents a *time-varying recursive digital filter*. This is seen more clearly when equations (16-4) and (16-31) are substituted into (17-11). The resulting equation can be rewritten as

$$\begin{aligned}\hat{\mathbf{x}}(k+1|k+1) &= [\mathbf{I} - \mathbf{K}(k+1)\mathbf{H}(k+1)]\mathbf{\Phi}(k+1,k)\hat{\mathbf{x}}(k|k) \\ &\quad + \mathbf{K}(k+1)\mathbf{z}(k+1) \\ &\quad + [\mathbf{I} - \mathbf{K}(k+1)\mathbf{H}(k+1)]\mathbf{\Psi}(k+1,k)\mathbf{u}(k)\end{aligned} \tag{17-22}$$

   for $k = 0, 1, \ldots$. This is a state equation for state vector $\hat{\mathbf{x}}$, whose time-varying plant matrix is $[\mathbf{I} - \mathbf{K}(k+1)\mathbf{H}(k+1)]\mathbf{\Phi}(k+1,k)$. Equation (17-22) is time-varying even if our dynamical system in equations (15-17) and (15-18) is time-invariant and stationary, because gain matrix $\mathbf{K}(k+1)$ is still time-varying in that case. It is possible,

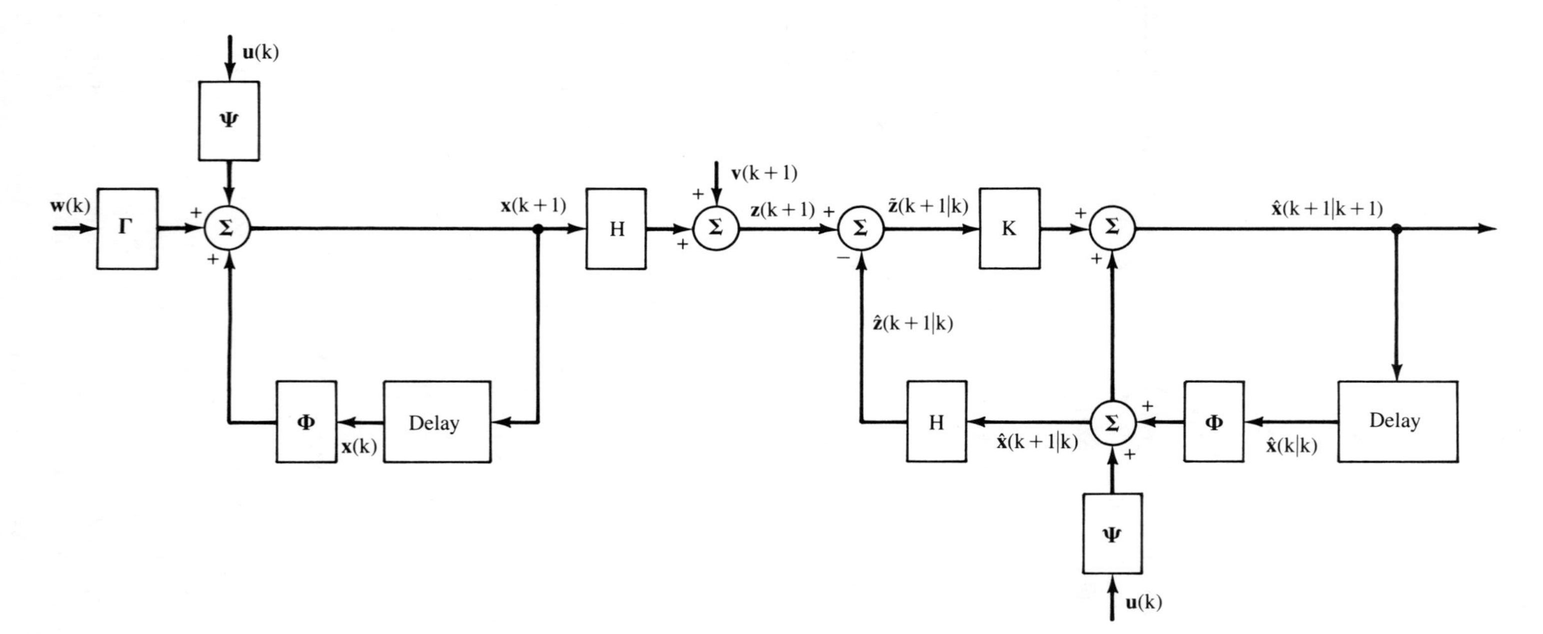

Figure 17-1 Interconnection of system and Kalman filter. (Mendel, 1983b, © 1983, Academic Press, Inc.).

however, for $\mathbf{K}(k+1)$ to reach a limiting value (i.e., a steady-state value, $\bar{\mathbf{K}}$), in which case (17-22) reduces to a recursive constant coefficient filter. We will have more to say about this important steady-state case in Lesson 19.

Equation (17-22) is in a *recursive filter form*, in that it relates the filtered estimate of $\mathbf{x}(k+1)$, $\hat{\mathbf{x}}(k+1|k+1)$, to the filtered estimate of $\mathbf{x}(k)$, $\hat{\mathbf{x}}(k|k)$. Using substitutions similar to those used in the derivation of (17-22), one can also obtain the following *recursive predictor form* of the Kalman filter (details left as an exercise),

$$\hat{\mathbf{x}}(k+1|k) = \mathbf{\Phi}(k+1,k)[\mathbf{I} - \mathbf{K}(k)\mathbf{H}(k)]\hat{\mathbf{x}}(k|k-1) + \mathbf{\Phi}(k+1,k)\mathbf{K}(k)\mathbf{z}(k) + \mathbf{\Psi}(k+1,k)\mathbf{u}(k) \tag{17-23}$$

Observe that in (17-23) the predicted estimate of $\mathbf{x}(k+1)$, $\hat{\mathbf{x}}(k+1|k)$, is related to the predicted estimate of $\mathbf{x}(k)$, $\hat{\mathbf{x}}(k|k-1)$. Interestingly enough, the recursive predictor (17-23), and not the recursive filter (17-22), plays an important role in mean-squared smoothing, as we shall see in Lesson 21.

The structures of (17-22) and (17-23) are summarized in Figure 17-2. This figure supports the claim made in Lesson 1 that our recursive estimators are nothing more than time-varying digital filters that operate on random (and also deterministic) inputs.

4. Embedded within the recursive Kalman filter equations is another set of recursive equations—(17-12), (17-13) and (17-14). Because $\mathbf{P}(0|0)$ initializes these calculations, these equations must be ordered as follows: $\mathbf{P}(k|k) \rightarrow \mathbf{P}(k+1|k) \rightarrow \mathbf{K}(k+1) \rightarrow \mathbf{P}(k+1|k+1) \rightarrow$ etc.

   By combining these three equations it is possible to get a matrix recursive equation for $\mathbf{P}(k+1|k)$ as a function of $\mathbf{P}(k|k-1)$, or a similar equation for $\mathbf{P}(k+1|k+1)$ as a function of $\mathbf{P}(k|k)$. These equations are nonlinear and are known as *matrix Riccati equations*. For example, the matrix Riccati equation for $\mathbf{P}(k+1|k)$ is

$$\mathbf{P}(k+1|k) = \mathbf{\Phi}\mathbf{P}(k|k-1)\{\mathbf{I} - \mathbf{H}'[\mathbf{H}\mathbf{P}(k|k-1)\mathbf{H}' + \mathbf{R}]^{-1}\mathbf{H}\mathbf{P}(k|k-1)\}\mathbf{\Phi}' + \mathbf{\Gamma}\mathbf{Q}\mathbf{\Gamma}' \tag{17-24}$$

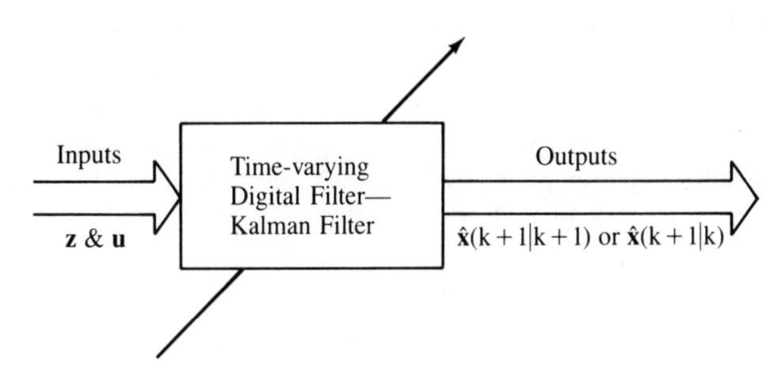

Figure 17-2 Input-output interpretation of the Kalman filter.

where we have omitted the temporal arguments on $\mathbf{\Phi}$, $\mathbf{\Gamma}$, $\mathbf{H}$, $\mathbf{Q}$ and $\mathbf{R}$ for notational simplicity [their correct arguments are $\mathbf{\Phi}(k+1,k)$, $\mathbf{\Gamma}(k+1,k)$, $\mathbf{H}(k)$, $\mathbf{Q}(k)$, and $\mathbf{R}(k)$]. The matrix Riccati equation for $\mathbf{P}(k+1|k+1)$ is obtained by substituting (17-12) into (17-14), and then (17-13) into the resulting equation. We leave its derivation as an exercise.

5. A measure of recursive predictor performance is provided by matrix $\mathbf{P}(k+1|k)$. This covariance matrix can be calculated prior to any processing of real data, using its matrix Riccati Equation (17-24) or Equations (17-13), (17-14), and (17-12). A measure of recursive filter performance is provided by matrix $\mathbf{P}(k+1|k+1)$, and this covariance matrix can also be calculated prior to any processing of real data. Note that $\mathbf{P}(k+1|k+1) \neq \mathbf{P}(k+1|k)$. These calculations are often referred to as *performance analyses*. It is indeed interesting that the Kalman filter utilizes a measure of its mean-squared error during its real-time operation.
6. Two formulas are available for computing $\mathbf{P}(k+1|k+1)$, namely (17-14), which is known as the *standard form*, and (17-9), which is known as the *stabilized form*. Although the stabilized form requires more computations than the standard form, it is much less sensitive to numerical errors from the prior calculation of gain matrix $\mathbf{K}(k+1)$ than is the standard form. In fact, one can show that first-order errors in the calculation of $\mathbf{K}(k+1)$ propagate as first-order errors in the calculation of $\mathbf{P}(k+1|k+1)$, when the standard form is used, but only as second-order errors in the calculation of $\mathbf{P}(k+1|k+1)$ when the stabilized form is used. This is why (17-9) is called the stabilized form [for detailed derivations, see Aoki (1967), Jazwinski (1970), or Mendel (1973)].
7. On the subject of computation, the calculation of $\mathbf{P}(k+1|k)$ is the most costly one for the Kalman filter, because of the term $\mathbf{\Phi}(k+1,k)\mathbf{P}(k|k)\mathbf{\Phi}'(k+1,k)$, which entails two multiplications of two $n \times n$ matrices [i.e., $\mathbf{P}(k|k)\mathbf{\Phi}'(k+1,k)$ and $\mathbf{\Phi}(k+1,k)(\mathbf{P}(k|k)\mathbf{\Phi}'(k+1,k))$]. Total computation for the two matrix multiplications is on the order of $2n^3$ multiplications and $2n^3$ additions [for more detailed computation counts, including storage requirements, for all of the Kalman filter equations, see Mendel (1971), Gura and Bierman (1971), and Bierman (1973a)].

   One must be very careful to code the standard or stabilized forms for $\mathbf{P}(k+1|k+1)$ so that $n \times n$ matrices are never multiplied. We leave it to the reader to show that the standard algorithm can be coded in such a manner that it only requires on the order of $\frac{1}{2}mn^2$ multiplications, whereas the stabilized algorithm can be coded in such a manner that it only requires on the order of $\frac{5}{2}mn^2$ multiplications. In many applications, system order n is larger than number of measurements m, so that $mn^2 \ll n^3$. Usually, computation is most sensitive to system order; so, *whenever possible*, use low-order (but adequate) models.

8. Because of the equivalence between mean-squared, best-linear unbiased, and weighted-least squares filtered estimates of our state vector $\mathbf{x}(k)$ (see Lesson 14), we must realize that our Kalman filter equations are just a recursive solution to a system of normal equations (see Lesson 3). Other implementations of the Kalman filter that solve the normal equations using stable algorithms from numerical linear algebra (see, e.g., Bierman, 1977) and involve orthogonal transformations, have better numerical properties than (17-11)–(17-14). We reiterate, however, that because this is a book on estimation *theory*, theoretical formulas such as those in (17-11)–(17-14) are appropriate.
9. In Lesson 4 we developed two forms for a recursive least-squares estimator, namely, the covariance and information forms. Compare $\mathbf{K}(k+1)$ in (17-12) with $\mathbf{K}_W(k+1)$ in (4-25) to see they have the same structure; hence, our formulation of the Kalman filter is often known as the *covariance formulation.* We leave it to the reader to show that $\mathbf{K}(k+1)$ can also be computed as

$$\mathbf{K}(k+1) = \mathbf{P}(k+1|k+1)\mathbf{H}'(k+1)\mathbf{R}^{-1}(k+1) \qquad (17\text{-}25)$$

where $\mathbf{P}(k+1|k+1)$ is computed as

$$\mathbf{P}^{-1}(k+1|k+1) = \mathbf{P}^{-1}(k+1|k) + \mathbf{H}'(k+1)\mathbf{R}^{-1}(k+1)\mathbf{H}(k+1) \qquad (17\text{-}26)$$

When these equations are used along with (17-11) and (17-13) we have the *information formulation* of the Kalman filter. Of course, the ordering of the computations in these two formulations of the Kalman filter are different. See Lessons 4 and 9 for related discussions.
10. In Lesson 14 we showed that $\hat{\mathbf{x}}_{MAP}(k_1|N) = \hat{\mathbf{x}}_{MS}(k_1|N)$; hence,

$$\hat{\mathbf{x}}_{MAP}(k|k) = \hat{\mathbf{x}}_{MS}(k|k) \qquad (17\text{-}27)$$

This means that the Kalman filter also gives MAP estimates of the state vector $\mathbf{x}(k)$ for our basic state-variable model.
11. At the end of the introduction section in this lesson we mentioned that $\mathbf{z}(k+1)$ and $\tilde{\mathbf{z}}(k+1|k)$ are *causally invertible.* This means that we can compute one from the other using a causal (i.e., realizable) system. For example, when the measurements are available, then $\tilde{\mathbf{z}}(k+1|k)$ can be obtained from Equations (17-23) and (16-31), which we repeat here for the convenience of the reader:

$$\hat{\mathbf{x}}(k+1|k) = \mathbf{\Phi}(k+1,k)[\mathbf{I} - \mathbf{K}(k)\mathbf{H}(k)]\hat{\mathbf{x}}(k|k-1) + \mathbf{\Psi}(k+1,k)\mathbf{u}(k) + \mathbf{\Phi}(k+1,k)\mathbf{K}(k)\mathbf{z}(k) \qquad (17\text{-}28)$$

and

$$\tilde{\mathbf{z}}(k+1|k) = -\mathbf{H}(k+1)\hat{\mathbf{x}}(k+1|k) + \mathbf{z}(k+1) \qquad k = 0, 1, \ldots \qquad (17\text{-}29)$$

We refer to (17-28) and (17-29) in the rest of this book as the *Kalman innovations system*. It is initialized by $\hat{\mathbf{x}}(0|-1) = \mathbf{0}$.

On the other hand, if the innovations are given a priori, then $\mathbf{z}(k+1)$ can be obtained from

$$\hat{\mathbf{x}}(k+1|k) = \mathbf{\Phi}(k+1,k)\hat{\mathbf{x}}(k|k-1) + \mathbf{\Psi}(k+1,k)\mathbf{u}(k) + \mathbf{\Phi}(k+1,k)\mathbf{K}(k)\tilde{\mathbf{z}}(k|k-1) \tag{17-30}$$

and

$$\mathbf{z}(k+1) = \mathbf{H}(k+1)\hat{\mathbf{x}}(k+1|k) + \tilde{\mathbf{z}}(k+1|k) \qquad k = 0, 1, \ldots \tag{17-31}$$

Equation (17-30) was obtained by rearranging the terms in (17-28), and (17-31) was obtained by solving (17-29) for $\mathbf{z}(k+1)$. Equation (17-30) is again initialized by $\hat{\mathbf{x}}(0|-1) = \mathbf{0}$.

Note that (17-30) and (17-31) are equivalent to our basic state-variable model in (15-17) and (15-18), from an input-output point of view. Consequently, model (17-30) and (17-31) is often the starting point for important problems such as the stochastic realization problem (e.g., Faurre, 1976).

PROBLEMS

17-1. Prove that $\tilde{\mathbf{x}}(k+1|k+1)$ is zero mean, Gaussian and first-order Markov.

17-2. Derive the recursive predictor form of the Kalman filter, given in (17-23).

17-3. Derive the matrix Riccati equation for $\mathbf{P}(k+1|k+1)$.

17-4. Show that gain matrix $\mathbf{K}(k+1)$ can also be computed using (17-25).

17-5. Suppose a small error $\delta\mathbf{K}$ is made in the computation of the Kalman filter gain $\mathbf{K}(k+1)$.

(a) Show that when $\mathbf{P}(k+1|k+1)$ is computed from the "standard form" equation, then to first-order terms,

$$\delta\mathbf{P}(k+1|k+1) = -\delta\mathbf{K}(k+1)\mathbf{H}(k+1)\mathbf{P}(k+1|k)$$

(b) Show that when $\mathbf{P}(k+1|k+1)$ is computed from the "stabilized form" equation, then to first-order terms

$$\delta\mathbf{P}(k+1|k+1) \simeq \mathbf{0}$$

17-6. Consider the basic scalar system $x(k+1) = \phi x(k) + w(k)$ and $z(k+1) = x(k+1) + v(k+1)$.

(a) Show that $p(k+1|k) \geq q$, which means that the variance of the system disturbance sets the performance limit on prediction accuracy.

(b) Show that $0 \leq K(k+1) \leq 1$.

(c) Show that $0 \leq p(k+1|k+1) \leq r$.

17-7. An RC filter with time constant τ is excited by Gaussian white noise, and the output is measured every T seconds. The output at the sample times obeys the equation $x(k) = e^{-T/\tau}x(k-1) + w(k-1)$, where $\mathbf{E}\{x(0)\} = 1$, $\mathbf{E}\{x^2(0)\} = 2$, $q = 2$, and $T = \tau = 0.1$ sec. The measurements are described by $z(k) = x(k) + v(k)$, where $v(k)$ is a white but *non-Gaussian* noise sequence for which $\mathbf{E}\{v(k)\} = 0$ and $\mathbf{E}\{v^2(k)\} = 4$. Additionally, $w(k)$ and $v(k)$ are uncorrelated. Measurements $z(1) = 1.5$ and $z(2) = 3.0$.

(a) Find the best linear estimate of $x(1)$ based on $z(1)$.

(b) Find the best linear estimate of $x(2)$ based on $z(1)$ and $z(2)$.

17-8. Table 17-1 lists multiplications and additions for all the basic matrix operations used in a Kalman filter. Using the formulas for the Kalman filter given in Theorem 17-1, establish the number of multiplications and additions required to compute $\hat{\mathbf{x}}(k+1|k)$, $\mathbf{P}(k+1|k)$, $\mathbf{K}(k+1)$, $\hat{\mathbf{x}}(k+1|k+1)$, and $\mathbf{P}(k+1|k+1)$.

TABLE 17-1 Operation Characteristics

Name	Function	Multiplications	Additions
Matrix addition	$\mathbf{C}_{MN} = \mathbf{A}_{MN} + \mathbf{B}_{MN}$	—	MN
Matrix subtraction	$\mathbf{C}_{MN} = \mathbf{A}_{MN} - \mathbf{B}_{MN}$	—	MN
Matrix multiply	$\mathbf{C}_{ML} = \mathbf{A}_{MN}\mathbf{B}_{NL}$	MNL	$ML(N-1)$
Matrix transpose multiply	$\mathbf{C}_{MN} = \mathbf{A}_{ML}(\mathbf{B}'_{NL})$	MNL	$ML(N-1)$
Matrix inversion	$\mathbf{A}_{NN} \rightarrow \mathbf{A}_{NN}^{-1}$	αN^3	βN^3
Scalar-vector product	$\mathbf{C}_{N1} = \rho\mathbf{A}_{N1}$	N	—

17-9. Show that the standard algorithm for computing $\mathbf{P}(k+1|k+1)$ only requires on the order of $\frac{1}{2}mn^2$ multiplications, whereas the stabilized algorithm requires on the order of $\frac{5}{2}mn^2$ multiplications (use Table 17-1). Note that this last result requires a very clever coding of the stabilized algorithm.

Lesson 18

State Estimation: Filtering Examples

INTRODUCTION

In this lesson (which is an excellent one for self-study) we present five examples, which illustrate some interesting numerical and theoretical aspects of Kalman filtering.

EXAMPLES

Example 18-1

In Lesson 17 we learned that the Kalman filter is a dynamical feedback system. Its gain matrix and predicted- and filtering-error covariance matrices comprise a matrix feedback system operating within the Kalman filter. Of course, these matrices can be calculated prior to processing of data, and such calculations constitute a *performance analysis* of the Kalman filter. Here we examine the results of these calculations for two second-order systems, $H_1(z) = 1/(z^2 - 1.32z + 0.875)$ and $H_2(z) = 1/(z^2 - 1.36z + 0.923)$. The second system is less damped than the first. Impulse responses of both systems are depicted in Figure 18-1.

In Figure 18-1 we also depict $p_{11}(k|k)$, $p_{22}(k|k)$, $K_1(k)$ and $K_2(k)$ versus k for both systems. In both cases $\mathbf{P}(0|0)$ was set equal to the zero matrix. For system 1, $q = 1$ and $r = 5$, whereas for system 2 $q = 1$ and $r = 20$. Observe that the error variances and Kalman gains exhibit a transient response as well as a steady-state response; i.e., after a certain value of k ($k \cong 10$ for system-1 and $k \cong 15$ for system-2) $p_{11}(k|k)$, $p_{22}(k|k)$, $K_1(k)$, and $K_2(k)$ reaching limiting values. These limiting values do not depend on $\mathbf{P}(0|0)$, as can be seen from Figure 18-2. The Kalman filter is initially influenced by its initial conditions, but eventually ignores them, paying much greater attention to model

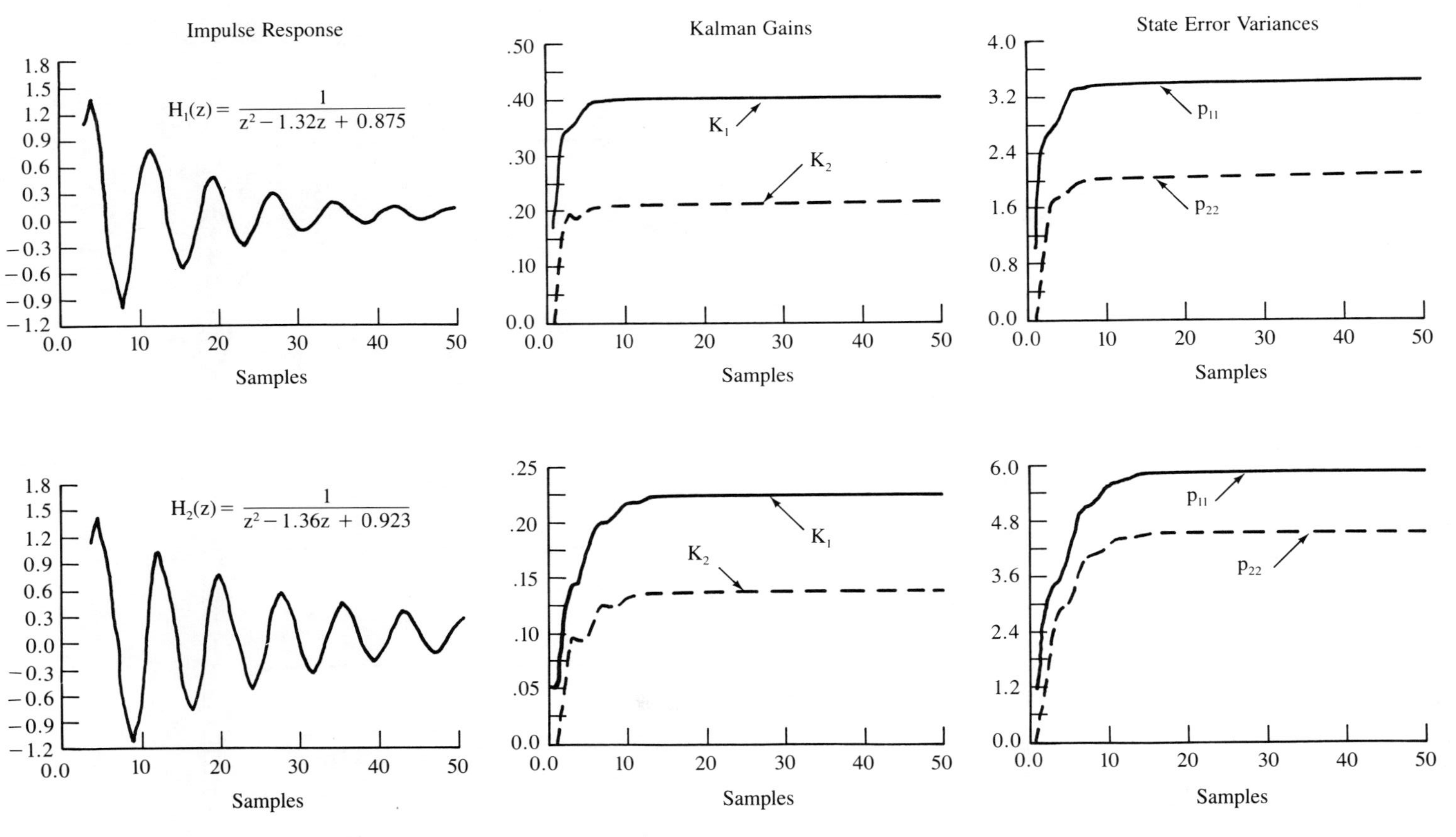

Figure 18-1 Kalman gains and filtering error variances for two second-order systems.

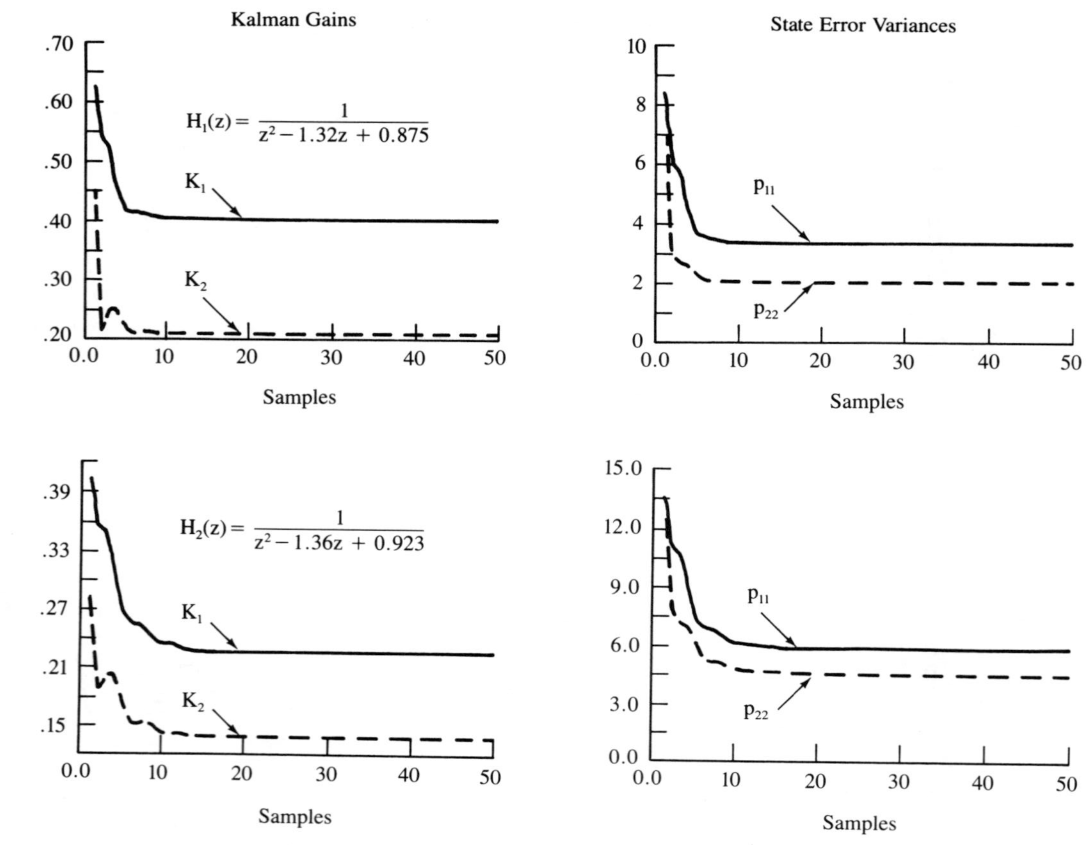

Figure 18-2 Same as Figure 18-1, except for a different choice of $\mathbf{P}(0|0)$.

parameters and the measurements. The relatively large steady-state values for p_{11} and p_{22} are due to the large values of r. □

Example 18-2

A state estimate is implicitly conditioned on knowing the true values for all system parameters (i.e., $\mathbf{\Phi}$, $\mathbf{\Gamma}$, $\mathbf{\Psi}$, $\mathbf{H}$, $\mathbf{Q}$, and $\mathbf{R}$). Sometimes we do not know these values exactly; hence, it is important to learn how *sensitive* (i.e., robust) the Kalman filter is to parameter errors. Many references which treat Kalman filter sensitivity issues can be found in Mendel and Gieseking (1971) under category 2*f*, "State Estimation: Sensitivity Considerations."

Let θ denote any parameter that may appear in $\mathbf{\Phi}$, $\mathbf{\Gamma}$, $\mathbf{\Psi}$, $\mathbf{H}$, $\mathbf{Q}$, or $\mathbf{R}$. In order to determine the sensitivity of the Kalman filter to *small* variations in θ one computes $\partial \hat{\mathbf{x}}(k+1|k+1)/\partial\theta$, whereas, for large variations in θ one computes $\Delta\hat{\mathbf{x}}(k+1|k+1)/\Delta\theta$. An analysis of $\Delta\hat{\mathbf{x}}(k+1|k+1)/\Delta\theta$, for example, reveals the interesting chain of events, that: $\Delta\hat{\mathbf{x}}(k+1|k+1)/\Delta\theta$ depends on $\Delta\mathbf{K}(k+1)/\Delta\theta$, which in turn depends on $\Delta\mathbf{P}(k+1|k)/\Delta\theta$, which in turn depends on $\Delta\mathbf{P}(k|k)/\Delta\theta$. Hence, for each variable parameter, θ, we have a *Kalman filter sensitivity system* comprised of equations from which we compute $\Delta\hat{\mathbf{x}}(k+1|k+1)/\Delta\theta$ (see also, Lesson 26). An alternative to using these equations is to perform a computer perturbation study. For example,

$$\frac{\Delta\mathbf{K}(k+1)}{\Delta\theta} = \frac{\mathbf{K}(k+1)\Big|_{\theta=\theta_N+\Delta\theta} - K(k+1)\Big|_{\theta=\theta_N}}{\Delta\theta} \tag{18-1}$$

where θ_N denotes a nominal value of θ.

We define *sensitivity* as the ratio of the percentage change in a function [e.g., $\mathbf{K}(k+1)$] to the percentage change in a parameter, θ. For example,

$$S_\theta^{K_{ij}(k+1)} = \frac{\dfrac{\Delta K_{ij}(k+1)}{K_{ij}(k+1)}}{\dfrac{\Delta\theta}{\theta}} \tag{18-2}$$

denotes the sensitivity of the *ij*th element of matrix $\mathbf{K}(k+1)$ with respect to parameter θ. All other sensitivities, such as $S_\theta^{P_{ij}(k|k)}$, are defined similarly.

Here we present some numerical sensitivity results for the simple first-order system

$$x(k+1) = a\,x(k) + b\,w(k) \tag{18-3}$$

$$z(k) = h\,x(k) + n(k) \tag{18-4}$$

where $a_N = 0.7$, $b_N = 1.0$, $h_N = 0.5$, $q_N = 0.2$, and $r_N = 0.1$. Figure 18-3 depicts $S_a^{K(k+1)}$, $S_b^{K(k+1)}$, $S_h^{K(k+1)}$, $S_q^{K(k+1)}$, and $S_r^{K(k+1)}$ for parameter variations of $\pm 5\%$, $\pm 10\%$, $\pm 20\%$, and $\pm 50\%$ about nominal values, a_N, b_N, h_N, q_N, and r_N.

Observe that the sensitivity functions vary with time and that they all reach steady-state values. Table 18-1 summarizes the steady-state sensitivity coefficients $\overline{S}_\theta^{K(k+1)}$, $\overline{S}_\theta^{P(k|k)}$ and $\overline{S}_\theta^{P(k+1|k)}$.

Some conclusions that can be drawn from these numerical results are: (1) $K(k+1)$, $P(k|k)$ and $P(k+1|k)$ are most sensitive to changes in parameter b, and

(2) $\overline{S}_\theta^{K(k+1)} = \overline{S}_\theta^{P(k|k)}$ for $\theta = a, b$, and q. This last observation could have been foreseen, because of our alternate equation for $K(k+1)$ [Equation (17-25)].

$$\mathbf{K}(k+1) = \mathbf{P}(k+1|k+1)\mathbf{H}(k+1)\mathbf{R}^{-1}(k+1) \tag{18-5}$$

This expression shows that if h and r are fixed then $\mathbf{K}(k+1)$ varies exactly the same way as $\mathbf{P}(k+1|k+1)$.

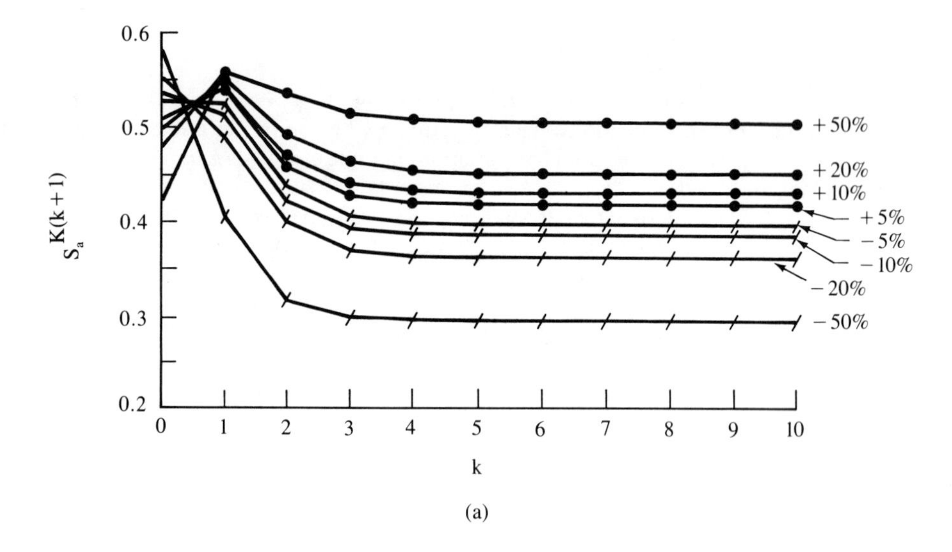

(a)

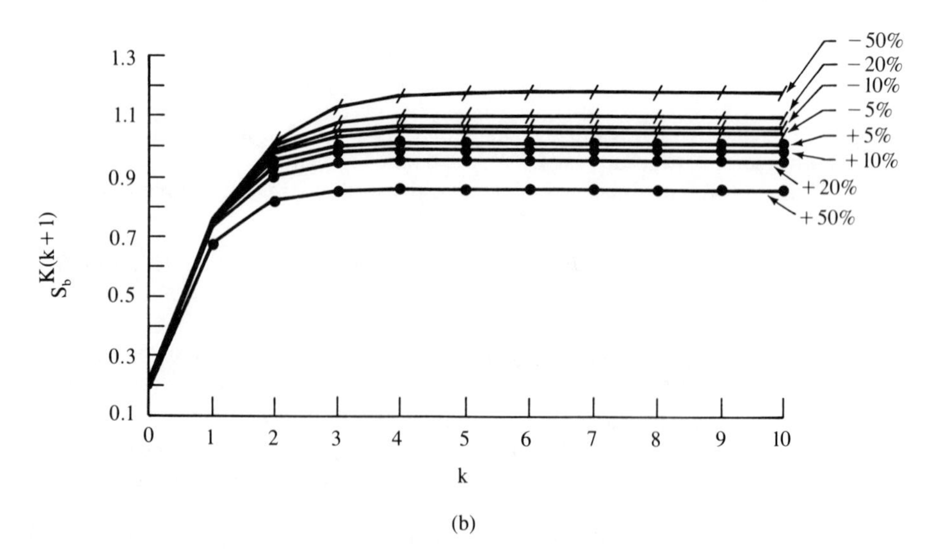

(b)

Figure 18-3 Sensitivity plots.

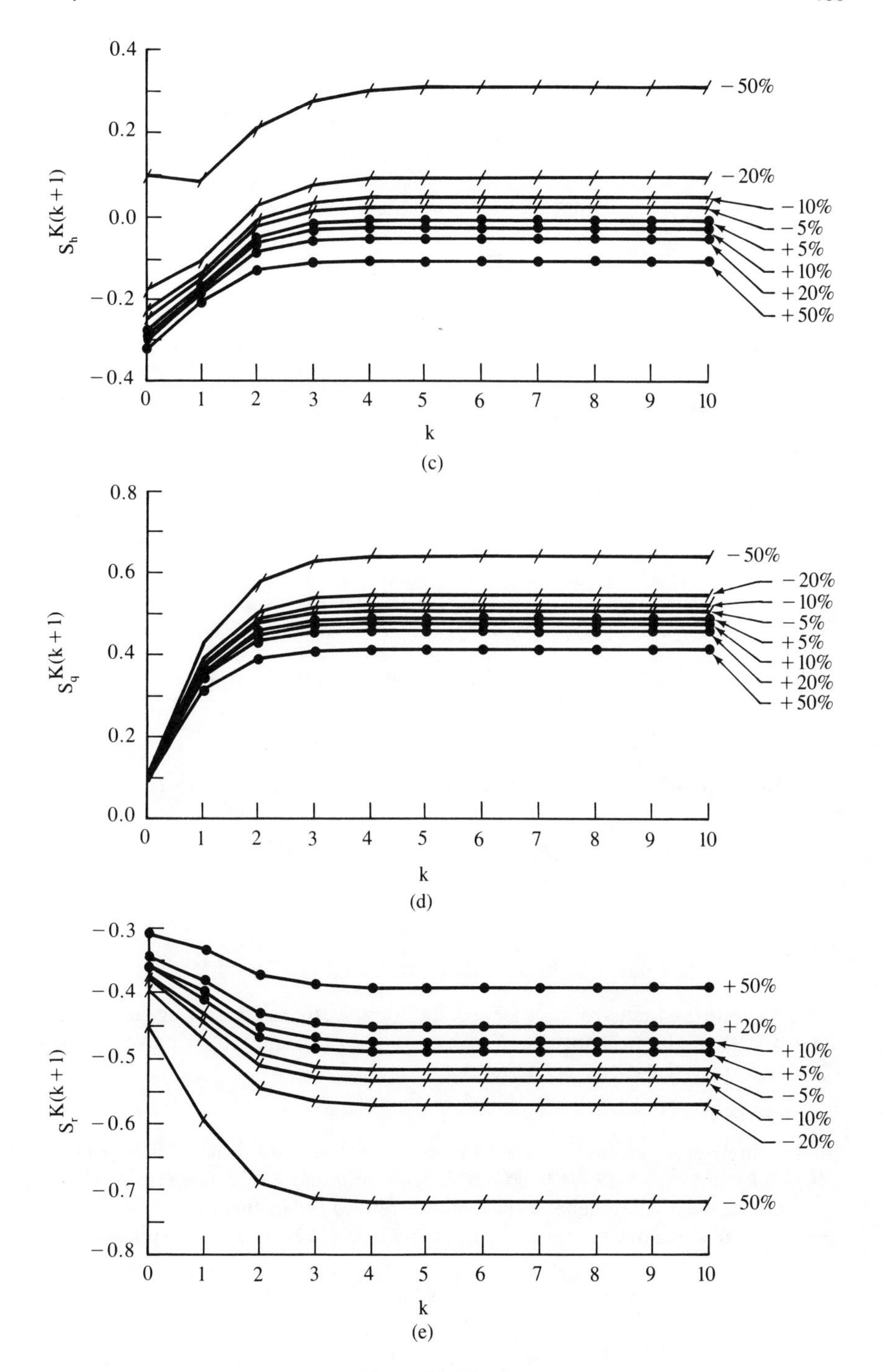

Figure 18-3 Cont.

TABLE 18-1 **Steady State Sensitivity Coefficients**

(a) $\overline{S}_\theta^{K(k+1)}$	Percentage Change in θ							
θ	+50	+20	+10	+5	−5	−10	−20	−50
a	.506	.451	.430	.419	.396	.384	.361	.294
b	.838	.937	.972	.989	1.025	1.042	1.077	1.158
h	−.108	−.053	−.026	−.010	.026	.047	.096	.317
q	.417	.465	.483	.493	.515	.526	.551	.648
r	−.393	−.452	−.476	−.490	−.518	−.534	−.570	−.717
(b) $\overline{S}_\theta^{P(k\|k)}$	Percentage Change in θ							
θ	+50	+20	+10	+5	−5	−10	−20	−50
a	.506	.451	.430	.419	.396	.384	.361	.294
b	.838	.937	.972	.989	1.025	1.042	1.077	1.158
h	−.739	−.877	−.932	−.962	−1.025	−1.059	−1.130	−1.367
q	.417	.465	.483	.493	.515	.526	.551	.648
r	.410	.457	.476	.486	.507	.519	.544	.642
(c) $\overline{S}_\theta^{P(k+1\|k)}$	Percentage Change in θ							
θ	+50	+20	+10	+5	−5	−10	−20	−50
a	1.047	.820	.754	.723	.664	.637	.585	.453
b	2.022	1.836	1.775	1.745	1.684	1.653	1.591	1.402
h	−.213	−.253	−.268	−.277	−.295	−.305	−.325	−.393
q	.832	.846	.851	.854	.860	.864	.871	.899
r	.118	.131	.137	.140	.146	.149	.157	.185

Here is how to use the results in Table 18-1. From Equation (18-2), for example, we see that

$$\frac{\Delta K_{ij}(k+1)}{K_{ij}(k+1)} = \left(\frac{\Delta\theta}{\theta}\right) S_\theta^{K_{ij}(k+1)} \tag{18-6}$$

or

$$\%\text{ Change in } K_{ij}(k+1) = (\%\text{ Change in } \theta) \times S_\theta^{K_{ij}(k+1)} \tag{18-7}$$

From Table 18-1 and this formula we see, for example, that a 20% change in a produces a (20)(.451) = 9.02% change in K, whereas a 20% change in h only produces a (20)(−.053) = −1.06% change in K, etc. □

Example 18-3

In the single-channel case, when $\mathbf{w}(k)$, $\mathbf{z}(k)$ and $\mathbf{v}(k)$ are scalars, then $\mathbf{K}(k+1)$, $\mathbf{P}(k+1|k)$ and $\mathbf{P}(k+1|k+1)$ do not depend on q and r separately. Instead, as we demonstrate next, they depend only on the ratio q/r. In this case, $\mathbf{H} = \mathbf{h}'$, $\boldsymbol{\Gamma} = \boldsymbol{\gamma}$, $\mathbf{Q} = q$ and $\mathbf{R} = r$, and Equations (17-12), (17-13) and (17-14) can be expressed as

$$\mathbf{K}(k+1) = \frac{\mathbf{P}(k+1|k)}{r}\mathbf{h}\left[\mathbf{h}'\frac{\mathbf{P}(k+1|k)}{r}\mathbf{h} + 1\right]^{-1} \tag{18-8}$$

$$\frac{\mathbf{P}(k+1|k)}{r} = \boldsymbol{\Phi}\frac{\mathbf{P}(k|k)}{\mathrm{r}}\boldsymbol{\Phi}' + \frac{q}{r}\boldsymbol{\gamma}\boldsymbol{\gamma}' \tag{18-9}$$

and

$$\frac{\mathbf{P}(k+1|k+1)}{r} = [\mathbf{I} - \mathbf{K}(k+1)\mathbf{h}']\frac{\mathbf{P}(k+1|k)}{r} \tag{18-10}$$

Observe that, given ratio q/r, we can compute $\mathbf{P}(k+1|k)/r$, $\mathbf{K}(k+1)$, and $\mathbf{P}(k+1|k+1)/r$. We refer to Equations (18-8), (18-9), and (18-10) as the *scaled* Kalman filter equations.

Ratio q/r can be viewed as a *filter tuning parameter*. Recall that, in Lesson 15, we showed q/r is related to signal-to-noise ratio; thus, using (15-50) for example, *we can also view signal-to-noise ratio as a (single-channel) Kalman filter tuning parameter.* Suppose, for example, that data quality is quite poor, so that signal-to-noise ratio, as measured by $\overline{\text{SNR}}$, is very small. Then q/r will also be very small, because $q/r = \overline{\text{SNR}}/\mathbf{h}'(\overline{\mathbf{P}}_{\mathbf{x}}/q)\mathbf{h}$. In this case the Kalman filter rejects the low-quality data, i.e., the Kalman gain matrix approaches the zero matrix, because

$$\frac{\mathbf{P}(k+1|k)}{r} \simeq \boldsymbol{\Phi}\frac{\mathbf{P}(k|k)}{r}\boldsymbol{\Phi}' \rightarrow \mathbf{0} \tag{18-11}$$

The Kalman filter is therefore quite sensitive to signal-to-noise ratio, as indeed are most digital filters. Its dependence on signal-to-noise ratio is complicated and nonlinear. Although signal-to-noise ratio (or q/r) enters quite simply into the equation for $\mathbf{P}(k+1|k)/r$, it is transformed in a nonlinear manner via (18-8) and (18-10). □

Example 18-4

A recursive unbiased minimum-variance estimator (BLUE) of a random parameter vector $\boldsymbol{\theta}$ can be obtained from the Kalman filter equations in Theorem 17-1 by setting $\mathbf{x}(k) = \boldsymbol{\theta}$, $\boldsymbol{\Phi}(k+1,k) = \mathbf{I}$, $\boldsymbol{\Gamma}(k+1,k) = \mathbf{0}$, $\boldsymbol{\Psi}(k+1,k) = \mathbf{0}$, and $\mathbf{Q}(k) = \mathbf{0}$. Under these conditions we see that $\mathbf{w}(k) = \mathbf{0}$ for all k, and

$$\mathbf{x}(k+1) = \mathbf{x}(k)$$

which means, of course, that $\mathbf{x}(k)$ is a vector of constants, namely $\boldsymbol{\theta}$. The Kalman filter equations reduce to

$$\hat{\boldsymbol{\theta}}(k+1|k+1) = \hat{\boldsymbol{\theta}}(k|k) + \mathbf{K}(k+1)[\mathbf{z}(k+1) - \mathbf{H}(k+1)\hat{\boldsymbol{\theta}}(k|k)], \tag{18-12}$$

$$\mathbf{P}(k+1|k) = \mathbf{P}(k|k)$$

$$\mathbf{K}(k+1) = \mathbf{P}(k|k)\mathbf{H}'(k+1)[\mathbf{H}(k+1)\mathbf{P}(k|k)\mathbf{H}'(k+1) + \mathbf{R}(k+1)]^{-1} \tag{18-13}$$

and

$$\mathbf{P}(k+1|k+1) = [\mathbf{I} - \mathbf{K}(k+1)\mathbf{H}(k+1)]\mathbf{P}(k|k) \tag{18-14}$$

Note that it is no longer necessary to distinguish between filtered and predicted quantities, because $\hat{\boldsymbol{\theta}}(k+1|k) = \hat{\boldsymbol{\theta}}(k|k)$ and $\mathbf{P}(k+1|k) = \mathbf{P}(k|k)$; hence, the notation $\hat{\boldsymbol{\theta}}(k|k)$ can be simplified to $\hat{\boldsymbol{\theta}}(k)$, for example. Equations (18-12), (18-13), and (18-14) were obtained earlier in Lesson 9 (see Theorem 9-7) for the case of scalar measurements. □

Example 18-5

This example illustrates the *divergence phenomenon*, which often occurs when either process noise or measurement noise or both are small. We shall see that the Kalman filter locks onto wrong values for the state, but believes them to be the true values, i.e., it "learns" the wrong state too well.

Our example is adapted from Jazwinski (1970, pp. 302–303). We begin with the following simple first-order system

$$x(k+1) = x(k) + b \tag{18-15}$$

$$z(k+1) = x(k+1) + v(k+1) \tag{18-16}$$

where b is a very small bias, so small that, when we design our Kalman filter, we choose to neglect it. Our Kalman filter is based on the following model

$$x_m(k+1) = x_m(k) \tag{18-17}$$

$$z(k+1) = x_m(k+1) + v(k+1) \tag{18-18}$$

Using this model we estimate $x(k)$ and $\hat{x}_m(k|k)$, where it is straightforward to show that

$$\hat{x}_m(k+1|k+1) = \hat{x}_m(k|k) + \overbrace{\frac{p(0)}{(k+1)p(0)+r}}^{K(k+1)}[z(k+1) - \hat{x}_m(k|k)] \tag{18-19}$$

Observe that, as $k \to \infty$, $K(k+1) \to 0$ so that $\hat{x}_m(k+1|k+1) \to \hat{x}_m(k|k)$. The Kalman filter is rejecting the new measurements because it believes (18-17) to be the true model for $x(k)$; but, of course, it is not the true model.

The Kalman filter computes the error variance, $p_m(k|k)$, between $\hat{x}_m(k|k)$ and $x_m(k)$. The true error variance is associated with $\tilde{x}(k|k)$, where

$$\tilde{x}(k|k) = x(k) - \hat{x}_m(k|k) \tag{18-20}$$

We leave it to the reader to show that

$$\tilde{x}(k|k) = \frac{r}{kp(0)+r}\tilde{x}(0|0) - \frac{p(0)}{kp(0)+r}\sum_{i=1}^{k} v(i) + \frac{[(k-1)k/2]p(0)+kr}{kp(0)+r}b \tag{18-21}$$

As $k \to \infty$, $\tilde{x}(k|k) \to \infty$ because the third term on the right-hand side of (18-21) diverges to infinity. This term contains the bias b that was neglected in the model used by the Kalman filter. Note also that $\tilde{x}_m(k|k) = x_m(k) - \hat{x}_m(k|k) \to 0$ as $k \to \infty$; thus, the Kalman filter has locked onto the wrong state and is unaware that the true error variance is diverging.

A number of different remedies have been proposed for controlling divergence effects, including:

1. adding fictitious process noise,
2. finite-memory filtering, and
3. fading memory filtering.

Fictitious process noise, which appears in the state equation, can be used to account for neglected modeling effects that enter into the state equation (e.g., truncation of second- and higher-order effects when a nonlinear state equation is linearized, as described in Lesson 24). This process noise introduces $\mathbf{Q}$ into the Kalman filter equations. Observe, in our first-order example, that $\mathbf{Q}$ does not appear in the equations for $\hat{x}_m(k+1|k+1)$ or $\tilde{x}(k|k)$, because state equation (18-17) contains no process noise.

Divergence is a large-sample property of the Kalman filter. Finite-memory and fading memory filtering control divergence by not letting the Kalman filter get into its "large sample" regime. Finite-memory filtering (Jazwinski, 1970) uses a finite window of measurements (of fixed length W) to estimate $x(k)$. As we move from $t = k_1$ to $t = k_1 + 1$, we must account for two effects, namely, the new measurement at $t = k_1 + 1$ and a discarded measurement at $t = k_1 - W$. Fading-memory filtering, due to Sorenson and Sacks (1971) exponentially ages the measurements, weighting the recent measurement most heavily and past measurements much less heavily. It is analogous to weighted least squares, as described in Lesson 3.

Fading-memory filtering seems to be the most successful and popular way to control divergence effects. □

PROBLEMS

18-1. Derive the equations for $\hat{x}_m(k+1|k+1)$ and $\tilde{x}(k|k)$, in (18-19) and (18-21), respectively.

18-2. In Lesson 5 we described cross-sectional processing for weighted least-squares estimates. Cross-sectional (also known as sequential) processing can be performed in Kalman filtering. Suppose $\mathbf{z}(k+1) = \text{col}(\mathbf{z}_1(k+1), \mathbf{z}_2(k+1), \ldots, \mathbf{z}_q(k+1))$, where $\mathbf{z}_i(k+1) = \mathbf{H}_i\mathbf{x}(k+1) + \mathbf{v}_i(k+1)$, $\mathbf{v}_i(k+1)$ are mutually uncorrelated for $i = 1, 2, \ldots$, $\mathbf{z}_i(k+1)$ is $m_i \times 1$, and $m_1 + m_2 + \cdots + m_q = m$. Let $\hat{\mathbf{x}}_i(k+1|k+1)$ be a "corrected" estimate of $\mathbf{x}(k+1)$ that is associated with processing $\mathbf{z}_i(k+1)$.

(a) Using the Fundamental Theorem of Estimation Theory, prove that a cross-sectional structure for the corrector equation of the Kalman filter is:

$$\begin{aligned}
\hat{\mathbf{x}}_1(k+1|k+1) &= \hat{\mathbf{x}}(k+1|k) + \mathbf{E}\{\mathbf{x}(k+1)|\tilde{\mathbf{z}}_1(k+1|k)\} \\
\hat{\mathbf{x}}_2(k+1|k+1) &= \hat{\mathbf{x}}_1(k+1|k+1) + \mathbf{E}\{\mathbf{x}(k+1)|\tilde{\mathbf{z}}_2(k+1|k)\} \\
&\vdots \\
\hat{\mathbf{x}}_q(k+1|k+1) &= \hat{\mathbf{x}}_{q-1}(k+1|k+1) + \mathbf{E}\{\mathbf{x}(k+1)|\tilde{\mathbf{z}}_q(k+1|k)\} \\
&= \hat{\mathbf{x}}(k+1|k+1)
\end{aligned}$$

(b) Provide equations for computing $\mathbf{E}\{\mathbf{x}(k+1)|\tilde{\mathbf{z}}_i(k+1|k)\}$.

18-3. (Project) Choose a second-order system and perform a thorough sensitivity study of its associated Kalman filter. Do this for various nominal values and for both small and large variations of the system's parameters. *You will need a computer for this project.* Present the results both graphically and tabularly, as in Example 18-2. Draw as many conclusions as possible.

Lesson 19

State Estimation: Steady-State Kalman Filter and Its Relationship to a Digital Wiener Filter

INTRODUCTION

In this lesson we study the steady-state Kalman filter from different points of view, and we then show how it is related to a digital Wiener filter.

STEADY-STATE KALMAN FILTER

For time-invariant and stationary systems, if $\lim_{k\to\infty} \mathbf{P}(k+1|k) = \overline{\mathbf{P}}$ exists, then $\lim_{k\to\infty} \mathbf{K}(k) \to \overline{\mathbf{K}}$ and the Kalman filter (17-11) becomes a constant coefficient filter. Because $\mathbf{P}(k+1|k)$ and $\mathbf{P}(k|k)$ are intimately related, then, if $\overline{\mathbf{P}}$ exists, $\lim_{k\to\infty} \mathbf{P}(k|k) \triangleq \overline{\mathbf{P}}_1$ also exists. We have already observed limiting behaviors for $\mathbf{K}(k+1)$ and $\mathbf{P}(k|k)$ in Example 18-1. The following theorem, which is adopted from Anderson and Moore (1979, pg. 77), tells us when $\overline{\mathbf{P}}$ exists, and assures us that the steady-state Kalman filter will be asymptotically stable.

Theorem 19-1 (Steady-State Kalman Filter). *If our dynamical model in Equations (15-17) and (15-18) is time-invariant, stationary, and asymptotically stable (i.e., all the eigenvalues of $\mathbf{\Phi}$ lie inside the unit circle), then:*

a. *For any nonnegative-definite symmetric initial condition $\mathbf{P}(0|-1)$, one has $\lim_{k\to\infty} \mathbf{P}(k+1|k) = \overline{\mathbf{P}}$ with $\overline{\mathbf{P}}$ independent of $\mathbf{P}(0|-1)$ and satisfying the following steady-state version of Equation (17-24):*

$$\overline{\mathbf{P}} = \mathbf{\Phi}\overline{\mathbf{P}}[\mathbf{I} - \mathbf{H}'(\mathbf{H}\overline{\mathbf{P}}\mathbf{H}' + \mathbf{R})^{-1}\mathbf{H}\overline{\mathbf{P}}]\mathbf{\Phi}' + \mathbf{\Gamma}\mathbf{Q}\mathbf{\Gamma}' \tag{19-1}$$

Equation (19-1) is often referred to either as a steady-state or algebraic Riccati equation.

b. *The eigenvalues of the steady-state Kalman filter,* $\lambda[\mathbf{\Phi} - \overline{\mathbf{K}}\mathbf{H}\mathbf{\Phi}]$, *all lie within the unit circle, so that the filter is asymptotically stable; i.e.,*

$$|\lambda[\mathbf{\Phi} - \overline{\mathbf{K}}\mathbf{H}\mathbf{\Phi}]| < 1 \tag{19-2}$$

If our dynamical model in Equations (15-17) and (15-18) is time-invariant and stationary, but is not necessarily asymptotically stable, then points (a) and (b) still hold as long as the system is completely stabilizable and detectable. □

A proof of this theorem is beyond the scope of this textbook. It can be found in Anderson and Moore, pp. 78–82 (1979). For definitions of the system-theoretic terms *stabilizable* and *detectable*, the reader should consult a textbook on linear systems, such as Kailath (1980) or Chen (1970). By *completely detectable* and *completely stabilizable*, we mean that $(\mathbf{\Phi},\mathbf{H})$ is completely detectable and $(\mathbf{\Phi},\mathbf{\Gamma}\mathbf{Q}_1)$ is completely stabilizable, where $\mathbf{Q} = \mathbf{Q}_1\mathbf{Q}_1'$. Additionally, any asymptotically stable model is always completely stabilizable and detectable.

Probably the most interesting case of a system that is not asymptotically stable, for which we want to design a steady-state Kalman filter, is one that has a pole on the unit circle.

Example 19-1

In this example (which is similar to Example 5.4 in Meditch, 1969, pp. 189–190) we consider the scalar system

$$x(k+1) = x(k) + w(k) \tag{19-3}$$

$$z(k+1) = x(k+1) + v(k+1) \tag{19-4}$$

It has a pole on the unit circle. When $q = 20$ and $r = 5$, Equations (17-13), (17-12) and (17-14) reduce to:

$$p(k+1|k) = p(k|k) + 20 \tag{19-5}$$

$$K(k+1) = \frac{p(k|k) + 20}{p(k|k) + 25} \tag{19-6}$$

and

$$p(k+1|k+1) = \frac{5[p(k|k) + 20]}{p(k|k) + 25} = 5\,K(k+1) \tag{19-7}$$

Starting with $p(0|0) = p_x(0) = 50$, it is a relatively simple matter to compute $p(k+1|k)$, $K(k+1)$ and $p(k+1|k+1)$ for $k = 0, 1, \ldots$. The results for the first few iterations are given in Table 19-1.

TABLE 19-1 Kalman Filter Quantities

k	$p(k\|k-1)$	$K(k)$	$p(k\|k)$
0	...	...	50
1	70	0.933	4.67
2	24.67	0.831	4.16
3	24.16	0.829	4.14
4	24.14	0.828	4.14

The steady-state value of $p(k|k)$ is obtained by setting $p(k+1|k+1) = p(k|k) = \overline{p}_1$ in the last of the above three relations to obtain

$$\overline{p}_1^2 + 20\overline{p}_1 - 100 = 0 \tag{19-8}$$

Because $\overline{p}_1$ is a variance it must be nonnegative; hence, only the solution $\overline{p}_1 = 4.14$ is valid. Comparing this result with $p(3|3)$ we see that the filter is in the steady state to within the indicated computational accuracy after processing just three measurements. In steady-state, the Kalman filter equation is

$$\begin{aligned}\hat{x}(k+1|k+1) &= \hat{x}(k|k) + 0.828[z(k+1) - \hat{x}(k|k)] \\ &= 0.172\,\hat{x}(k|k) + 0.828\,z(k+1)\end{aligned} \tag{19-9}$$

Observe that the filter's pole at 0.172 lies inside the unit circle. □

Many ways have been reported for solving the algebraic Riccati equation (19-1) [see Laub (1979), for example], ranging from direct iteration of the matrix Riccati Equation (17-24) until $\mathbf{P}(k+1|k)$ does not change appreciably from $\mathbf{P}(k|k-1)$, to solving the nonlinear algebraic Riccati equation via an iterative Newton-Raphson procedure, to solving that equation in one shot by the Schur method. Iterative methods are quite sensitive to error accumulation. The one-shot Schur method possesses a high degree of numerical integrity, and appears to be one of the most successful ways for obtaining $\overline{\mathbf{P}}$. For details about this method, see Laub (1979).

In summary, then, to design a steady-state Kalman filter:

1. given $(\mathbf{\Phi},\mathbf{\Gamma},\mathbf{\Psi},\mathbf{H},\mathbf{Q},\mathbf{R})$, compute $\overline{\mathbf{P}}$, the positive definite solution of (19-1);
2. compute $\overline{\mathbf{K}}$, as

$$\overline{\mathbf{K}} = \overline{\mathbf{P}}\mathbf{H}'(\mathbf{H}\overline{\mathbf{P}}\mathbf{H}' + \mathbf{R})^{-1} \tag{19-10}$$

3. use (19-10) in

$$\begin{aligned}\hat{\mathbf{x}}(k+1|k+1) &= \mathbf{\Phi}\hat{\mathbf{x}}(k|k) + \mathbf{\Psi}\mathbf{u}(k) + \overline{\mathbf{K}}\tilde{\mathbf{z}}(k+1|k) \\ &= (\mathbf{I} - \overline{\mathbf{K}}\mathbf{H})\mathbf{\Phi}\hat{\mathbf{x}}(k|k) + \overline{\mathbf{K}}\mathbf{z}(k+1) \\ &\quad + (\mathbf{I} - \overline{\mathbf{K}}\mathbf{H})\mathbf{\Psi}\mathbf{u}(k)\end{aligned} \tag{19-11}$$

SINGLE-CHANNEL STEADY-STATE KALMAN FILTER

The steady-state Kalman filter can be viewed outside of the context of estimation theory as a recursive digital filter. As such, it is sometimes useful to be able to compute its impulse response, transfer function, and frequency response. In this section we restrict our attention to the single-channel steady-state Kalman filter. From (19-11) we observe that this filter is excited by two inputs, $\mathbf{z}(k+1)$ and $\mathbf{u}(k)$ [in the single-channel case, $z(k+1)$ and $u(k)$]. In this section we shall only be interested in transfer functions which are associated with the effect of $z(k+1)$ on the filter; hence, we set $u(k) = 0$. Additionally, we shall view the signal component of measurement $z(k)$ as our desired output. The signal component of $z(k)$ is $\mathbf{h}'\mathbf{x}(k)$.

Let $H_f(z)$ denote the z-transform of the impulse response of the steady-state filter; i.e.,

$$H_f(z) = \mathfrak{Z}\{\hat{z}(k|k) \text{ when } z(k) = \delta(k)\} \tag{19-12}$$

This transfer function is found from the following *steady-state filter system*:

$$\hat{\mathbf{x}}(k|k) = [\mathbf{I} - \overline{\mathbf{K}}\mathbf{h}']\mathbf{\Phi}\hat{\mathbf{x}}(k-1|k-1) + \overline{\mathbf{K}}\delta(k) \tag{19-13}$$

$$\hat{\mathbf{z}}(k|k) = \mathbf{h}'\hat{\mathbf{x}}(k|k) \tag{19-14}$$

Taking the z-transform of (19-13) and (19-14), it follows that

$$H_f(z) = \mathbf{h}'(\mathbf{I} - \mathbf{\Phi}_f z^{-1})^{-1}\overline{\mathbf{K}} \tag{19-15}$$

where

$$\mathbf{\Phi}_f = [\mathbf{I} - \overline{\mathbf{K}}\mathbf{h}']\mathbf{\Phi} \tag{19-16}$$

Equation (19-15) can also be written as

$$H_f(z) = h_f(0) + h_f(1)z^{-1} + h_f(2)z^{-2} + \cdots \tag{19-17}$$

where the filter's coefficients (i.e., Markov parameters), $h_f(j)$, are

$$h_f(j) = \mathbf{h}'\mathbf{\Phi}_f^j\overline{\mathbf{K}} \tag{19-18}$$

for $j = 0, 1, \ldots$.

In our study of mean-squared smoothing, in Lessons 20, 21, and 22, we will see that the steady-state predictor system plays an important role. The *steady-state predictor*, obtained from (17-23), is given by

$$\hat{\mathbf{x}}(k+1|k) = \mathbf{\Phi}_p\hat{\mathbf{x}}(k|k-1) + \boldsymbol{\gamma}_p z(k) \tag{19-19}$$

where

$$\mathbf{\Phi}_p = \mathbf{\Phi}(\mathbf{I} - \overline{\mathbf{K}}\mathbf{h}') \tag{19-20}$$

and

$$\boldsymbol{\gamma}_p = \mathbf{\Phi}\overline{\mathbf{K}} \tag{19-21}$$

Let $H_p(z)$ denote the z-transform of the impulse response of the steady-state predictor, i.e.,

$$H_p(z) = \mathfrak{Z}\{\hat{z}(k|k-1) \text{ when } z(k) = \delta(k)\} \tag{19-22}$$

This transfer function is found from the following *steady-state predictor system*:

$$\hat{\mathbf{x}}(k+1|k) = \mathbf{\Phi}_p \hat{\mathbf{x}}(k|k-1) + \boldsymbol{\gamma}_p \delta(k) \tag{19-23}$$

$$\hat{\mathbf{z}}(k|k-1) = \mathbf{h}'\hat{\mathbf{x}}(k|k-1) \tag{19-24}$$

Taking the z-transform of these equations, we find that

$$H_p(z) = \mathbf{h}'(\mathbf{I} - \mathbf{\Phi}_p z^{-1})^{-1} z^{-1} \boldsymbol{\gamma}_p \tag{19-25}$$

which can also be written as

$$H_p(z) = h_p(1)z^{-1} + h_p(2)z^{-2} + \cdots \tag{19-26}$$

where the predictor's coefficients, $h_p(j)$, are

$$h_p(j) = \mathbf{h}'\mathbf{\Phi}_p^{j-1} \boldsymbol{\gamma}_p \tag{19-27}$$

for $j = 1, 2, \ldots$.

Although $H_f(z)$ and $H_p(z)$ contain an infinite number of terms, they can usually be truncated, because both are associated with asymptotically stable filters.

Example 19-2

In this example we examine the impulse response of the steady-state predictor for the first-order system

$$x(k+1) = \frac{1}{\sqrt{2}} x(k) + w(k) \tag{19-28}$$

$$z(k+1) = \frac{1}{\sqrt{2}} x(k+1) + v(k+1) \tag{19-29}$$

In Example 15-2 of Lesson 15, we showed that ratio q/r is proportional to signal-to-noise ratio, $\overline{\text{SNR}}$. For this example, $\overline{\text{SNR}} = q/r$. Our numerical results, given below, are for $\overline{\text{SNR}} = 20, 5, 1$ and use the scaled steady-state prediction-error variance $\overline{p}/r$, which can be solved for from (19-1) when it is written as:

$$\frac{\overline{p}}{r} = \frac{1}{2}\frac{\overline{p}}{r}\left[1 - \frac{1}{2}\left(\frac{1}{2}\frac{\overline{p}}{r} + 1\right)^{-1}\frac{\overline{p}}{r}\right] + \overline{\text{SNR}} \tag{19-30}$$

Note that we could just as well have solved (19-1) for $\overline{p}/q$ but the structure of its equation is more complicated than (19-30). The positive solution of (19-30) is

$$\frac{\overline{p}}{r} = \frac{1}{2}(\overline{\text{SNR}} - 1) + \frac{1}{2}\sqrt{(\overline{\text{SNR}} - 1)^2 + 8\,\overline{\text{SNR}}} \tag{19-31}$$

Additionally,

$$\overline{K} = \frac{1}{\sqrt{2}}\frac{\overline{p}}{r}\left(\frac{1}{2}\frac{\overline{p}}{r} + 1\right)^{-1} \tag{19-32}$$

Table 19-2 summarizes $\overline{p}/r$, $\overline{K}$, ϕ_p and γ_p, quantities which are needed to compute the impulse response $h_p(k)$ for $k \geq 0$, which is depicted in Figure 19-1. Observe that all three responses peak at $j = 1$; however, the decay time for $\overline{\text{SNR}} = 20$ is quicker than

TABLE 19-2 Steady-State Predictor Quantities

$\overline{\text{SNR}}$	$\overline{p}/r$	$\overline{K}$	ϕ_p	γ_p
20	20.913	1.287	0.064	0.910
5	5.742	1.049	0.183	0.742
1	1.414	0.586	0.414	0.414

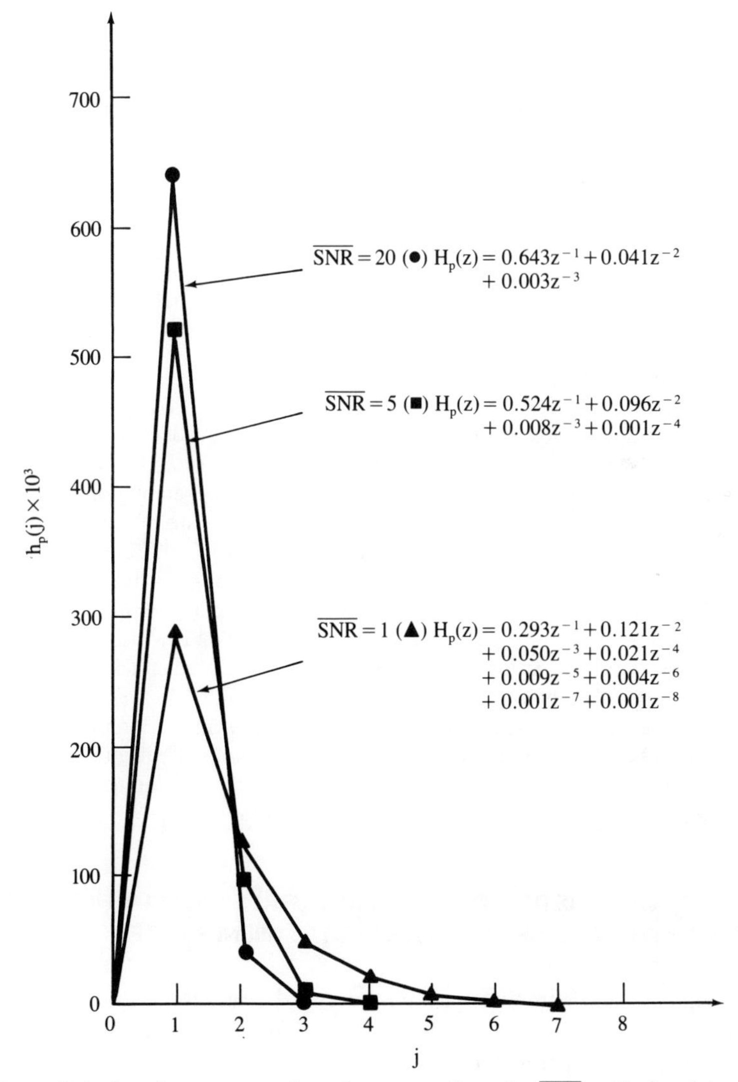

Figure 19-1 Impulse response of steady-state predictor for $\overline{SNR} = 20$, 5 and 1. $H_p(z)$ is shown to three significant figures (Mendel, 1981, © 1981, IEEE).

the decay times for lower $\overline{\text{SNR}}$ values, and the peak amplitude is larger for higher $\overline{\text{SNR}}$ values. The steady-state predictor tends to reject measurements which have a low signal-to-noise ratio (the same is true for the steady-state filter). □

Example 19-3

In this example we present impulse response and frequency response plots for the steady-state predictors associated with the systems

$$H_1(z) = \frac{1}{z^2 - z + 0.5} \tag{19-33}$$

and

$$H_2(z) = \frac{-0.688z^3 + 1.651z^2 - 1.221z + 0.25}{z^4 - 2.586z^3 + 2.489z^2 - 1.033z + 0.168} \tag{19-34}$$

Matrices $\mathbf{\Phi}$, $\boldsymbol{\gamma}$ and $\mathbf{h}$ for these systems are:

$$\mathbf{\Phi}_1 = \begin{pmatrix} 0 & 1 \\ -1/2 & 1 \end{pmatrix} \quad \boldsymbol{\gamma}_1 = \begin{pmatrix} 0 \\ 1 \end{pmatrix} \quad \mathbf{h}_1 = \begin{pmatrix} 1 \\ 0 \end{pmatrix}$$

and

$$\mathbf{\Phi}_2 = \begin{pmatrix} 0 & 1 & 0 & 0 \\ 0 & 0 & 1 & 0 \\ 0 & 0 & 0 & 1 \\ -0.168 & +1.033 & -2.489 & +2.586 \end{pmatrix} \quad \boldsymbol{\gamma}_2 = \begin{pmatrix} 0 \\ 0 \\ 0 \\ 1 \end{pmatrix} \quad \mathbf{h}_2 = \begin{pmatrix} 0.250 \\ -1.221 \\ 1.651 \\ -0.688 \end{pmatrix}$$

Figure 19-2 depicts $h_1(k)$, $|H_1(j\omega)|$ (in db) and $\angle H_1(j\omega)$ as well as $h_{p1}(k)$, $|H_{p1}(j\omega)|$, and $\angle H_{p1}(j\omega)$ for $\overline{\text{SNR}} = 1$, 5, and 20. Figure 19-3 depicts comparable quantities for the second system. Observe that, as signal-to-noise ratio decreases, the steady-state predictor rejects the measurements; for the amplitudes of $h_{p1}(k)$ and $h_{p2}(k)$ become smaller as $\overline{\text{SNR}}$ becomes smaller. It also appears, from examination of $|H_{p1}(j\omega)|$ and $|H_{p2}(j\omega)|$, that at high signal-to-noise ratios the steady-state predictor behaves like a high-pass filter for system-1 and a bandpass filter for system-2. On the other hand, at low signal-to-noise ratios it appears to behave like a band-pass filter.

The steady-state predictor appears to be quite dependent on system dynamics at high signal-to-noise ratios, but is much less dependent on the system dynamics at low signal-to-noise ratios. At low signal-to-noise ratios the predictor is rejecting the measurements regardless of system dynamics.

Just because the steady-state predictor behaves like a high-pass filter at high signal-to-noise ratios does not mean that it passes a lot of noise through it, because *high signal-to-noise ratio* means that measurement noise level is quite low. Of course, a spurious burst of noise would pass through this filter quite easily. □

RELATIONSHIPS BETWEEN THE STEADY-STATE KALMAN FILTER AND A FINITE IMPULSE RESPONSE DIGITAL WIENER FILTER

The steady-state Kalman filter is a recursive digital filter with filter coefficients equal to $h_f(j)$, $j = 0, 1, \ldots$ [see (19-18)]. Quite often $h_f(j) \rightarrow 0$ for $j \geq J$, so that $H_f(z)$ can be truncated, i.e.,

$$H_f(z) \simeq h_f(0) + h_f(1)z^{-1} + \cdots + h_f(J)z^{-J} \tag{19-35}$$

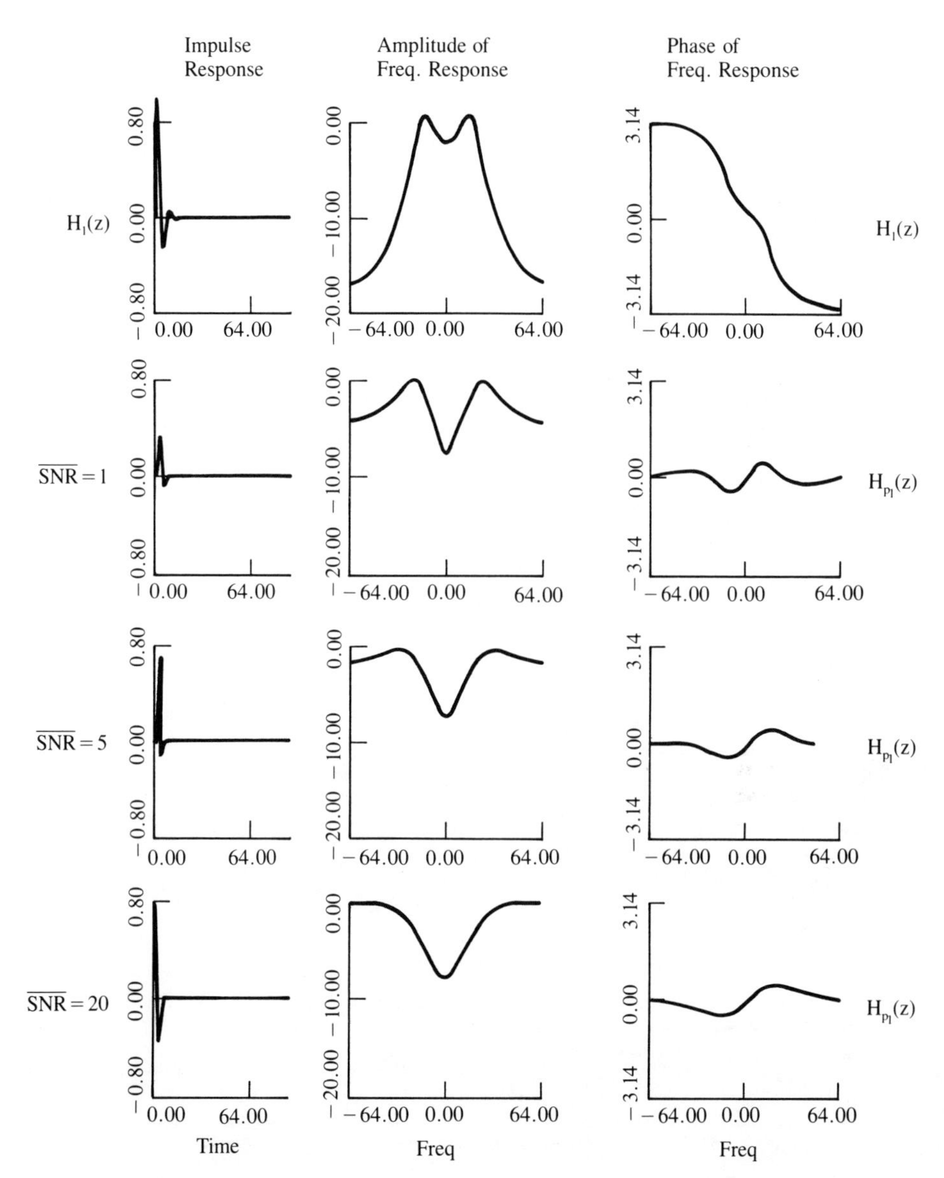

Figure 19-2 Impulse and frequency responses for system $H_1(z)$ and its associated steady-state predictor.

The *truncated steady-state Kalman filter* can then be implemented as a finite-impulse response (FIR) digital filter.

There is a more direct way for designing a FIR minimum mean-squared error filter, i.e., a *digital Wiener filter*, as we describe next.

Consider the situation depicted in Figure 19-4. We wish to design digital filter F(z)'s coefficients $f(0), f(1), \ldots, f(\eta)$ so that the filter's output, $y(k)$, is

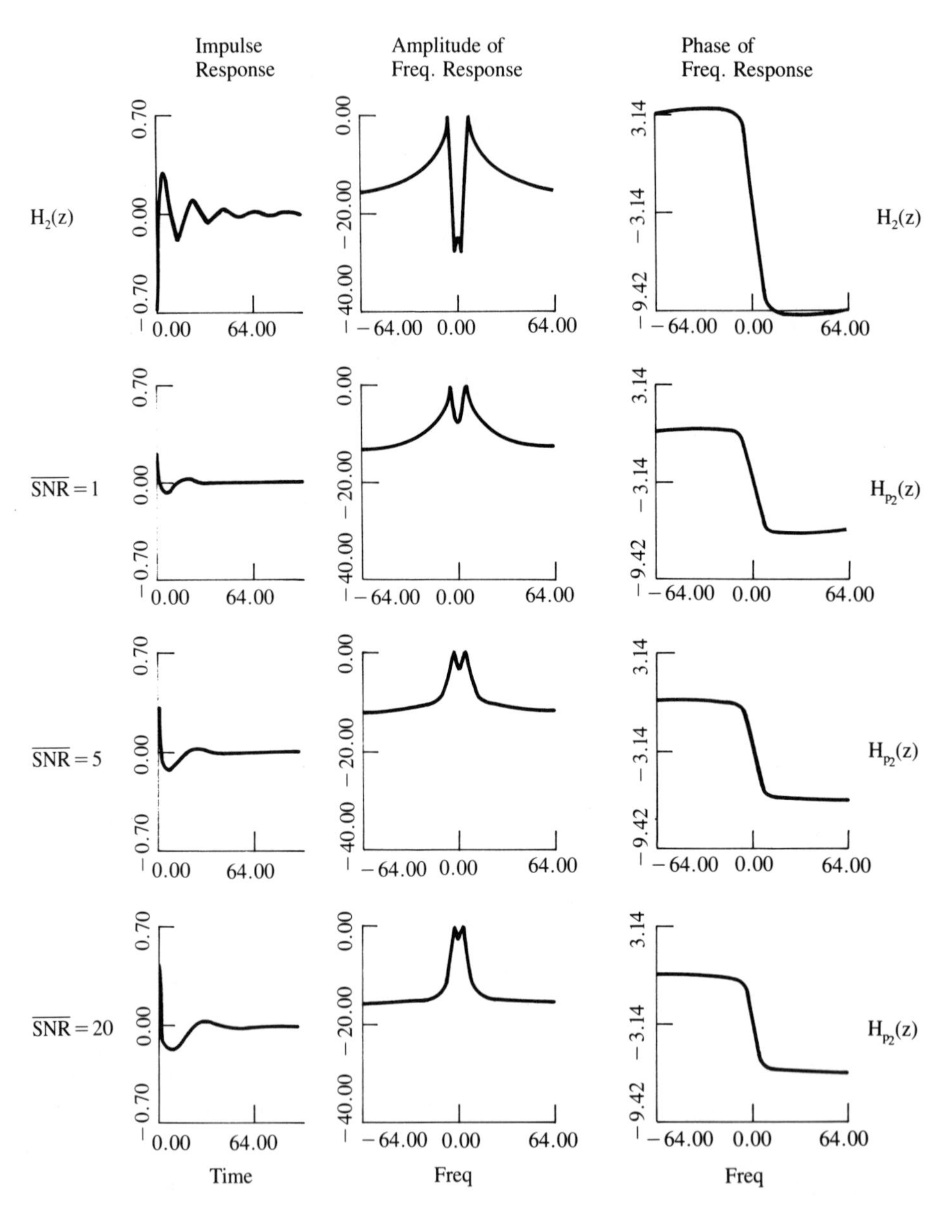

Figure 19-3 Impulse and frequency responses for system $H_2(z)$ and its associated steady-state predictor.

close, in some sense, to a desired signal $d(k)$. In a digital Wiener filter design, $f(0), f(1), \ldots, f(\eta)$ are obtained by minimizing the following mean-squared error

$$I(\mathbf{f}) = \mathbf{E}\{[d(k) - y(k)]^2\} = \mathbf{E}\{e^2(k)\} \tag{19-36}$$

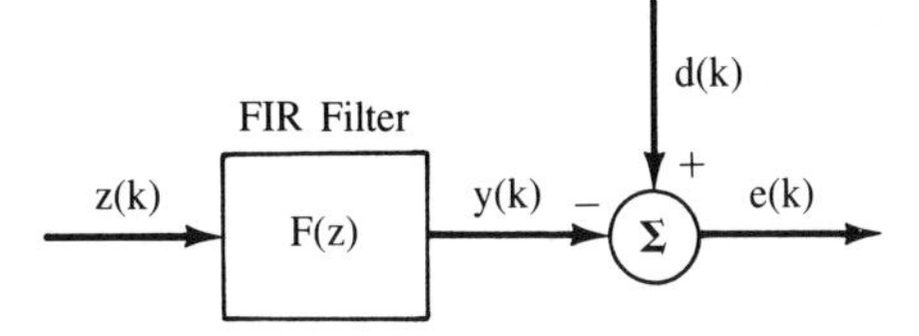

Figure 19-4 Starting point for design of FIR digital Wiener filter.

Using the fact that

$$y(k) = f(k)*z(k) = \sum_{i=0}^{\eta} f(i)z(k-i) \tag{19-37}$$

we see that

$$I(\mathbf{f}) = \mathbf{E}\{[d(k) - \sum_{i=0}^{\eta} f(i)z(k-i)]^2\} \tag{19-38}$$

The filter coefficients that minimize $I(\mathbf{f})$ are found by setting $\partial I(\mathbf{f})/\partial f(j) = 0$ for $j = 0, 1, \ldots, \eta$. We leave it as an exercise for the reader to show that doing this results in the following linear system of $\eta + 1$ equations in the $\eta + 1$ unknown filter coefficients,

$$\sum_{i=0}^{\eta} f(i)\phi_{zz}(i-j) = \phi_{zd}(j) \qquad j = 0, 1, \ldots, \eta \tag{19-39}$$

where $\phi_{zz}(\cdot)$ is the auto-correlation function of filter input $z(k)$, and $\phi_{zd}(\cdot)$ is the cross-correlation function between $z(k)$ and $d(k)$. Equations (19-39) are known as the *discrete-time Wiener-Hopf equations.* They are a system of normal equations, and can be solved in many different ways, the fastest of which is by the Levinson Algorithm (Treitel and Robinson, 1966).

The minimum mean-squared error, $I^*(\mathbf{f})$, can be shown to be given by

$$I^*(\mathbf{f}) = \phi_{dd}(0) - \sum_{i=0}^{\eta} f(i)\phi_{zd}(i) \tag{19-40}$$

One property of the digital Wiener filter is that $I^*(\mathbf{f})$ becomes smaller as η, the number of filter coefficients, increases. In general, $I^*(\mathbf{f})$ approaches a nonzero limiting value, a value that is often reached for modest values of η.

In order to relate this FIR Wiener filter to the truncated steady-state Kalman filter, we must first assume a signal-plus-noise model for $z(k)$, i.e., (see Figure 19-5),

$$z(k) = s(k) + v(k) = h(k)*w(k) + v(k) \tag{19-41}$$

where $h(k)$ is the impulse response of a linear time-invariant system, and, as in our basic state-variable model (Lesson 15), $w(k)$ and $v(k)$ are mutually uncorrelated (stationary) white noise processes with variances q and r, respectively. We must also specify an explicit form for "desired signal" $d(k)$. We shall require that

$$d(k) = s(k) = h(k)*w(k) \tag{19-42}$$

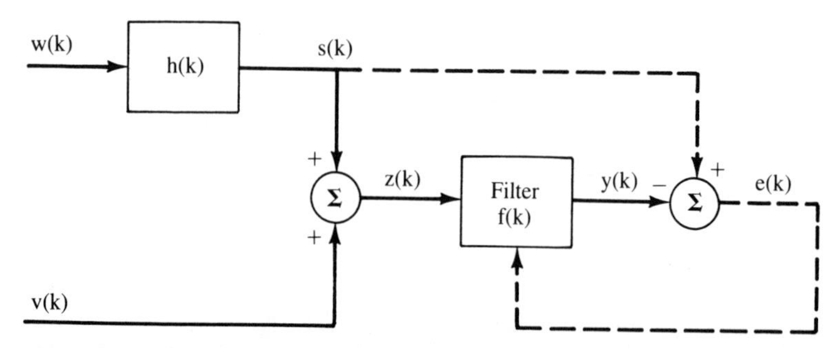

Figure 19-5 Signal-plus-noise model incorporated into the design of a FIR digital Wiener filter. The dashed lines denote paths that exist only during the design stage; these paths do not exist in the implementation of the filter.

which means, of course, that we want the output of the FIR digital Wiener filter to be as close as possible to signal $s(k)$.

Using (19-41) and (19-42), it is another straightforward exercise to compute $\phi_{zz}(i-j)$ and $\phi_{zd}(j)$. The resulting discrete-time Wiener-Hopf equations are

$$\sum_{i=0}^{\eta} f(i)\left[\frac{q}{r}\phi_{hh}(j-i)+\delta(i-j)\right]=\frac{q}{r}\phi_{hh}(j) \qquad j=0,1,\ldots,\eta \tag{19-43}$$

where

$$\phi_{hh}(i)=\sum_{l=0}^{\infty} h(l)h(l+i) \tag{19-44}$$

Observe, that, as in the case of a single-channel Kalman filter, our single channel Wiener filter only depends on the ratio q/r (and subsequently on $\overline{\text{SNR}}$).

Theorem 19-2. *The steady-state Kalman filter is an infinite-length digital Wiener filter. The truncated steady-state Kalman filter is a FIR digital Wiener filter.*

Proof (heuristic). The digital Wiener filter has constant coefficients as does the steady-state Kalman filter. Both filters minimize error variances. The infinite-length Wiener filter has the smallest error variance of all Wiener filters, as does the steady-state Kalman filter; hence, the steady-state Kalman filter is an infinite length digital Wiener filter.

The second part of the theorem is proved in a similar manner. □

Using Theorem 19-2, we suggest the following *procedure for designing a recursive Wiener filter* (i.e., a steady-state Kalman filter):

1. obtain a state-variable representation of $h(k)$;
2. determine the steady-state Kalman gain matrix, $\overline{\mathbf{K}}$, as described above;
3. implement the steady-state Kalman filter (19-11);
4. compute the estimate of desired signal $s(k)$, as

$$\hat{s}(k|k) = \mathbf{h}'\hat{\mathbf{x}}(k|k) \tag{19-45}$$

COMPARISONS OF KALMAN AND WIENER FILTERS

We conclude this lesson with a brief comparison of Kalman and Wiener filters. First of all, the filter designs use different types of modeling information. Auto-correlation information is used for FIR digital Wiener filtering, whereas a combination of auto-correlation and difference equation information is used in Kalman filtering. In order to compute $\phi_{hh}(l)$, $h(k)$ must be known; it is always possible to go directly from $h(k)$ to the parameters in a state-variable model [i.e., $(\mathbf{\Phi}, \boldsymbol{\gamma}, \mathbf{h})$], using an approximate realization procedure (e.g., Kung, 1978).

Because the Kalman filter is recursive, it is an infinite-length filter; hence, unlike the FIR Wiener filter, where filter length is a design variable, the filter's length is not a design variable in the Kalman filter.

Our derivation of the digital Wiener filter was for a single-channel model, whereas the derivation of the Kalman filter was for a more general multichannel model (i.e., the case of vector inputs and outputs). For a derivation of a multichannel digital Wiener filter, see Treitel (1970).

There is a conceptual difference between Kalman and FIR Wiener filtering. A Wiener filter acts directly on the measurements to provide a signal $\hat{s}(k|k)$ that is "close" to $s(k)$. The Kalman filter does not do this directly. It provides us with such a signal in two steps. In the first step, a signal is obtained that is "close" to $\mathbf{x}(k)$, and in the second step, a signal is obtained that is "close" to $s(k)$. The first step provides the optimal filtered value of $\mathbf{x}(k)$, $\hat{\mathbf{x}}(k|k)$; the second step provides the optimal filtered value of $s(k)$, because $\hat{s}(k|k) = \mathbf{h}'\hat{\mathbf{x}}(k|k)$.

Finally, we can picture a diagram similar to the one in Figure 19-5 for the Kalman filter, except that the dashed lines are solid ones. In the Kalman filter the filter coefficients are usually time-varying and are affected in real time by a measure of error $e(k)$ (i.e., by error-covariance matrices), whereas Wiener filter coefficients are constant and have only been affected by a measure of $e(k)$ during their design.

In short, a Kalman filter is a generalization of the FIR digital Wiener filter to include time-varying filter coefficients which incorporate the effects of error in an active manner. Additionally, a Kalman filter is applicable to either time-varying or nonstationary systems or both, whereas the digital Wiener filter is not.

PROBLEMS

19-1. Derive the discrete-time Wiener-Hopf equations given in (19-39).

19-2. Prove that the minimum mean-squared error, $I^*(\mathbf{f})$, given in (19-40), becomes smaller as η, the number of filter coefficients, increases.

19-3. Consider the basic state-variable model, $x(k+1) = \dfrac{2}{\sqrt{3}}\, x(k) + w(k)$ and $z(k+1) = x(k+1) + v(k+1)$, where $q = 2$, $r = 1$, $m_x(0) = 0$ and $\mathbf{E}\{x^2(0)\}=2$.

(a) Specify $\hat{x}(0|0)$ and $p(0|0)$.

(b) Give the recursive predictor for this system.

(c) Obtain the steady-state predictor.

(d) Suppose $z(5)$ is not provided (i.e., there is a gap in the measurements at $k = 5$). How does this affect $p(6|5)$ and $p(100|99)$?

19-4. Consider the basic scalar system $x(k+1) = \phi x(k) + w(k)$ and $z(k+1) = x(k+1) + v(k+1)$. Assume that $q = 0$ and let $\lim_{k \to \infty} p(k|k) = \bar{p}_1$.

(a) Show that $\bar{p}_1 = 0$ and $\bar{p}_1 = (\phi^2 - 1)r / \phi^2$.

(b) Which of the two solutions in (a) is the correct one when $\phi^2 < 1$?

Lesson 20

State Estimation: Smoothing

THREE TYPES OF SMOOTHERS

Recall that in smoothing we obtain mean-squared estimates of state vector $\mathbf{x}(k)$, $\hat{\mathbf{x}}(k|j)$ for which $k < j$. From the Fundamental Theorem of Estimation Theory, Theorem 13-1, we know that the structure of the mean-squared smoother is

$$\hat{\mathbf{x}}(k|j) = \mathbf{E}\{\mathbf{x}(k)|\mathfrak{Z}(j)\} \qquad \text{where} \qquad k < j \tag{20-1}$$

In this lesson we shall develop recursive smoothing algorithms that are comparable to our recursive Kalman filter algorithm.

Smoothing is much more complicated than filtering, primarily because we are using future measurements to obtain estimates at earlier time points. Because there is some flexibility associated with how we choose to process future measurements, it is convenient to distinguish between three types of smoothing (Meditch, 1969, pp. 204–208): fixed-interval, fixed-point, and fixed-lag.

The *fixed-interval smoother* is $\hat{\mathbf{x}}(k|N)$, $k = 0, 1, \ldots, N-1$, where N is a fixed positive integer. The situation here is as follows: with an experiment completed, we have measurements available over the fixed interval $1 \leq k \leq N$. For each time point within this interval we wish to obtain the optimal estimate of state vector $\mathbf{x}(k)$, which is based on *all* the available measurement data $\{\mathbf{z}(j), j = 1, 2, \ldots, N\}$. Fixed-interval smoothing is very useful in signal processing, where the processing is done after all the data is collected. It cannot be carried out on-line during an experiment, as filtering can be. Because all the available data is used, one cannot hope to do better (by other forms of smoothing) than by fixed-interval smoothing.

The *fixed-point* smoothed estimate is $\hat{\mathbf{x}}(k|j), j = k + 1, k + 2, \ldots$ where k is a fixed positive integer. Suppose we want to improve our estimate of a state at a specific time by making use of future measurements. Let this time be $\bar{k}$. Then fixed-point smoothed estimates of $\mathbf{x}(\bar{k})$ will be $\hat{\mathbf{x}}(\bar{k}|\bar{k} + 1)$, $\hat{\mathbf{x}}(\bar{k}|\bar{k} + 2), \ldots$, etc. The last possible fixed-point estimate is $\hat{\mathbf{x}}(\bar{k}|N)$, which is the same as the fixed-interval estimate of $\mathbf{x}(\bar{k})$. Fixed-point smoothing can be carried out on-line, if desired, but the calculation of $\hat{\mathbf{x}}(\bar{k}|\bar{k} + d)$ is subject to a delay of dT sec.

The *fixed-lag* smoothed estimate is $\hat{\mathbf{x}}(k|k + L), k = 0, 1, \ldots$, where L is a fixed positive integer. In this case, the point at which we seek the estimate of the system's state lags the time point of the most recent measurement by a fixed interval of time, L; i.e., $t_{k+L} - t_k = L$ which is a positive constant for all $k = 0, 1, \ldots$. This type of estimator can be used where a constant lag between measurements and state estimates is permissible. Fixed-lag smoothing can be carried out on-line, if desired, but the calculation of $\hat{\mathbf{x}}(k|k + L)$ is subject to an LT sec delay.

APPROACHES FOR DERIVING SMOOTHERS

The literature on smoothing is filled with many different approaches for deriving recursive smoothers. By augmenting suitably defined states to (15-17) and (15-18), one can reduce the derivation of smoothing formulas to a Kalman filter for the augmented state-variable model (Anderson and Moore, 1979). The "filtered" estimates of the newly-introduced states turn out to be equivalent to smoothed values of $\mathbf{x}(k)$. We shall examine this augmentation approach in Lesson 21. A second approach is to use the orthogonality principle to derive a discrete-time Wiener-Hopf equation, which can then be used to establish the smoothing formulas. We do not treat this approach in this book.

A third approach, the one we shall follow in this lesson, is based on the causal invertibility between the innovations process $\tilde{\mathbf{z}}(k + j|k + j - 1)$ and measurement $\mathbf{z}(k + j)$, and repeated applications of Theorem 12-4.

A SUMMARY OF IMPORTANT FORMULAS

The following formulas, which have been derived in earlier lessons, are used so frequently in this lesson, as well as in Lesson 21, that we collect them here for the convenience of the reader:

$$\tilde{\mathbf{z}}(k+1|k) = \mathbf{z}(k+1) - \mathbf{H}(k+1)\hat{\mathbf{x}}(k+1|k) \tag{20-2}$$

$$\tilde{\mathbf{z}}(k+1|k) = \mathbf{H}(k+1)\tilde{\mathbf{x}}(k+1|k) + \mathbf{v}(k+1) \tag{20-3}$$

$$\mathbf{P}_{\tilde{z}\tilde{z}}(k+1|k) = \mathbf{H}(k+1)\mathbf{P}(k+1|k)\mathbf{H}'(k+1) + \mathbf{R}(k+1) \tag{20-4}$$

$$\tilde{\mathbf{x}}(k+1|k) = \mathbf{\Phi}(k+1,k)\tilde{\mathbf{x}}(k|k) + \mathbf{\Gamma}(k+1,k)\mathbf{w}(k) \tag{20-5}$$

and

$$\tilde{\mathbf{x}}(k+1|k+1) = [\mathbf{I} - \mathbf{K}(k+1)\mathbf{H}(k+1)]\tilde{\mathbf{x}}(k+1|k) - \mathbf{K}(k+1)\mathbf{v}(k+1) \tag{20-6}$$

SINGLE-STAGE SMOOTHER

As in our study of prediction, it is useful first to develop a single-stage smoother and then to obtain more general smoothers.

Theorem 20-1. *The single-stage mean-squared smoothed estimator of* $\mathbf{x}(k)$, $\hat{\mathbf{x}}(k|k+1)$, *is given by the expression*

$$\hat{\mathbf{x}}(k|k+1) = \hat{\mathbf{x}}(k|k) + \mathbf{M}(k|k+1)\tilde{\mathbf{z}}(k+1|k) \tag{20-7}$$

where single-stage smoother gain matrix, $\mathbf{M}(k|k+1)$, *is*

$$\mathbf{M}(k|k+1) = \mathbf{P}(k|k)\mathbf{\Phi}'(k+1,k)\mathbf{H}'(k+1) [\mathbf{H}(k+1)\mathbf{P}(k+1|k)\mathbf{H}'(k+1) + \mathbf{R}(k+1)]^{-1} \tag{20-8}$$

Proof. From the Fundamental Theorem of Estimation Theory and Theorem 12-4, we know that

$$\begin{aligned}\hat{\mathbf{x}}(k|k+1) &= \mathbf{E}\{\mathbf{x}(k)|\mathfrak{Z}(k+1)\} \\ &= \mathbf{E}\{\mathbf{x}(k)|\mathfrak{Z}(k),\mathbf{z}(k+1)\} \\ &= \mathbf{E}\{\mathbf{x}(k)|\mathfrak{Z}(k),\tilde{\mathbf{z}}(k+1|k)\} \\ &= \mathbf{E}\{\mathbf{x}(k)|\mathfrak{Z}(k)\} + \mathbf{E}\{\mathbf{x}(k)|\tilde{\mathbf{z}}(k+1|k)\} - \mathbf{m}_{\mathbf{x}}(k)\end{aligned} \tag{20-9}$$

which can also be expressed as (see Corollary 13-1)

$$\hat{\mathbf{x}}(k|k+1) = \hat{\mathbf{x}}(k|k) + \mathbf{P}_{\mathbf{x}\tilde{\mathbf{z}}}(k,k+1|k)\mathbf{P}_{\tilde{\mathbf{z}}\tilde{\mathbf{z}}}^{-1}(k+1|k)\tilde{\mathbf{z}}(k+1|k) \tag{20-10}$$

Defining single-stage smoother gain matrix $\mathbf{M}(k|k+1)$ as

$$\mathbf{M}(k|k+1) = \mathbf{P}_{\mathbf{x}\tilde{\mathbf{z}}}(k,k+1|k)\mathbf{P}_{\tilde{\mathbf{z}}\tilde{\mathbf{z}}}^{-1}(k+1|k) \tag{20-11}$$

(20-10) reduces to (20-7).

Next, we must show that $\mathbf{M}(k|k+1)$ can also be expressed as in (20-8). We already have an expression for $\mathbf{P}_{\tilde{\mathbf{z}}\tilde{\mathbf{z}}}(k+1|k)$, namely (20-4); hence, we must compute $\mathbf{P}_{x\tilde{z}}(k,k+1|k)$. To do this, we make use of (20-3), (20-5) and the orthogonality principle, as follows:

$$\begin{aligned}\mathbf{P}_{x\tilde{z}}(k,k+1|k) &= \mathbf{E}\{\mathbf{x}(k)\tilde{\mathbf{z}}'(k+1|k)\}\\ &= \mathbf{E}\{\mathbf{x}(k)[\mathbf{H}(k+1)\tilde{\mathbf{x}}(k+1|k)+\mathbf{v}(k+1)]'\}\\ &= \mathbf{E}\{\mathbf{x}(k)\tilde{\mathbf{x}}'(k+1|k)\}\mathbf{H}'(k+1)\\ &= \mathbf{E}\{\mathbf{x}(k)[\mathbf{\Phi}(k+1,k)\tilde{\mathbf{x}}(k|k)\\ &\quad +\mathbf{\Gamma}(k+1,k)\mathbf{w}(k)]'\}\mathbf{H}'(k+1)\\ &= \mathbf{E}\{\mathbf{x}(k)\tilde{\mathbf{x}}'(k|k)\}\mathbf{\Phi}'(k+1,k)\mathbf{H}'(k+1)\\ &= \mathbf{E}\{[\hat{\mathbf{x}}(k|k)+\tilde{\mathbf{x}}(k|k)]\tilde{\mathbf{x}}'(k|k)\}\mathbf{\Phi}'(k+1,k)\mathbf{H}'(k+1)\\ &= \mathbf{P}(k|k)\mathbf{\Phi}'(k+1,k)\mathbf{H}'(k+1)\end{aligned} \tag{20-12}$$

Substituting (20-12) and (20-4) into (20-11), we obtain (20-8). □

For future reference, we record the following fact,

$$\mathbf{E}\{\mathbf{x}(k)\tilde{\mathbf{x}}'(k+1|k)\} = \mathbf{P}(k|k)\mathbf{\Phi}'(k+1,k) \tag{20-13}$$

This is obtained by comparing the third and last lines of (20-12).

Observe that the structure of the single-stage smoother is quite similar to that of the Kalman filter. The Kalman filter obtains $\hat{\mathbf{x}}(k+1|k+1)$ by adding a correction that depends on the most recent innovations, $\tilde{\mathbf{z}}(k+1|k)$, to the *predicted* value of $\mathbf{x}(k+1)$. The single-stage smoother, on the other hand, obtains $\hat{\mathbf{x}}(k|k+1)$ by adding a correction that also depends on $\tilde{\mathbf{z}}(k+1|k)$, to the *filtered* value of $\mathbf{x}(k)$. We see that filtered estimates are required to obtain smoothed estimates.

Corollary 20-1. *Kalman gain matrix* $\mathbf{K}(k+1)$ *is a factor in* $\mathbf{M}(k|k+1)$, *i.e.,*

$$\mathbf{M}(k|k+1) = \mathbf{A}(k)\mathbf{K}(k+1) \tag{20-14}$$

where

$$\mathbf{A}(k) \triangleq \mathbf{P}(k|k)\mathbf{\Phi}'(k+1,k)\mathbf{P}^{-1}(k+1|k) \tag{20-15}$$

Proof. Using the fact that [Equation (17-12)]

$$\mathbf{K}(k+1) = \mathbf{P}(k+1|k)\mathbf{H}'(k+1)\,[\mathbf{H}(k+1)\mathbf{P}(k+1|k)\mathbf{H}'(k+1)+\mathbf{R}(k+1)]^{-1} \tag{20-16}$$

we see that

$$\mathbf{H}'(k+1)[\mathbf{H}(k+1)\mathbf{P}(k+1|k)\mathbf{H}'(k+1)+\mathbf{R}(k+1)]^{-1} = \mathbf{P}^{-1}(k+1|k)\mathbf{K}(k+1) \tag{20-17}$$

When (20-17) is substituted into (20-8), we obtain (20-14). □

Corollary 20-2. *Another way to express* $\hat{\mathbf{x}}(k|k+1)$ *is*

$$\hat{\mathbf{x}}(k|k+1) = \hat{\mathbf{x}}(k|k) + \mathbf{A}(k)[\hat{\mathbf{x}}(k+1|k+1) - \hat{\mathbf{x}}(k+1|k)] \tag{20-18}$$

Proof. Substitute (20-14) into (20-7), to see that

$$\hat{\mathbf{x}}(k|k+1) = \hat{\mathbf{x}}(k|k) + \mathbf{A}(k)\mathbf{K}(k+1)\tilde{\mathbf{z}}(k+1|k) \tag{20-19}$$

but [see (17-11)]

$$\mathbf{K}(k+1)\tilde{\mathbf{z}}(k+1|k) = \hat{\mathbf{x}}(k+1|k+1) - \hat{\mathbf{x}}(k+1|k) \tag{20-20}$$

Substitute (20-20) into (20-19) to obtain the desired result in (20-18). □

Formula (20-7) is useful for computational purposes, whereas (20-18) is most useful for theoretical purposes. These facts will become more clear when we examine double-stage smoothing in our next section.

Whereas the structure of the single-stage smoother is similar to that of the Kalman filter, we see that $\mathbf{M}(k|k+1)$ does not depend on single-stage smoothing error-covariance matrix $\mathbf{P}(k|k+1)$. Kalman gain $\mathbf{K}(k+1)$, of course, does depend on $\mathbf{P}(k+1|k)$ [or $\mathbf{P}(k|k)$]. In fact, $\mathbf{P}(k|k+1)$ does not appear at all in the smoothing equations and must be computed (if one desires to do so) separately. We address this calculation in Lesson 21.

DOUBLE-STAGE SMOOTHER

Instead of immediately generalizing the single-stage smoother to an N stage smoother, we first present results for the double-stage smoother. We will then be able to write down the general results (almost) by inspection of the single- and double-stage results.

Theorem 20-2. *The double-stage mean-squared smoothed estimator of* $\mathbf{x}(k)$, $\hat{\mathbf{x}}(k|k+2)$, *is given by the expression*

$$\hat{\mathbf{x}}(k|k+2) = \hat{\mathbf{x}}(k|k+1) + \mathbf{M}(k|k+2)\tilde{\mathbf{z}}(k+2|k+1) \tag{20-21}$$

where double-stage smoother gain matrix, $\mathbf{M}(k|k+2)$, *is*

$$\begin{aligned}\mathbf{M}(k|k+2) = \mathbf{P}(k|k)&\boldsymbol{\Phi}'(k+1,k)[\mathbf{I} - \mathbf{K}(k+1)\\ &\mathbf{H}(k+1)]'\boldsymbol{\Phi}'(k+2,k+1)\\ &\mathbf{H}'(k+2)[\mathbf{H}(k+2)\mathbf{P}(k+2|k+1)\\ &\mathbf{H}'(k+2) + \mathbf{R}(k+2)]^{-1}\end{aligned} \tag{20-22}$$

Proof. From the Fundamental Theorem of Estimation Theory, Theorem 12-4, and Corollary 13-1, we know that

$$\begin{aligned}\hat{\mathbf{x}}(k|k+2) &= \mathbf{E}\{\mathbf{x}(k)|\mathfrak{Z}(k+2)\}\\ &= \mathbf{E}\{\mathbf{x}(k)|\mathfrak{Z}(k+1), \mathbf{z}(k+2)\}\\ &= \mathbf{E}\{\mathbf{x}(k)|\mathfrak{Z}(k+1), \tilde{\mathbf{z}}(k+2|k+1)\}\\ &= \mathbf{E}\{\mathbf{x}(k)|\mathfrak{Z}(k+1)\} + \mathbf{E}\{\mathbf{x}(k)|\tilde{\mathbf{z}}(k+2|k+1)\} - \mathbf{m}_{\mathbf{x}}(k)\end{aligned} \tag{20-23}$$

which can also be expressed as

$$\hat{\mathbf{x}}(k|k+2) = \hat{\mathbf{x}}(k|k+1) + \mathbf{P}_{\mathbf{x}\tilde{\mathbf{z}}}(k,k+2|k+1)\mathbf{P}_{\tilde{\mathbf{z}}\tilde{\mathbf{z}}}^{-1}(k+2|k+1)\tilde{\mathbf{z}}(k+2|k+1) \tag{20-24}$$

Defining double-stage smoother gain matrix $\mathbf{M}(k|k+2)$ as

$$\mathbf{M}(k|k+2) = \mathbf{P}_{\mathbf{x}\tilde{\mathbf{z}}}(k,k+2|k+1)\mathbf{P}_{\tilde{\mathbf{z}}\tilde{\mathbf{z}}}^{-1}(k+2|k+1) \tag{20-25}$$

(20-24) reduces to (20-21).

In order to show that $\mathbf{M}(k|k+2)$ in (20-25) can be expressed as in (20-22), one proceeds as in our derivation of $\mathbf{M}(k|k+1)$ in (20-12); however, the details are lengthier because $\tilde{\mathbf{z}}(k+2|k+1)$ involves quantities that are two time units away from $\mathbf{x}(k)$, whereas $\tilde{\mathbf{z}}(k+1|k)$ involves quantities that are only one time unit away from $\mathbf{x}(k)$. Equation (20-13) is used during the derivation. We leave the detailed derivation of (20-22) as an exercise for the reader. □

Whereas (20-22) is a computationally useful formula, it is not useful from a theoretical viewpoint; i.e., when we examine $\mathbf{M}(k|k+1)$ in (20-8) and $\mathbf{M}(k|k+2)$ in (20-22), it is not at all obvious how to generalize these formulas to $\mathbf{M}(k|k+N)$, or even to $\mathbf{M}(k|k+3)$. The following result for $\mathbf{M}(k|k+2)$ is easily generalized.

Corollary 20-3. *Kalman gain matrix* $\mathbf{K}(k+2)$ *is a factor in* $\mathbf{M}(k|k+2)$, *i.e.*,

$$\mathbf{M}(k|k+2) = \mathbf{A}(k)\mathbf{A}(k+1)\mathbf{K}(k+2) \tag{20-26}$$

where $\mathbf{A}(k)$ *is defined in (20-15).*

Proof. Increment k to $k+1$ in (20-17) to see that

$$\mathbf{H}'(k+2)[\mathbf{H}(k+2)\mathbf{P}(k+2|k+1)\mathbf{H}'(k+2) + \mathbf{R}(k+2)]^{-1} = \mathbf{P}^{-1}(k+2|k+1)\mathbf{K}(k+2) \tag{20-27}$$

Next, note from (17-14), that

$$\mathbf{I} - \mathbf{K}(k+1)\mathbf{H}(k+1) = \mathbf{P}(k+1|k+1)\mathbf{P}^{-1}(k+1|k) \tag{20-28}$$

hence,

$$[\mathbf{I} - \mathbf{K}(k+1)\mathbf{H}(k+1)]' = \mathbf{P}^{-1}(k+1|k)\mathbf{P}(k+1|k+1) \tag{20-29}$$

Substitute (20-27) and (20-29) into (20-22) to see that

$$\begin{aligned}\mathbf{M}(k|k+2) = {} & \mathbf{P}(k|k)\mathbf{\Phi}'(k+1,k)\mathbf{P}^{-1}(k+1|k) \\ & \mathbf{P}(k+1|k+1)\mathbf{\Phi}'(k+2,k+1) \\ & \mathbf{P}^{-1}(k+2|k+1)\mathbf{K}(k+2)\end{aligned} \tag{20-30}$$

Using the definition of matrix $\mathbf{A}(k)$ in (20-15), we see that (20-30) can be expressed as in (20-26). □

Corollary 20-4. *Two other ways to express* $\hat{\mathbf{x}}(k|k+2)$ *are:*

$$\hat{\mathbf{x}}(k|k+2) = \hat{\mathbf{x}}(k|k+1) + \mathbf{A}(k)\mathbf{A}(k+1)[\hat{\mathbf{x}}(k+2|k+2) - \hat{\mathbf{x}}(k+2|k+1)] \tag{20-31}$$

and

$$\hat{\mathbf{x}}(k|k+2) = \hat{\mathbf{x}}(k|k) + \mathbf{A}(k)[\hat{\mathbf{x}}(k+1|k+2) - \hat{\mathbf{x}}(k+1|k)] \tag{20-32}$$

Proof. The derivation of (20-31) follows exactly the same path as the derivation of (20-18), and is therefore left as an exercise for the reader. Equation (20-31) is the starting place for the derivation of (20-32). Observe, from (20-18), that

$$\mathbf{A}(k+1)[\hat{\mathbf{x}}(k+2|k+2) - \hat{\mathbf{x}}(k+2|k+1)] = \hat{\mathbf{x}}(k+1|k+2) - \hat{\mathbf{x}}(k+1|k+1) \tag{20-33}$$

thus, (20-31) can be written as

$$\hat{\mathbf{x}}(k|k+2) = \hat{\mathbf{x}}(k|k+1) + \mathbf{A}(k)[\hat{\mathbf{x}}(k+1|k+2) - \hat{\mathbf{x}}(k+1|k+1)] \tag{20-34}$$

Substituting (20-18) into (20-34), we obtain the desired result in (20-32). □

The alternate forms we have obtained for both $\hat{\mathbf{x}}(k|k+1)$ and $\hat{\mathbf{x}}(k|k+2)$ will suggest how we can generalize our single- and double-stage smoothers to N-stage smoothers.

SINGLE- AND DOUBLE-STAGE SMOOTHERS AS GENERAL SMOOTHERS

At the beginning of this lesson we described three types of smoothers, namely: fixed-interval, fixed-point, and fixed-lag smoothers. Table 20-1 shows how our single- and double-stage smoothers fit into these three categories.

In order to obtain the fixed-interval smoother formulas, given for both the single- and double-stage smoothers, set $k+1 = N$ in (20-18) and $k+2 = N$ in (20-32), respectively. Doing this forces the left-hand side of both equations to be conditioned on data length N. Observe that, before we can compute $\hat{\mathbf{x}}(N-1|N)$ or $\hat{\mathbf{x}}(N-2|N)$, we must run a Kalman filter on all of the data in order to obtain $\hat{\mathbf{x}}(N|N)$. This last filtered state estimate initializes the backward running fixed-interval smoother. Observe, also, that we must compute $\hat{\mathbf{x}}(N-1|N)$ before we can compute $\hat{\mathbf{x}}(N-2|N)$. Clearly, the limitation of our results so far is that we can only perform fixed-interval smoothing for

TABLE 20-1 Smoothing Interrelationships

	Single-Stage	Double-Stage
Fixed-Interval	$\hat{\mathbf{x}}(N-1\|N) = \hat{\mathbf{x}}(N-1\|N-1) + \mathbf{A}(N-1)[\hat{\mathbf{x}}(N\|N) - \hat{\mathbf{x}}(N\|N-1)]$ 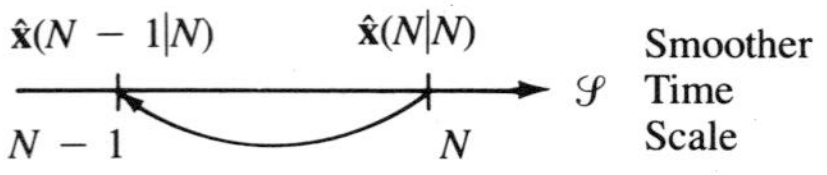 $\hat{\mathbf{x}}(N-1\|N)$ $\hat{\mathbf{x}}(N\|N)$ $\mathcal{S}$ Smoother Time Scale $N-1$ N Solution proceeds in reverse time, from N to $N-1$, where N is fixed.	$\hat{\mathbf{x}}(N-2\|N) = \hat{\mathbf{x}}(N-2\|N-2) + \mathbf{A}(N-2)[\hat{\mathbf{x}}(N-1\|N) - \hat{\mathbf{x}}(N-1\|N-2)]$ 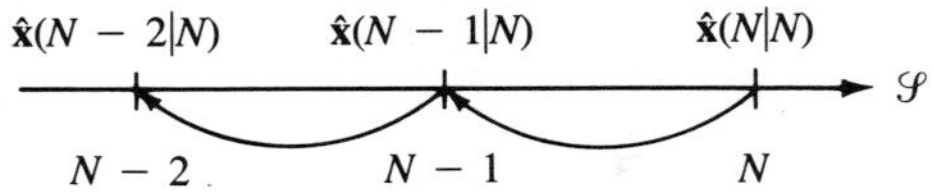 $\hat{\mathbf{x}}(N-2\|N)$ $\hat{\mathbf{x}}(N-1\|N)$ $\hat{\mathbf{x}}(N\|N)$ $\mathcal{S}$ $N-2$ $N-1$ N Solution proceeds in reverse time, from N to $N-1$, to $N-2$, where N is fixed.
Fixed-Point	$\hat{\mathbf{x}}(\bar{k}\|\bar{k}+1) = \hat{\mathbf{x}}(\bar{k}\|\bar{k}) + \mathbf{A}(\bar{k})[\hat{\mathbf{x}}(\bar{k}+1\|\bar{k}+1) - \hat{\mathbf{x}}(\bar{k}+1\|\bar{k})]$ k fixed at $\bar{k}$ $\hat{\mathbf{x}}(\bar{k}\|\bar{k})$ $\hat{\mathbf{x}}(\bar{k}+1\|\bar{k}+1)$ $\mathcal{F}$ Filter Time Scale $\bar{k}$ $\bar{k}+1$ $\hat{\mathbf{x}}(\bar{k}\|\bar{k}+1)$ $\mathcal{S}$ Smoother Time Scale $\bar{k}$ 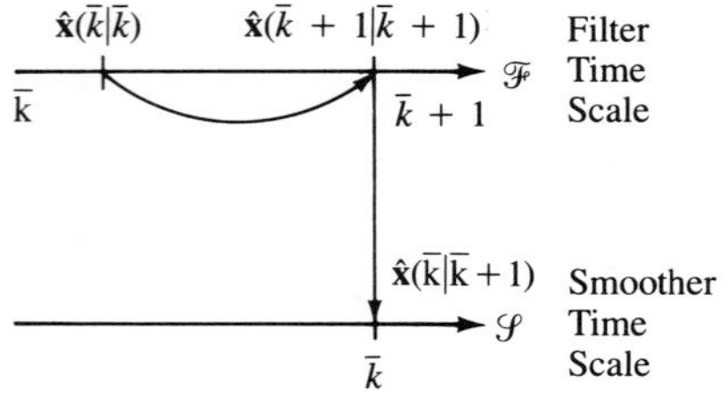 Solution proceeds in forward time from $\bar{k}$ to $\bar{k}+1$ on the filtering time scale and then back to $\bar{k}$ on the smoothing time scale. A one-unit time delay is present.	$\hat{\mathbf{x}}(\bar{k}\|\bar{k}+2) = \hat{\mathbf{x}}(\bar{k}\|\bar{k}+1) + \mathbf{A}(\bar{k})\mathbf{A}(\bar{k}+1)[\hat{\mathbf{x}}(\bar{k}+2\|\bar{k}+2) - \hat{\mathbf{x}}(\bar{k}+2\|\bar{k}+1)]$ k fixed at $\bar{k}$ 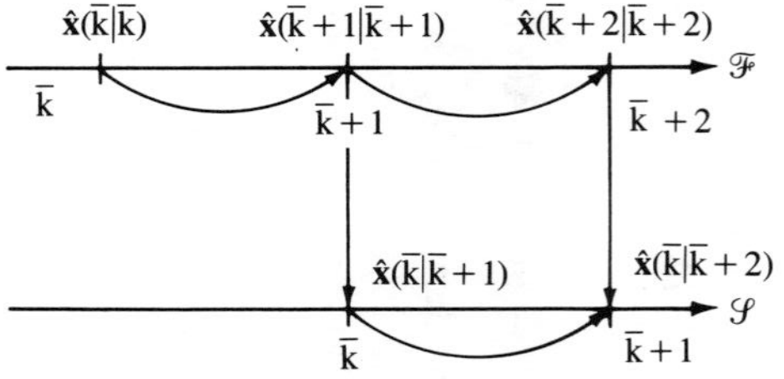 $\hat{\mathbf{x}}(\bar{k}\|\bar{k})$ $\hat{\mathbf{x}}(\bar{k}+1\|\bar{k}+1)$ $\hat{\mathbf{x}}(\bar{k}+2\|\bar{k}+2)$ $\mathcal{F}$ $\bar{k}$ $\bar{k}+1$ $\bar{k}+2$ $\hat{\mathbf{x}}(\bar{k}\|\bar{k}+1)$ $\hat{\mathbf{x}}(\bar{k}\|\bar{k}+2)$ $\mathcal{S}$ $\bar{k}$ $\bar{k}+1$ Solution proceeds in forward time. Results from single-stage smoother as well as optimal filter are required at $\mathcal{S}=\bar{k}+1$, whereas at $\mathcal{S}=\bar{k}$ only results from optimal filter are required. A two-unit time delay is present.

Table 20-1 Cont.

	Single-Stage	Double-Stage
Fixed-Lag	$\hat{\mathbf{x}}(k\|k+1) = \hat{\mathbf{x}}(k\|k) + \mathbf{A}(k)[\hat{\mathbf{x}}(k+1\|k+1) - \hat{\mathbf{x}}(k+1\|k)]$	$\hat{\mathbf{x}}(k\|k+2) = \hat{\mathbf{x}}(k\|k+1) + \mathbf{A}(k)\mathbf{A}(k+1)[\hat{\mathbf{x}}(k+2\|k+2) - \hat{\mathbf{x}}(k+2\|k+1)]$
	k variable	k variable
	Picture and discussion same as in Fixed-Point case, replacing $\bar{k}$ by k. Here we have a one-unit window from k to $k+1$.	Picture and discussion same as in Fixed-Point case, replacing $\bar{k}$ by k. Here we have a two-unit window from k to $k+2$.

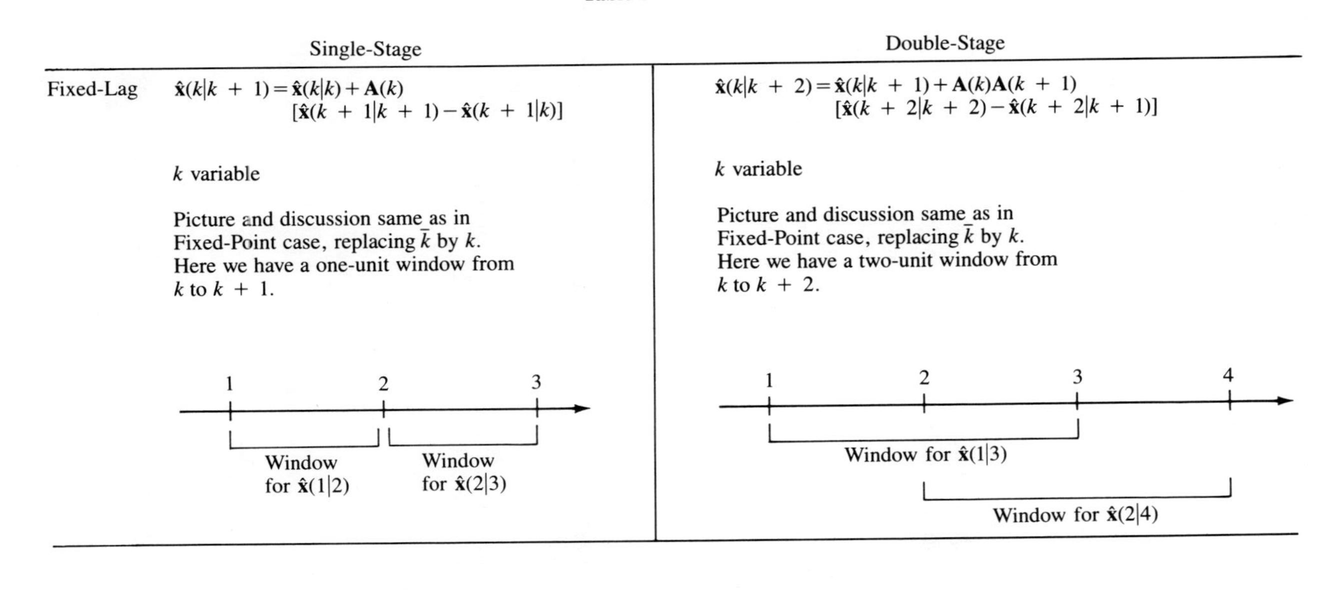

$k = N - 1$ and $N - 2$. More general results, that will permit us to perform fixed-interval smoothing for any $k < N$, are described in Lesson 21.

In order to obtain the fixed-point smoother formulas, given for both the single- and double-stage smoothers, set $k = \bar{k}$ in (20-18) and (20-31), respectively. Time point $\bar{k}$ represents the fixed point in our smoother formulas. As noted at the beginning of this lesson, fixed-point smoothing can be carried out on-line, but it is subject to a delay. From an information availability point of view, a one-unit time delay is present in the single-stage fixed-point smoother, because our smoothed estimate at $k = \bar{k}$ uses the filtered estimate computed at $k = \bar{k} + 1$. A two-unit time delay is present in the double-stage fixed-point smoother, because our smoothed estimate at $\mathcal{S} = \bar{k} + 1$ uses the filtered estimate computed at $k = \bar{k} + 1$ and $\bar{k} + 2$.

The fixed-lag smoother formulas look just like the fixed-point formulas, except that k is a variable in the former and is fixed in the latter. Observe that the windows of measurements used in the single-stage fixed-lag smoother are nonoverlapping, whereas they are overlapping in the case of the double-stage fixed-lag smoother. As in the case of fixed-point smoothing, fixed-interval smoothing can be carried out on line, subject, of course, to a fixed delay equal to the "lag" of the smoother.

PROBLEMS

20-1. Derive the formula for the double-stage smoother gain matrix $\mathbf{M}(k|k + 2)$, given in (20-22).

20-2. Derive the alternative expression for $\hat{\mathbf{x}}(k|k + 2)$ given in (20-31).

20-3. Using the Fundamental Theorem of Estimation Theory, derive expressions for $\hat{\mathbf{w}}(k|k + 1)$ and $\hat{\mathbf{w}}(k|k + 2)$. These represent single- and double-stage estimators of disturbance $\mathbf{w}(k)$.

Lesson 21

State-Estimation: Smoothing (General Results)

INTRODUCTION

In Lesson 20 we introduced three general types of smoothers, namely, fixed-interval, fixed-point, and fixed-lag smoothers. We also developed formulas for single-stage and double-stage smoothers and showed how these specialized smoothers fit into the three general categories. In this lesson we shall develop general formulas for fixed-interval, fixed-point, and fixed-lag smoothers.

FIXED-INTERVAL SMOOTHERS

In this section we develop a number of algorithms for the fixed-interval smoothed estimator of $\mathbf{x}(k)$, $\hat{\mathbf{x}}(k|N)$. Not all of them will be useful, because we will not be able to compute with some. Only those with which we can compute will be considered useful.

Recall that

$$\hat{\mathbf{x}}(k|N) = \mathbf{E}\{\mathbf{x}(k)|\mathfrak{Z}(N)\} \tag{21-1}$$

It is straightforward to proceed as we did in Lesson 20, by showing first that

$$\begin{aligned}\hat{\mathbf{x}}(k|N) &= \mathbf{E}\{\mathbf{x}(k)|\mathfrak{Z}(N-1),\tilde{\mathbf{z}}(N|N-1)\} \\ &= \mathbf{E}\{\mathbf{x}(k)|\mathfrak{Z}(N-1)\} + \mathbf{E}\{\mathbf{x}(k)|\tilde{\mathbf{z}}(N|N-1)\} - \mathbf{m}_{\mathbf{x}}(k) \\ &= \hat{\mathbf{x}}(k|N-1) + \mathbf{M}(k|N)\tilde{\mathbf{z}}(N|N-1)\end{aligned} \tag{21-2}$$

where

$$\mathbf{M}(k|N) = \mathbf{P}_{\mathbf{x}\tilde{\mathbf{z}}}(k, N|N-1)\mathbf{P}_{\tilde{\mathbf{z}}\tilde{\mathbf{z}}}^{-1}(N|N-1) \tag{21-3}$$

and then that

$$\mathbf{M}(k|N) = \mathbf{A}(k)\mathbf{A}(k+1)\ldots\mathbf{A}(N-1)\mathbf{K}(N) \tag{21-4}$$

and, finally, that other algorithms for $\hat{\mathbf{x}}(k|N)$ are

$$\hat{\mathbf{x}}(k|N) = \hat{\mathbf{x}}(k|N-1) + \left[\prod_{i=k}^{N-1}\mathbf{A}(i)\right][\hat{\mathbf{x}}(N|N) - \hat{\mathbf{x}}(N|N-1)] \tag{21-5}$$

and

$$\hat{\mathbf{x}}(k|N) = \hat{\mathbf{x}}(k|k) + \mathbf{A}(k)[\hat{\mathbf{x}}(k+1|N) - \hat{\mathbf{x}}(k+1|k)] \tag{21-6}$$

A detailed derivation of these results that uses a mathematical induction proof, can be found in Meditch (1969, pp. 216–220). The proof relies, in part, on our previously derived results in Lesson 20, for the single- and double-stage smoothers. The reader, however, ought to be able to obtain (21-4), (21-5), and (21-6) directly from the rules of formation that can be inferred from the Lesson 20 formulas for $\mathbf{M}(k|k+1)$ and $\mathbf{M}(k|k+2)$, and $\hat{\mathbf{x}}(k|k+1)$ and $\hat{\mathbf{x}}(k|k+2)$.

In order to compute $\hat{\mathbf{x}}(k|N)$ for specific values of k, all of the terms that appear on the right-hand side of an algorithm must be available. We have three possible algorithms for $\hat{\mathbf{x}}(k|N)$; however, as we explain next, only one is "useful."

The first value of k for which the right-hand side of (21-2) is fully available is $k = N-1$. By running a Kalman filter over all of the measurements, we are able to compute $\hat{\mathbf{x}}(N-1|N-1)$ and $\tilde{\mathbf{z}}(N|N-1)$, so that we can compute $\hat{\mathbf{x}}(N-1|N)$. We can now try to iterate (21-2) in the backward direction, to see if we can compute $\hat{\mathbf{x}}(N-2|N)$, $\hat{\mathbf{x}}(N-3|N)$, etc. Setting $k = N-2$ in (21-2), we obtain

$$\hat{\mathbf{x}}(N-2|N) = \hat{\mathbf{x}}(N-2|N-1) + \mathbf{M}(N-2|N)\tilde{\mathbf{z}}(N|N-1) \tag{21-7}$$

Unfortunately, $\hat{\mathbf{x}}(N-2|N-1)$, which is a single-stage smoothed estimate of $\mathbf{x}(N-2)$, has not been computed; hence, (21-7) is not useful for computing $\hat{\mathbf{x}}(N-2|N)$. *We, therefore, reject (21-2) as a useful fixed-interval smoother.*

A similar argument can be made against (21-5); thus, *we also reject (21-5) as a useful fixed-interval smoother.*

Equation (21-6) is a useful fixed-inerval smoother, because all of its terms on its right-hand side are available when we iterate it in the backward direction. For example,

$$\hat{\mathbf{x}}(N-1|N) = \hat{\mathbf{x}}(N-1|N-1) + \mathbf{A}(N-1)[\hat{\mathbf{x}}(N|N) - \hat{\mathbf{x}}(N|N-1)] \tag{21-8}$$

and

$$\hat{\mathbf{x}}(N-2|N) = \hat{\mathbf{x}}(N-2|N-2) + \mathbf{A}(N-2)[\hat{\mathbf{x}}(N-1|N) - \hat{\mathbf{x}}(N-1|N-2)] \tag{21-9}$$

Observe how $\hat{\mathbf{x}}(N-1|N)$ is used in the calculation of $\hat{\mathbf{x}}(N-2|N)$.

Equation (21-6) was developed by Rauch, et al. (1965).

Theorem 21-1. *A useful mean-squared fixed-interval smoothed estimator of* $\mathbf{x}(k)$, $\hat{\mathbf{x}}(k|N)$, *is given by the expression*

$$\hat{\mathbf{x}}(k|N) = \hat{\mathbf{x}}(k|k) + \mathbf{A}(k)[\hat{\mathbf{x}}(k+1|N) - \hat{\mathbf{x}}(k+1|k)] \tag{21-10}$$

where matrix $\mathbf{A}(k)$ *is defined in (20-15), and* $k = N-1, N-2, \ldots, 0$. *Additionally, the error-covariance matrix associated with* $\hat{\mathbf{x}}(k|N)$, $\mathbf{P}(k|N)$, *is given by*

$$\mathbf{P}(k|N) = \mathbf{P}(k|k) + \mathbf{A}(k)[\mathbf{P}(k+1|N) - \mathbf{P}(k+1|k)]\mathbf{A}'(k) \tag{21-11}$$

where $k = N-1, N-2, \ldots, 0$.

Proof. We have already derived (21-10); hence, we direct our attention at the derivation of the algorithm for $\mathbf{P}(k|N)$. Our derivation of (21-11) follows the one given in Meditch, 1969, pp. 222–224. To begin, we know that

$$\tilde{\mathbf{x}}(k|N) = \mathbf{x}(k) - \hat{\mathbf{x}}(k|N) \tag{21-12}$$

Substitute (21-10) into (21-12) to see that

$$\tilde{\mathbf{x}}(k|N) = \tilde{\mathbf{x}}(k|k) - \mathbf{A}(k)[\hat{\mathbf{x}}(k+1|N) - \hat{\mathbf{x}}(k+1|k)] \tag{21-13}$$

Collecting terms conditioned on N on the left-hand side, and terms conditioned on k on the right-hand side, (21-13) becomes

$$\tilde{\mathbf{x}}(k|N) + \mathbf{A}(k)\hat{\mathbf{x}}(k+1|N) = \tilde{\mathbf{x}}(k|k) + \mathbf{A}(k)\hat{\mathbf{x}}(k+1|k) \tag{21-14}$$

Treating (21-14) as an identity, we see that the covariance of its left-hand side must equal the covariance of its right-hand side; hence, [in order to obtain (21-15) we use the orthogonality principle to eliminate all the cross-product terms]

$$\mathbf{P}(k|N) + \mathbf{A}(k)\mathbf{P}_{\hat{\mathbf{x}}\hat{\mathbf{x}}}(k+1|N)\mathbf{A}'(k) = \mathbf{P}(k|k) + \mathbf{A}(k)\mathbf{P}_{\hat{\mathbf{x}}\hat{\mathbf{x}}}(k+1|k)\mathbf{A}'(k) \tag{21-15}$$

or

$$\mathbf{P}(k|N) = \mathbf{P}(k|k) + \mathbf{A}(k)[\mathbf{P}_{\hat{\mathbf{x}}\hat{\mathbf{x}}}(k+1|k) - \mathbf{P}_{\hat{\mathbf{x}}\hat{\mathbf{x}}}(k+1|N)]\mathbf{A}'(k) \tag{21-16}$$

We must now evaluate the two covariances $\mathbf{P}_{\hat{\mathbf{x}}\hat{\mathbf{x}}}(k+1|k)$ and $\mathbf{P}_{\hat{\mathbf{x}}\hat{\mathbf{x}}}(k+1|N)$.

Recall that $\mathbf{x}(k+1) = \hat{\mathbf{x}}(k+1|k) + \tilde{\mathbf{x}}(k+1|k)$; thus,

$$\mathbf{P}_{\mathbf{x}}(k+1) = \mathbf{P}_{\hat{\mathbf{x}}\hat{\mathbf{x}}}(k+1|k) + \mathbf{P}(k+1|k) \tag{21-17}$$

Additionally, $\mathbf{x}(k+1) = \hat{\mathbf{x}}(k+1|N) + \tilde{\mathbf{x}}(k+1|N)$, so that

$$\mathbf{P}_{\mathbf{x}}(k+1) = \mathbf{P}_{\hat{\mathbf{x}}\hat{\mathbf{x}}}(k+1|N) + \mathbf{P}(k+1|N) \tag{21-18}$$

Equating (21-17) and (21-18) we find that

$$\mathbf{P}_{\hat{\mathbf{x}}\hat{\mathbf{x}}}(k+1|k) - \mathbf{P}_{\hat{\mathbf{x}}\hat{\mathbf{x}}}(k+1|N) = \mathbf{P}(k+1|N) - \mathbf{P}(k+1|k) \tag{21-19}$$

Finally, substituting (21-19) into (21-16), we obtain the desired expression for $\mathbf{P}(k|N)$, (21-11). □

We leave proof of the fact that $\{\tilde{\mathbf{x}}(k|N),\ k = N, N-1, \ldots, 0\}$ is a zero-mean *second-order* Gauss-Markov process as an exercise for the reader.

Example 21-1

In order to illustrate fixed-interval smoothing and obtain at the same time a comparison of the relative accuracies of smoothing and filtering, we return to Example 19-1. To review briefly, we have the scalar system $x(k+1) = x(k) + w(k)$ with the scalar measurement $z(k+1) = x(k+1) + v(k+1)$ and $p(0) = 50$, $q = 20$, and $r = 5$. In this example (which is similar to Example 6.1 in Meditch, 1969, pg. 225) we choose $N = 4$ and compute quantities that are associated with $\hat{x}(k|4)$, where from (21-10)

$$\hat{x}(k|4) = \hat{x}(k|k) + A(k)[\hat{x}(k+1|4) - \hat{x}(k+1|k)] \tag{21-20}$$

$k = 3, 2, 1, 0$. Because $\Phi = 1$ and $p(k+1|k) = p(k|k) + 20$

$$A(k) = p(k|k)\Phi' p^{-1}(k+1|k) = \frac{p(k|k)}{p(k|k) + 20} \tag{21-21}$$

and, therefore,

$$p(k|4) = p(k|k) + \left[\frac{p(k|k)}{p(k|k) + 20}\right]^2 [p(k+1|4) - p(k+1|k)] \tag{21-22}$$

Utilizing these last two expressions, we compute $A(k)$ and $p(k|4)$ for $k = 3, 2, 1, 0$ and present them, along with the results summarized in Table 19-1, in Table 21-1. The three estimation error variances are given in adjacent columns for ease in comparison.

TABLE 21-1 Kalman Filter and Fixed-interval Smoother Quantities

k	$p(k\|k-1)$	$p(k\|k)$	$p(k\|4)$	$K(k)$	$A(k)$
0	...	50	16.31	...	0.714
1	70	4.67	3.92	0.933	0.189
2	24.67	4.16	3.56	0.831	0.172
3	24.16	4.14	3.74	0.829	0.171
4	24.14	4.14	4.14	0.828	0.171

Observe the large improvement (percentage wise) of $p(k|4)$ over $p(k|k)$. Improvement seems to get larger the farther away we get from the end of our data; thus, $p(0|4)$ is more than three times as small as $p(0|0)$. Of course, it should be, because

$p(0|0)$ is an *initial condition* that is data independent, whereas $p(0|4)$ is a result of processing $z(1)$, $z(2)$, $z(3)$, and $z(4)$. In essence, fixed-interval smoothing has let us look into the future and reflect the future back to time zero.

Finally, note that, for large values of k, $A(k)$ reaches a steady-state value, $\overline{A}$, where in this example

$$\overline{A} = \overline{p}_1/(\overline{p}_1 + 20) = 0.171 \tag{21-23}$$

This steady-state value is achieved for $k = 3$. □

Equation (21-10) requires the multiplication of 3 $n \times n$ matrices as well as a matrix inversion at each iteration; hence, it is somewhat limited for practical computing purposes. The following results, which are due to Bryson and Frazier (1963) and Bierman (1973b), represent the most practical way for computing $\hat{\mathbf{x}}(k|N)$ and also $\mathbf{P}(k|N)$.

Theorem 21-2. (a) *A most useful mean-squared fixed-interval smoothed estimator of* $\mathbf{x}(k)$, $\hat{\mathbf{x}}(k|N)$, *is*

$$\hat{\mathbf{x}}(k|N) = \hat{\mathbf{x}}(k|k-1) + \mathbf{P}(k|k-1)\mathbf{r}(k|N) \tag{21-24}$$

where $k = N-1, N-2, \ldots, 1$, *and* $n \times 1$ *vector* $\mathbf{r}$ *satisfies the backward-recursive equation*

$$\mathbf{r}(j|N) = \mathbf{\Phi}_p'(j+1,j)\mathbf{r}(j+1|N) + \mathbf{H}'(j)[\mathbf{H}(j)\mathbf{P}(j|j-1)\mathbf{H}'(j) + \mathbf{R}(j)]^{-1}\tilde{\mathbf{z}}(j|j-1) \tag{21-25}$$

where $j = N, N-1, \ldots, 1$ *and* $\mathbf{r}(N+1|N) = \mathbf{0}$.

(b) *The smoothing error-covariance matrix,* $\mathbf{P}(k|N)$, *is*

$$\mathbf{P}(k|N) = \mathbf{P}(k|k-1) - \mathbf{P}(k|k-1)\mathbf{S}(k|N)\mathbf{P}(k|k-1) \tag{21-26}$$

where $k = N-1, N-2, \ldots, 1$, *and* $n \times n$ *matrix* $\mathbf{S}(j|N)$, *which is the covariance matrix of* $\mathbf{r}(j|N)$, *satisfies the backward-recursive equation*

$$\mathbf{S}(j|N) = \mathbf{\Phi}_p'(j+1,j)\mathbf{S}(j+1|N)\mathbf{\Phi}_p(j+1,j) + \mathbf{H}'(j)[\mathbf{H}(j)\mathbf{P}(j|j-1)\mathbf{H}'(j) + \mathbf{R}(j)]^{-1}\mathbf{H}(j) \tag{21-27}$$

where $j = N, N-1, \ldots, 1$ *and* $\mathbf{S}(N+1|N) = \mathbf{0}$. *Matrix* $\mathbf{\Phi}_p$ *is defined in (21-33).*

Proof. (Mendel, 1983b, pp. 64–65). (a) Substitute the Kalman filter equation (17-11) for $\hat{\mathbf{x}}(k|k)$ into Equation (21-10), to show that

$$\hat{\mathbf{x}}(k|N) = \hat{\mathbf{x}}(k|k-1) + \mathbf{K}(k)\tilde{\mathbf{z}}(k|k-1) + \mathbf{A}(k)[\hat{\mathbf{x}}(k+1|N) - \hat{\mathbf{x}}(k+1|k)] \tag{21-28}$$

Residual state vector $\mathbf{r}(k|N)$ is defined as

$$\mathbf{r}(k|N) = \mathbf{P}^{-1}(k|k-1)[\hat{\mathbf{x}}(k|N) - \hat{\mathbf{x}}(k|k-1)] \tag{21-29}$$

Next, substitute $\mathbf{r}(k|N)$ and $\mathbf{r}(k+1|N)$, using (21-29), into (21-28), to show that

$$\mathbf{r}(k|N) = \mathbf{P}^{-1}(k|k-1)[\mathbf{K}(k)\tilde{\mathbf{z}}(k|k-1) + \mathbf{P}(k|k)\mathbf{\Phi}'(k+1,k)\mathbf{r}(k+1|N)] \tag{21-30}$$

From (17-12) and (17-13) and the symmetry of covariance matrices, we find that

$$\mathbf{P}^{-1}(k|k-1)\mathbf{K}(k) = \mathbf{H}'(k)[\mathbf{H}(k)\mathbf{P}(k|k-1)\mathbf{H}'(k) + \mathbf{R}(k)]^{-1} \tag{21-31}$$

and

$$\mathbf{P}^{-1}(k|k-1)\mathbf{P}(k|k) = [\mathbf{I} - \mathbf{K}(k)\mathbf{H}(k)]' \tag{21-32}$$

Substituting (21-31) and (21-32) into Equation (21-30), and defining

$$\mathbf{\Phi}_p(k+1,k) = \mathbf{\Phi}(\mathrm{k}+1,k)[\mathbf{I} - \mathbf{K}(k)\mathbf{H}(k)] \tag{21-33}$$

we obtain Equation (21-25). Setting $k = N + 1$ in (21-29), we establish $\mathbf{r}(N+1|N) = \mathbf{0}$. Finally, solving (21-29) for $\hat{\mathbf{x}}(k|N)$ we obtain Equation (21-24).

(b) The orthogonality principle in Corollary 13-3 leads us to conclude that

$$\mathbf{E}\{\tilde{\mathbf{x}}(k|N)\mathbf{r}'(k|N)\} = \mathbf{0} \tag{21-34}$$

because $\mathbf{r}(k|N)$ is simply a linear combination of all the observations $\mathbf{z}(1)$, $\mathbf{z}(2), \ldots, \mathbf{z}(N)$. From (21-24) we find that

$$\tilde{\mathbf{x}}(k|k-1) = \tilde{\mathbf{x}}(k|N) - \mathbf{P}(k|k-1)\mathbf{r}(k|N) \tag{21-35}$$

and, therefore, using (21-34), we find that

$$\mathbf{P}(k|k-1) = \mathbf{P}(k|N) - \mathbf{P}(k|k-1)\mathbf{S}(\mathrm{k}|N)\mathbf{P}(k|k-1) \tag{21-36}$$

where

$$\mathbf{S}(k|N) = \mathbf{E}\{\mathbf{r}(k|N)\mathbf{r}'(k|N)\} \tag{21-37}$$

is the covariance-matrix of $\mathbf{r}(k|N)$ [note that $\mathbf{r}(k|N)$ is zero mean]. Equation (21-36) is solved for $\mathbf{P}(k|N)$ to give the desired result in Equation (21-26).

Because the innovations process is uncorrelated, (21-27) follows from substitution of (21-25) into (21-37). Finally, $\mathbf{S}(N+1|N) = \mathbf{0}$ because $\mathbf{r}(N+1|N) = \mathbf{0}$. □

Equations (21-24) and (21-25) are very efficient; they require no matrix inversions or multiplications of $n \times n$ matrices. The calculation of $\mathbf{P}(k|N)$ does require multiplications of $n \times n$ matrices.

Matrix $\mathbf{\Phi}_p(k+1,k)$ in (21-33) is the plant matrix of the recursive predictor (Lesson 16). It is interesting that the recursive predictor and not the recursive filter plays the predominant role in fixed-interval smoothing. This is further borne out by the appearance of predictor quantities on the right-hand side of (21-24). Observe that (21-25) looks quite similar to a recursive predictor which is excited by the innovations—one which is running in a backward direction.

Finally, note that (21-24) can also be used for $k = N$, in which case its right-hand side reduces to $\hat{\mathbf{x}}(N|N-1) + \mathbf{K}(N)\tilde{\mathbf{z}}(N|N-1)$, which, of course, is $\hat{\mathbf{x}}(N|N)$.

FIXED-POINT SMOOTHING

A fixed-point smoother, $\hat{\mathbf{x}}(k|j)$ where $j = k+1, k+2, \ldots$, can be obtained in exactly the same manner as we obtained fixed-interval smoother (21-5). It is obtained from this equation by setting N equal to j and then letting $j = k+1$, $k+2, \ldots$; thus,

$$\hat{\mathbf{x}}(k|j) = \hat{\mathbf{x}}(k|j-1) + \mathbf{B}(j)[\hat{\mathbf{x}}(j|j) - \hat{\mathbf{x}}(j|j-1)] \tag{21-38}$$

where

$$\mathbf{B}(j) = \prod_{i=k}^{j-1} \mathbf{A}(i) \tag{21-39}$$

and $j = k+1, k+2, \ldots$. Additionally, one can show that the fixed-point smoothing error-covariance matrix, $\mathbf{P}(k|j)$, is computed from

$$\mathbf{P}(k|j) = \mathbf{P}(k|j-1) + \mathbf{B}(j)[\mathbf{P}(j|j) - \mathbf{P}(j|j-1)]\mathbf{B}'(j) \tag{21-40}$$

where $j = k+1, k+2, \ldots$.

Equation (21-38) is impractical from a computational viewpoint, because of the many multiplications of $n \times n$ matrices required first to form the $\mathbf{A}(i)$ matrices and then to form the $\mathbf{B}(j)$ matrices. Additionally, the inverse of matrix $\mathbf{P}(i+1|i)$ is needed in order to compute matrix $\mathbf{A}(i)$. The following results present a "fast" algorithm for computing $\hat{\mathbf{x}}(k|j)$. It is fast in the sense that no multiplications of $n \times n$ matrices are needed to implement it.

Theorem 21-3. *A most useful mean-squared fixed-point smoothed estimator of* $\mathbf{x}(k)$, $\hat{\mathbf{x}}(k|k+l)$ *where* $l = 1, 2, \ldots$, *is given by the expression*

$$\hat{\mathbf{x}}(k|k+l) = \hat{\mathbf{x}}(k|k+l-1) + \mathbf{N}_{\mathbf{x}}(k|k+l)[\mathbf{z}(k+l) - \mathbf{H}(k+l)\hat{\mathbf{x}}(k+l|k+l-1)] \tag{21-41}$$

where

$$\mathbf{N}_{\mathbf{x}}(k|k+l) = \mathcal{D}_{\mathbf{x}}(k,l)\mathbf{H}'(k+l)[\mathbf{H}(k+l)\mathbf{P}(k+l|k+l-1)\mathbf{H}'(k+l) + \mathbf{R}(k+l)]^{-1} \tag{21-42}$$

and

$$\mathcal{D}_{\mathbf{x}}(k,l) = \mathcal{D}_{\mathbf{x}}(k,l-1)[\mathbf{I} - \mathbf{K}(k+l-1)\mathbf{H}(k+l-1)]'\mathbf{\Phi}'(k+l,k+l-1) \tag{21-43}$$

Equations (21-41) and (21-43) are initialized by $\hat{\mathbf{x}}(\mathrm{k}|\mathrm{k})$ *and* $\mathcal{D}_{\mathbf{x}}(\mathrm{k},1) = \mathbf{P}(\mathrm{k}|\mathrm{k})\mathbf{\Phi}'(\mathrm{k}+1,\mathrm{k})$, *respectively. Additionally,*

$$\mathbf{P}(k|k+l) = \mathbf{P}(k|k+l-1) - \mathbf{N}_{\mathbf{x}}(k|k+l)[\mathbf{H}(k+l)\mathbf{P}(k+l|k+l-1)\mathbf{H}'(k+l) + \mathbf{R}(k+l)]\mathbf{N}_{\mathbf{x}}'(k|k+l) \qquad (21\text{-}44)$$

which is initialized by $\mathbf{P}(\mathrm{k}|\mathrm{k})$. □

We leave the proof of this useful theorem, which is similar to a result given by Fraser (1967), as an exercise for the reader.

Example 21-2

Here we consider the problem of fixed-point smoothing to obtain a refined estimate of the initial condition for the system described in Example 21-1. Recall that $p(0|0) = 50$ and that by fixed-interval smoothing we had obtained the result $p(0|4) = 16.31$, which is a significant reduction in the uncertainty associated with the initial condition.

Using Equation (21-40) or (21-44) we compute $p(0|1)$, $p(0|2)$, and $p(0|3)$ to be 16.69, 16.32, and 16.31, respectively. Observe that a major reduction in the smoothing error variance occurs as soon as the first measurement is incorporated, and that the improvement in accuracy thereafter is relatively modest. This seems to be a general trait of fixed-point smoothing. □

Another way to derive fixed-point smoothing formulas is by the following *state augmentation procedure* (Anderson and Moore, 1979). We assume that for $k \geq j$

$$\mathbf{x}_a(k) \triangleq \mathbf{x}(j) \qquad (21\text{-}45)$$

The state equation for state vector $\mathbf{x}_a(k)$ is

$$\mathbf{x}_a(k+1) = \mathbf{x}_a(k) \qquad k \geq j \qquad (21\text{-}46)$$

It is initialized at $k = j$ by (21-45). Augmenting (21-46) to our basic state-variable model in (15-17) and (15-18), we obtain the following *augmented basic state-variable model*:

$$\begin{pmatrix} \mathbf{x}(k+1) \\ \mathbf{x}_a(k+1) \end{pmatrix} = \begin{pmatrix} \mathbf{\Phi}(k+1,k) & \mathbf{0} \\ \mathbf{0} & \mathbf{I} \end{pmatrix} \begin{pmatrix} \mathbf{x}(k) \\ \mathbf{x}_a(k) \end{pmatrix} + \begin{pmatrix} \mathbf{\Psi}(k+1,k) \\ \mathbf{0} \end{pmatrix} \mathbf{u}(k) + \begin{pmatrix} \mathbf{\Gamma}(k+1,k) \\ \mathbf{0} \end{pmatrix} \mathbf{w}(k) \qquad (21\text{-}47)$$

and

$$\mathbf{z}(k+1) = (\mathbf{H}(k+1)\ \mathbf{0}) \begin{pmatrix} \mathbf{x}(k) \\ \mathbf{x}_a(k) \end{pmatrix} + \mathbf{v}(k+1) \qquad (21\text{-}48)$$

The following two-step procedure can be used to obtain an algorithm for $\hat{\mathbf{x}}(j|k)$:

1. Write down the Kalman filter equations for the augmented basic state-

variable model. Anderson and Moore (1979) give these equations for the recursive predictor; i.e., they find

$$\mathbf{E}\left\{\begin{pmatrix}\mathbf{x}(k+1)\\ \mathbf{x}_a(k+1)\end{pmatrix}\middle|\mathfrak{X}(k)\right\} = \begin{pmatrix}\hat{\mathbf{x}}(k+1|k)\\ \hat{\mathbf{x}}_a(k+1|k)\end{pmatrix} = \begin{pmatrix}\hat{\mathbf{x}}(k+1|k)\\ \hat{\mathbf{x}}(j|k)\end{pmatrix} \quad (21\text{-}49)$$

where $k \geq j$. The last equality in (21-49) makes use of (21-46) and (21-45). Observe that $\hat{\mathbf{x}}(j|k)$, the fixed-point smoother of $\mathbf{x}(j)$, has been found as the second component of the recursive predictor for the augmented model.

2. The Kalman filter (or recursive predictor) equations are partitioned in order to obtain the explicit structure of the algorithm for $\hat{\mathbf{x}}(j|k)$. We leave the details of this two-step procedure as an exercise for the reader.

FIXED-LAG SMOOTHING

The earliest attempts to obtain a fixed-lag smoother $\hat{\mathbf{x}}(k|k+L)$ led to an algorithm (e.g., Meditch, 1969), which was later shown to be unstable (Kelly and Anderson, 1971). The following state augmentation procedure leads to a stable fixed-interval smoother for $\hat{\mathbf{x}}(k-L|k)$.

We introduce $L+1$ state vectors, as follows: $\mathbf{x}_1(k+1) = \mathbf{x}(k)$, $\mathbf{x}_2(k+1) = \mathbf{x}(k-1)$, $\mathbf{x}_3(k+1) = \mathbf{x}(k-2), \ldots,$ $\mathbf{x}_{L+1}(k+1) = \mathbf{x}(k-L)$ [i.e., $\mathbf{x}_i(k+1) = \mathbf{x}(k+1-i)$, $i = 1, 2, \ldots, L+1$]. The state equations for these $L+1$ state vectors are

$$\left.\begin{aligned}\mathbf{x}_1(k+1) &= \mathbf{x}(k)\\ \mathbf{x}_2(k+1) &= \mathbf{x}_1(k)\\ \mathbf{x}_3(k+1) &= \mathbf{x}_2(k)\\ &\cdots\\ \mathbf{x}_{L+1}(k+1) &= \mathbf{x}_L(k)\end{aligned}\right\} \quad (21\text{-}50)$$

Augmenting (21-50) to our basic state-variable model in (15-17) and (15-18), we obtain yet another *augmented basic state-variable model*:

$$\begin{pmatrix}\mathbf{x}(k+1)\\ \mathbf{x}_1(k+1)\\ \mathbf{x}_2(k+1)\\ \vdots\\ \mathbf{x}_{L+1}(k+1)\end{pmatrix} = \begin{pmatrix}\boldsymbol{\Phi}(k+1,k) & \mathbf{0} & \cdots & \mathbf{0} & \mathbf{0}\\ \mathbf{I} & \mathbf{0} & \cdots & \mathbf{0} & \mathbf{0}\\ \mathbf{0} & \mathbf{I} & \cdots & \mathbf{0} & \mathbf{0}\\ \vdots & \vdots & \ddots & \vdots & \vdots\\ \mathbf{0} & \mathbf{0} & \cdots & \mathbf{I} & \mathbf{0}\end{pmatrix}\begin{pmatrix}\mathbf{x}(k)\\ \mathbf{x}_1(k)\\ \mathbf{x}_2(k)\\ \vdots\\ \mathbf{x}_{L+1}(k)\end{pmatrix}$$

$$+ \begin{pmatrix}\boldsymbol{\Psi}(k+1,k)\\ \mathbf{0}\\ \mathbf{0}\\ \vdots\\ \mathbf{0}\end{pmatrix}\mathbf{u}(k) + \begin{pmatrix}\boldsymbol{\Gamma}(k+1,k)\\ \mathbf{0}\\ \mathbf{0}\\ \vdots\\ \mathbf{0}\end{pmatrix}\mathbf{w}(k) \quad (21\text{-}51)$$

The following two-step procedure can be used to obtain an algorithm for $\hat{\mathbf{x}}(k - L|k)$:

1. Write down the Kalman filter equations for the augmented basic state-variable model. Anderson and Moore (1979) give these equations for the recursive predictor; i.e., they find

$$
\begin{aligned}
\mathbf{E}\{\text{col}\,(\mathbf{x}(k+1), \mathbf{x}_1(k+1), \ldots, \mathbf{x}_{L+1}(k+1)|\mathfrak{X}(k)\} \\
= \text{col}\,(\hat{\mathbf{x}}(k+1|k), \hat{\mathbf{x}}_1(k+1|k), \ldots, \hat{\mathbf{x}}_{L+1}(k+1|k)) \\
= \text{col}\,(\hat{\mathbf{x}}(k+1|k), \hat{\mathbf{x}}(k|k), \hat{\mathbf{x}}(k-1|k), \ldots, \hat{\mathbf{x}}(k-L|k))
\end{aligned}
\tag{21-52}
$$

   The last equality in (21-52) makes use of the fact that $\mathbf{x}_i(k+1) = \mathbf{x}(k+1-i)$, $i = 1, 2, \ldots, L+1$.
2. The Kalman filter (or recursive predictor) equations are partitioned in order to obtain the explicit structure of the algorithm for $\hat{\mathbf{x}}(k-L|k)$. The detailed derivation of the algorithm for $\hat{\mathbf{x}}(k-L|k)$ is left as an exercise for the reader (it can be found in Anderson and Moore, 1979, pp. 177–181).

Some aspects of this fixed-lag smoother are:

1. It is numerically stable, because its stability is determined by the stability of the recursive predictor (i.e., no new feedback loops are introduced into the predictor as a result of the augmentation procedure);
2. In order to compute $\hat{\mathbf{x}}(k-L|k)$, we must also compute the $L-1$ fixed-lag estimates, $\hat{\mathbf{x}}(k-1|k)$, $\hat{\mathbf{x}}(k-2|k), \ldots, \hat{\mathbf{x}}(k-L+1|k)$; this may be costly to do from a computational point of view; and
3. Computation can be reduced by careful coding of the partitioned recursive predictor equations.

PROBLEMS

21-1. Derive the formula for $\hat{\mathbf{x}}(k|N)$ in (21-5) using mathematical induction. Then derive $\hat{\mathbf{x}}(k|N)$ in (21-6).

21-2. Prove that $\{\tilde{\mathbf{x}}(k|N), k = N, N-1, \ldots, 0\}$ is a zero-mean, *second-order* Gauss-Markov process.

21-3. Derive the formula for the fixed-point smoothing error-covariance matrix, $\mathbf{P}(k|j)$, given in (21-40).

21-4. Prove Theorem 21-3, which gives formulas for a most useful mean-squared fixed-point smoother of $\mathbf{x}(k)$, $\hat{\mathbf{x}}(k|k+l)$, $l = 1, 2, \ldots$.

21-5. Using the two-step procedure described at the end of the section entitled Fixed-Point Smoothing, derive the resulting fixed-point smoother equations.

21-6. Using the two-step procedure described at the end of the section entitled

Fixed-Lag Smoothing, derive the resulting fixed-lag smoother equations. Show, by means of a block diagram, that this smoother is stable.

21-7. (Meditch, 1969, Exercise 6-13, pg. 245). Consider the scalar system $x(k+1) = 2^{-k}x(k) + w(k)$, $z(k+1) = x(k+1)$, $k = 0, 1, \ldots$, where $x(0)$ has mean zero and variance σ_0^2, and $w(k)$, $k = 0, 1, \ldots$ is a zero mean Gaussian white sequence which is independent of $x(0)$ and has a variance equal to q.

(a) Assuming that optimal fixed-point smoothing is to be employed to determine $x(0|j)$, $j = 1, 2, \ldots$, what is the equation for the appropriate smoothing filter?

(b) What is the limiting value of $p(0|j)$ as $j \to \infty$?

(c) How does this value compare with $p(0|0)$?

Lesson 22

State Estimation: Smoothing Applications

INTRODUCTION

In this lesson we present some applications that illustrate interesting numerical and theoretical aspects of fixed-interval smoothing. These applications are taken from the field of digital signal processing.

MINIMUM-VARIANCE DECONVOLUTION (MVD)

Here, as in Examples 2-6 and 14-1, we begin with the convolutional model

$$z(k) = \sum_{i=1}^{k} \mu(i)h(k-i) + \nu(k), \qquad k = 1, 2, \ldots, N \tag{22-1}$$

Recall that deconvolution is the signal processing procedure for removing the effects of $h(j)$ and $\nu(j)$ from the measurements so that one is left with an estimate of $\mu(j)$. Here we shall obtain a useful algorithm for a mean-squared fixed-interval estimator of $\mu(j)$.

To begin, we must convert (22-1) into an equivalent state-variable model.

Theorem 22-1 (Mendel, 1983a, pp. 13–14). *The single-channel state-variable model*

$$\mathbf{x}(k+1) = \mathbf{\Phi}\mathbf{x}(k) + \boldsymbol{\gamma}\mu(k) \tag{22-2}$$

$$z(k) = \mathbf{h}'\mathbf{x}(k) + \nu(k) \tag{22-3}$$

is equivalent to the convolutional sum model in (22-1) when $\mathbf{x}(0) = \mathbf{0}$, $\mu(0) = 0$, $h(0) = 0$, *and*

$$h(l) = \mathbf{h}'\mathbf{\Phi}^{l-1}\boldsymbol{\gamma}, \qquad l = 1, 2, \ldots \tag{22-4}$$

Proof. Iterate (22-2) and substitute the results into (22-3). Compare the resulting equation with (22-1) to see that, under the conditions $\mathbf{x}(0) = \mathbf{0}$, $\mu(0) = 0$ and $h(0) = 0$, they are the same. □

The condition $\mathbf{x}(0) = \mathbf{0}$ merely initializes our state-variable model. The condition $\mu(0) = 0$ means there is no input at time zero. The coefficients in (22-4) represent sampled values of the impulse response. If we are given impulse response data $\{h(1), h(2), \ldots, h(L)\}$ then we can determine matrices $\mathbf{\Phi}$, $\boldsymbol{\gamma}$ and $\mathbf{h}$ as well as system order n by applying an approximate realization procedure, such as Kung's (1978), to $\{h(1), h(2), \ldots, h(L)\}$. Additionally, if $h(0) \neq 0$ it is simple to modify Theorem 22-1.

In Example 14-1 we obtained a rather unwieldy formula for $\hat{\boldsymbol{\mu}}_{MS}(N)$. Note that, in terms of our conditioning notation, the elements of $\hat{\boldsymbol{\mu}}_{MS}(N)$ are $\hat{\mu}_{MS}(k|N)$, $k = 1, 2, \ldots, N$. We now obtain a very useful algorithm for $\hat{\mu}_{MS}(k|N)$. For notational convenience, we shorten $\hat{\mu}_{MS}$ to $\hat{\mu}$.

Theorem 22-2 (Mendel, 1983a, pp. 68–70)

a. *A two-pass fixed-interval smoother for* $\mu(k)$ *is*

$$\hat{\mu}(k|N) = q(k)\boldsymbol{\gamma}'\mathbf{r}(k+1|N) \tag{22-5}$$

where $k = N-1, N-2, \ldots, 1$.

b. *The smoothing error variance,* $\sigma^2_{\tilde{\mu}}(k|N)$, *is*

$$\sigma^2_{\tilde{\mu}}(k|N) = q(k) - q(k)\boldsymbol{\gamma}'\mathbf{S}(k+1|N)\boldsymbol{\gamma}q(k) \tag{22-6}$$

where $k = N-1, N-2, \ldots, 1$. *In these formulas* $\mathbf{r}(k|N)$ *and* $\mathbf{S}(k|N)$ *are computed using (21-25) and (21-27), respectively, and* $\mathbf{E}\{\mu^2(k)\} = q(k)$ [here $q(k)$ denotes the variance of $\mu(k)$, and should not be confused with the event sequence; which appears in the product model for $\mu(k)$].

Proof a. To begin, we apply the fundamental theorem of estimation theory, Theorem 13-1, to (22-2). We operate on both sides of that equation with $\mathbf{E}\{\cdot|\mathfrak{Z}(N)\}$, to show that

$$\boldsymbol{\gamma}\hat{\mu}(k|N) = \hat{\mathbf{x}}(k+1|N) - \mathbf{\Phi}\hat{\mathbf{x}}(k|N) \tag{22-7}$$

By performing appropriate manipulations on this equation we can derive (22-5) as follows. Substitute $\hat{\mathbf{x}}(k|N)$ and $\hat{\mathbf{x}}(k+1|N)$ from Equation (21-24) into Equation (22-7), to see that

$$\begin{aligned}\boldsymbol{\gamma}\hat{\mu}(k|N) &= \hat{\mathbf{x}}(k+1|k) + \mathbf{P}(k+1|k)\mathbf{r}(k+1|N) \\ &\quad - \boldsymbol{\Phi}[\hat{\mathbf{x}}(k|k-1) + \mathbf{P}(k|k-1)\mathbf{r}(k|N)] \\ &= \hat{\mathbf{x}}(k+1|k) - \boldsymbol{\Phi}\hat{\mathbf{x}}(k|k-1) \\ &\quad + \mathbf{P}(k+1|k)\mathbf{r}(k+1|N) - \boldsymbol{\Phi}\mathbf{P}(k|k-1)\mathbf{r}(k|N)\end{aligned} \tag{22-8}$$

Applying (17-11) and (16-4) to the state-variable model in (22-2) and (22-3), it is straightforward to show that

$$\hat{\mathbf{x}}(k+1|k) = \boldsymbol{\Phi}\hat{\mathbf{x}}(k|k-1) + \boldsymbol{\Phi}\mathbf{K}(k)\tilde{z}(k|k-1) \tag{22-9}$$

hence, (22-8) reduces to

$$\begin{aligned}\boldsymbol{\gamma}\hat{\mu}(k|N) = \boldsymbol{\Phi}\mathbf{K}(k)\tilde{z}(k|k-1) &+ \mathbf{P}(k+1|k)\mathbf{r}(k+1|N) \\ &- \boldsymbol{\Phi}\mathbf{P}(k|k-1)\mathbf{r}(k|N)\end{aligned} \tag{22-10}$$

Next, substitute (21-25) into (22-10), to show that

$$\begin{aligned}\boldsymbol{\gamma}\hat{\mu}(k|N) &= \boldsymbol{\Phi}\mathbf{K}(k)\tilde{z}(k|k-1) + \mathbf{P}(k+1|k)\mathbf{r}(k+1|N) \\ &\quad - \boldsymbol{\Phi}\mathbf{P}(k|k-1)\boldsymbol{\Phi}_p'(k+1,k)\mathbf{r}(k+1|N) \\ &\quad - \boldsymbol{\Phi}\mathbf{P}(k|k-1)\mathbf{h}'[\mathbf{h}'\mathbf{P}(k|k-1)\mathbf{h} + r]^{-1}\tilde{z}(k|k-1)\end{aligned} \tag{22-11}$$

Making use of Equation (17-12) for $\mathbf{K}(k)$, we find that the first and last terms in Equation (22-11) are identical; hence,

$$\begin{aligned}\boldsymbol{\gamma}\hat{\mu}(k|N) = \mathbf{P}(k+1|k)\mathbf{r}(k+1|N) &- \boldsymbol{\Phi}\mathbf{P}(k|k-1) \\ &\boldsymbol{\Phi}_p'(k+1,k)\mathbf{r}(k+1|N)\end{aligned} \tag{22-12}$$

Combine Equations (17-13) and (17-14) to see that

$$\mathbf{P}(k+1|k) = \boldsymbol{\Phi}\mathbf{P}(k|k-1)\boldsymbol{\Phi}_p'(k+1,k) + \boldsymbol{\gamma}q(k)\boldsymbol{\gamma}' \tag{22-13}$$

Finally, substitute (22-13) into Equation (22-12) to observe that

$$\boldsymbol{\gamma}\hat{\mu}(k|N) = \boldsymbol{\gamma}q(k)\boldsymbol{\gamma}'\mathbf{r}(k+1|N)$$

which has the unique solution given by

$$\hat{\mu}(k|N) = q(k)\boldsymbol{\gamma}'\mathbf{r}(k+1|N) \tag{22-14}$$

which is Equation (22-5).

b. To derive Equation (22-6) we use (22-5) and the definition of estimation error $\tilde{\mu}(k|N)$,

$$\tilde{\mu}(k|N) = \mu(k) - \hat{\mu}(k|N) \tag{22-15}$$

to form

$$\mu(k) = \tilde{\mu}(k|N) + q(k)\boldsymbol{\gamma}'\mathbf{r}(k+1|N) \tag{22-16}$$

Taking the variance of both sides of (22-16), and using the orthogonality condition

$$\mathbf{E}\{\tilde{\mu}(k|N)\mathbf{r}(k+1|N)\} = \mathbf{0} \tag{22-17}$$

we see that

$$q(k) = \sigma_\mu^2(k|N) + q(k)\boldsymbol{\gamma}'\mathbf{S}(k+1|N)\boldsymbol{\gamma}q(k) \tag{22-18}$$

which is equivalent to (22-6). □

Observe, from (22-5) and (22-6), that $\hat{\mu}(k|N)$ and $\sigma_\mu^2(k|N)$ are easily computed once $\mathbf{r}(k|N)$ and $\mathbf{S}(k|N)$ have been computed.

The extension of Theorem 22-2 to a time-varying state-variable model is straightforward and can be found in Mendel (1983).

Example 22-1

In this example we compute $\hat{\mu}(k|N)$, first for a broadband channel IR, $h_1(k)$, and then for a narrower-band channel IR, $h_2(k)$. The transfer functions of these channel models are

$$H_1(z) = \frac{-0.76286z^3 + 1.5884z^2 - 0.82356z + 0.000222419}{z^4 - 2.2633z^3 + 1.77734z^2 - 0.49803z + 0.045546} \tag{22-19}$$

and

$$H_2(z) = \frac{0.0378417z^3 - 0.0306517z}{z^4 - 3.4016497z^3 + 4.5113732z^2 - 2.7553363z + 0.6561} \tag{22-20}$$

respectively. Plots of these IR's and their squared amplitude spectra are depicted in Figures 22-1 and 22-2.

Measurements, $z(k)$ ($k = 1, 2, \ldots, 250$, where $T = 3$ msec), were generated by convolving each of these IR's with a sparse spike train (i.e., a Bernoulli-Gaussian sequence) and then adding measurement noise to the results. These measurements, which, of course, represent the starting point for deconvolution, are depicted in Figure 22-3.

Figure 22-4 depicts $\hat{\mu}(k|N)$. Observe that much better results are obtained for the broadband channel than for the narrower-band channel, even though data quality, as measured by $\overline{\text{SNR}}$, is much lower in the former case. The MVD results for the narrower-band channel appear "smeared out," whereas the MVD results for the broadband channel are quite sharp. We provide a theoretical explanation for this effect below.

Observe, also, that $\hat{\mu}(k|N)$ tends to undershoot $\mu(k)$. See Chi and Mendel (1984) for a theoretical explanation about why this occurs. □

STEADY-STATE MVD FILTER

For a time-invariant IR and stationary noises, the Kalman gain matrix, as well as the error-covariance matrices, will reach steady-state values. When this occurs, both the Kalman innovations filter and anticausal μ-filter [(22-5) and (21-25)] become time invariant, and we then refer to the MVD filter as a *steady-state MVD filter*. In this section we examine an important property of this steady-state filter.

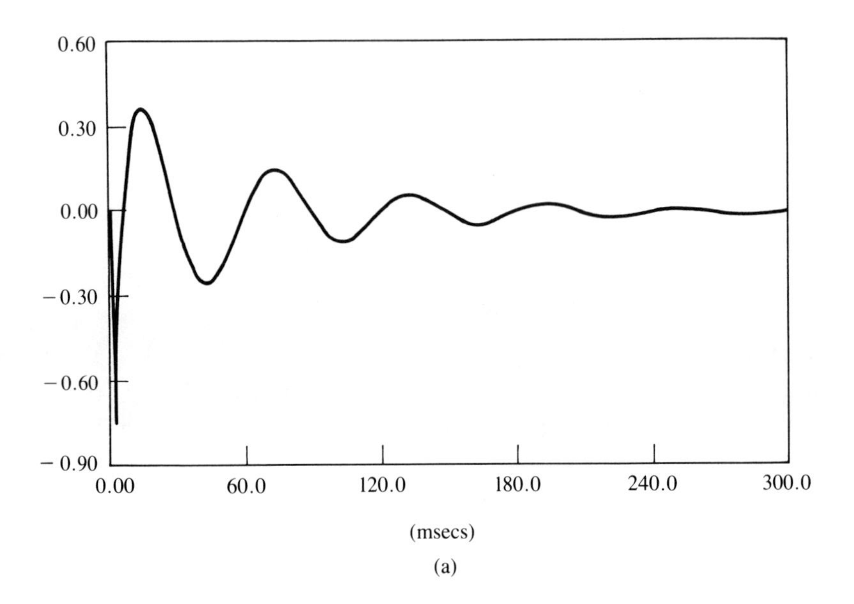

(a)

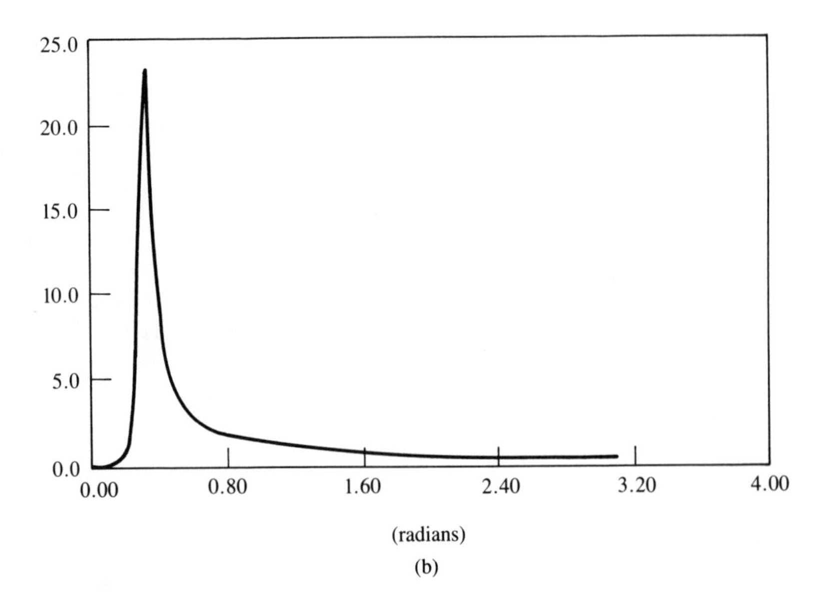

(b)

Figure 22-1 (a) Fourth-order broad-band channel IR, and (b) its squared amplitude spectrum (Chi, 1983).

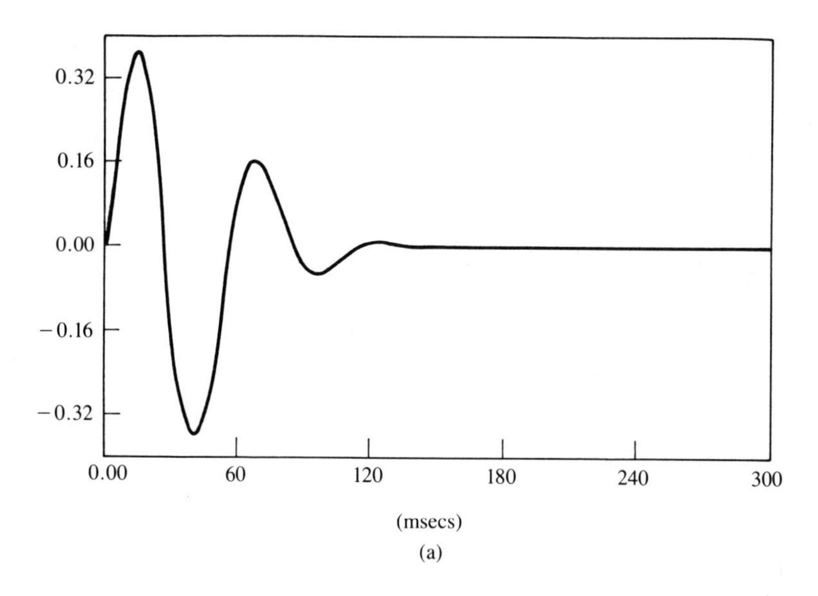

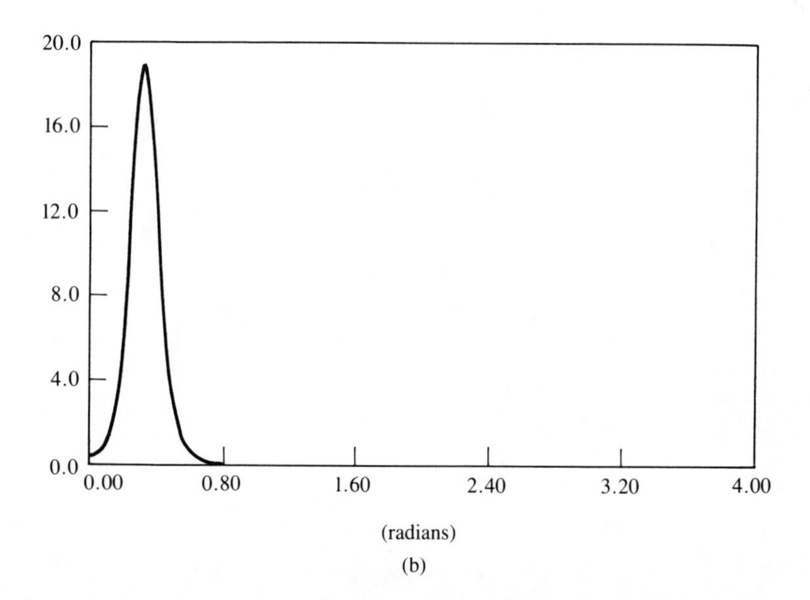

Figure 22-2 (a) Fourth-order narrower-band channel, IR, and (b) its squared amplitude spectrum (Chi and Mendel, 1984, © 1984, IEEE).

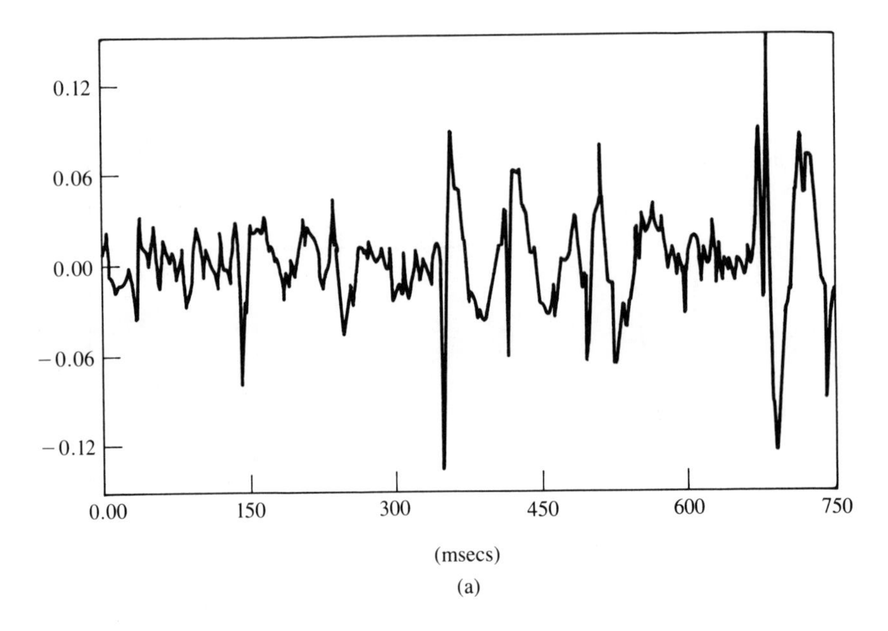

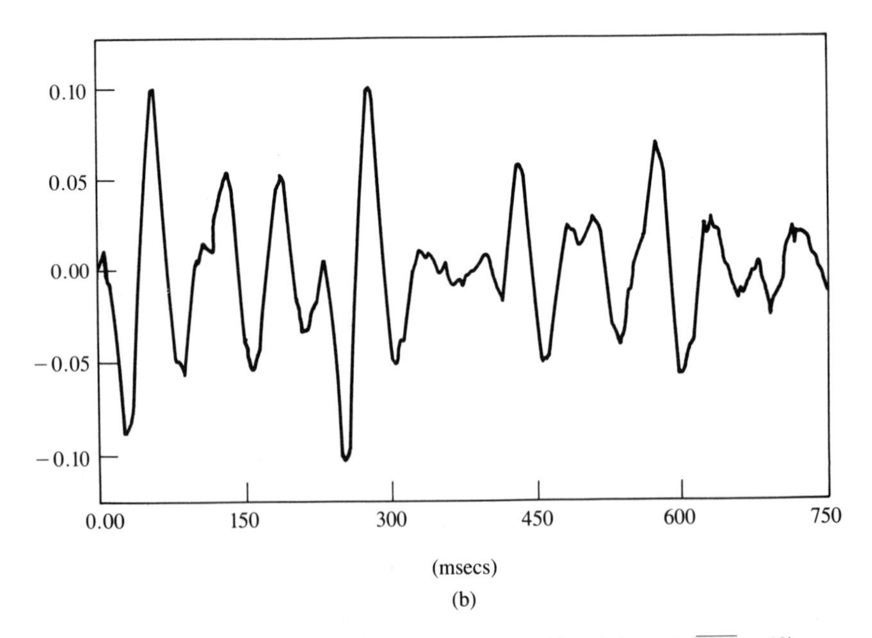

Figure 22-3 Measurements associated with (a) broad-band channel ($\overline{SNR} = 10$) and (b) narrower-band channel ($\overline{SNR} = 100$), (Chi and Mendel, 1984, © 1984, IEEE).

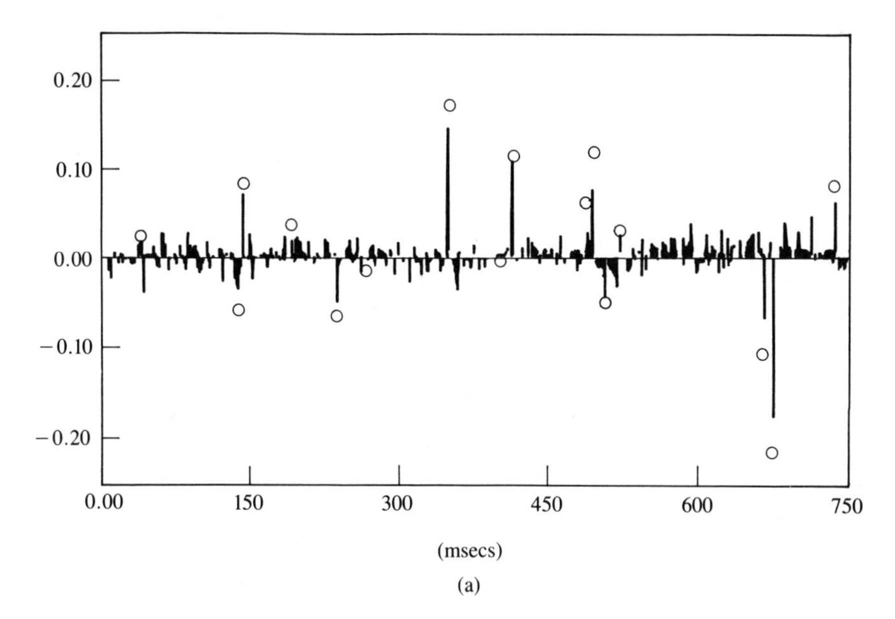

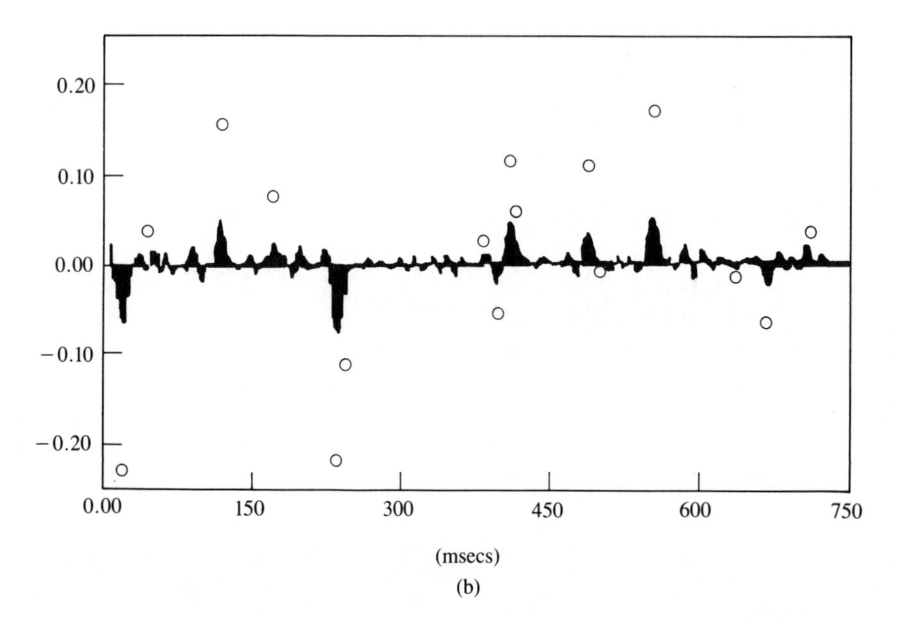

Figure 22-4 $\hat{\mu}(k|N)$ for (a) broadband channel ($\overline{SNR}$ = 10) and (b) narrower-band channel ($\overline{SNR}$ = 100). Circles depict true $\mu(k)$ and bars depict estimate of $\mu(k)$, (Chi and Mendel, 1984, © 1984, IEEE).

Let $h_i(k)$ and $h_\mu(k)$ denote the IR's of the steady-state Kalman innovations and anticausal μ-filters, respectively. Then,

$$\begin{aligned}\hat{\mu}(k|N) &= h_\mu(k)*\tilde{z}(k|k-1)\\ &= h_\mu(k)*h_i(k)*z(k)\\ &= h_\mu(k)*h_i(k)*h(k)*\mu(k)\\ &\quad + h_\mu(k)*h_i(k)*v(k)\end{aligned} \tag{22-21}$$

which can also be expressed as

$$\hat{\mu}(k|N) = \hat{\mu}_s(k|N) + n(k|N) \tag{22-22}$$

where the *signal component* of $\hat{\mu}(k|N)$, $\hat{\mu}_s(k|N)$, is

$$\hat{\mu}_s(k|N) = h_\mu(k)*h_i(k)*h(k)*\mu(k) \tag{22-23}$$

and the *noise component* of $\hat{\mu}(k|N)$, $n(k|N)$, is

$$n(k|N) = h_\mu(k)*h_i(k)*v(k) \tag{22-24}$$

We shall refer to $h_\mu(k)*h_i(k)$ as the IR of the MVD filter, $h_{\text{MV}}(k)$, i.e.,

$$h_{\text{MV}}(k) = h_\mu(k)*h_i(k) \tag{22-25}$$

The following result has been proven by Chi and Mendel (1984) for the slightly modified model $\mathbf{x}(k+1) = \mathbf{\Phi x}(k) + \boldsymbol{\gamma}\mu(k+1)$ and $z(k) = \mathbf{h}'\mathbf{x}(k) + v(k)$ [because of the $\mu(k+1)$ input instead of the $\mu(k)$ input, $h(0) \neq 0$].

Theorem 22-3. *In the stationary case:*

a. *The Fourier transform of* $h_{\text{MV}}(k)$ *is*

$$H_{\text{MV}}(\omega) = \frac{qH^*(\omega)}{q|H(\omega)|^2 + r} \tag{22-26}$$

where $H^*(\omega)$ *denotes the complex conjugate of* $H(\omega)$; *and*

b. *the signal component of* $\hat{\mu}(k|N)$, $\hat{\mu}_s(k|N)$, *is given by*

$$\hat{\mu}_s(k|N) = R(k)*\mu(k) \tag{22-27}$$

where $R(k)$ *is the auto-correlation function*

$$R(k) = \frac{q}{\eta}[h(k)*h_i(k)]*[h(-k)*h_i(-k)] \tag{22-28}$$

in which

$$\eta = \mathbf{h}'\bar{\mathbf{P}}\mathbf{h} + r \tag{22-29}$$

additionally,

$$R(\omega) = \frac{q|H(\omega)|^2}{q|H(\omega)|^2 + r} \quad \square \tag{22-30}$$

We leave the proof of this theorem as an exercise for the reader. Observe that part (b) of the theorem means that $\hat{\mu}_s(k|N)$ *is a zero-phase wave-shaped version of* $\mu(k)$. Observe, also, that $R(\omega)$ can be written as

$$R(\omega) = \frac{|H(\omega)|^2 q/r}{1 + |H(\omega)|^2 q/r} \tag{22-31}$$

which demonstrates that q/r, and subsequently $\overline{\text{SNR}}$, is an MVD filter tuning parameter. As $q/r \to \infty$, $R(\omega) \to 1$ so that $R(k) \to \delta(k)$; thus, for high signal-to-noise ratios $\hat{\mu}_s(k|N) \to \mu(k)$. Additionally, when $|H(\omega)|^2 q/r >> 1$, $R(\omega) \to 1$, and once again $R(k) \to \delta(k)$. Broadband IRs often satisfy this condition. In general, however, $\hat{\mu}_s(k|N)$ is a smeared-out version of $\mu(k)$; however, the nature of the smearing is quite dependent on the bandwidth of $h(k)$ and $\overline{\text{SNR}}$.

Example 22-2

This example is a continuation of Example 22-1. Figure 22-5 depicts $R(k)$ for both the broadband and narrower-band IRs, $h_1(k)$ and $h_2(k)$, respectively. As predicted by (22-31), $R_1(k)$ is much spikier than $R_2(k)$, which explains why the MVD results for the broadband IR are quite sharp, whereas the MVD results for the narrower-band IR are smeared out. Note, also, the difference in peak amplitudes for $R_1(k)$ and $R_2(k)$. This explains why $\hat{\mu}(k|N)$ underestimates the true values of $\mu(k)$ by such large amounts in the narrower-band case (see Figs. 22-4a and b). □

RELATIONSHIP BETWEEN STEADY-STATE MVD FILTER AND AN INFINITE IMPULSE RESPONSE DIGITAL WIENER DECONVOLUTION FILTER

We have seen that an MVD filter is a cascade of a causal Kalman innovations filter and an anticausal μ-filter; hence, it is a noncausal filter. Its impulse response extends from $k = -\infty$ to $k = +\infty$, and the IR of the steady-state MVD filter is given in the time-domain by $h_{\text{MV}}(k)$ in (22-25), or in the frequency domain by $H_{\text{MV}}(\omega)$ in (22-26).

There is a more direct way for designing an IIR minimum mean-squared error deconvolution filter, i.e., an *IIR digital Wiener deconvolution filter*, as we describe next.

We return to the situation depicted in Figure 19-4, but now we assume that: Filter $F(z)$ is an IIR filter, with coefficients $\{f(j), j = 0, \pm 1, \pm 2, \ldots\}$;

$$d(k) = \mu(k) \tag{22-32}$$

where $\mu(k)$ is a white noise sequence; $\mu(k)$, $v(k)$, and $n(k)$ are stationary; and, $\mu(k)$ and $v(k)$ are uncorrelated.
In this case, (19-39) becomes

$$\sum_{i=-\infty}^{\infty} f(i)\phi_{zz}(i-j) = \phi_{z\mu}(j), \qquad j = 0, \pm 1, \pm 2, \ldots \tag{22-33}$$

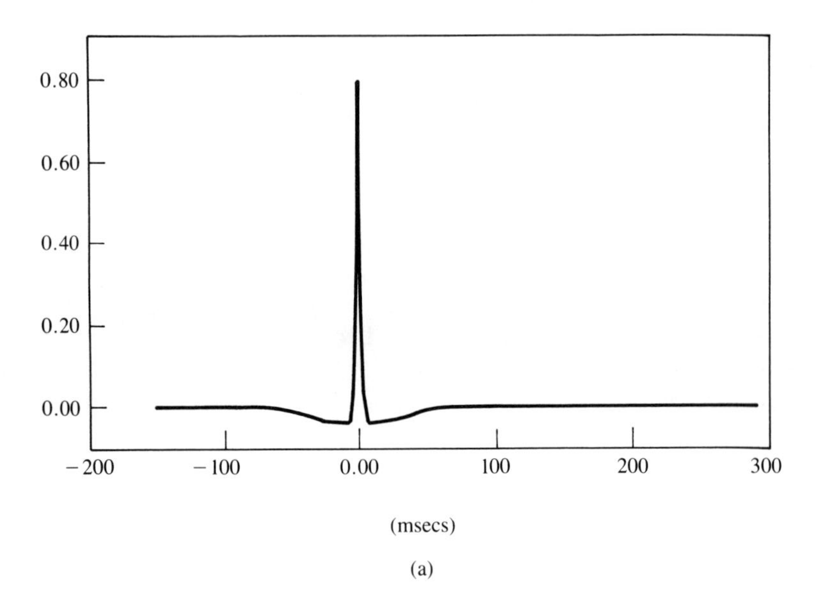

(a)

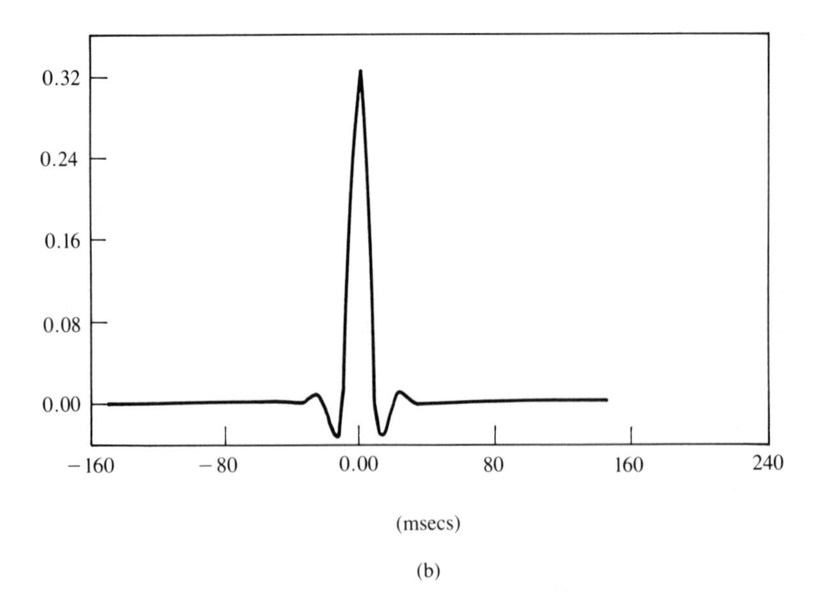

(b)

Figure 22-5 $R(k)$ for (a) broadband channel ($\overline{SNR} = 100$) and (b) narrower-band channel ($\overline{SNR} = 100$), (Chi and Mendel, 1984, © 1984, IEEE).

Using (22-1), the whiteness of $\mu(k)$ and the assumptions that $\mu(k)$ and $v(k)$ are uncorrelated and stationary, it is straightforward to show that

$$\phi_{z\mu}(j) = qh(-j) \tag{22-34}$$

Substituting (22-34) into (22-33), we have

$$\sum_{i=-\infty}^{\infty} f(i)\phi_{zz}(i-j) = qh(-j), \qquad j = 0, \pm 1, \pm 2, \ldots \tag{22-35}$$

Taking the discrete-time Fourier transform of (22-35), we see that

$$F(\omega)\Phi_{zz}(\omega) = qH^*(\omega) \tag{22-36}$$

but, from (22-1), we also know that

$$\Phi_{zz}(\omega) = q|H(\omega)|^2 + r \tag{22-37}$$

Substituting (22-37) into (22-36), we determine $F(\omega)$, as

$$F(\omega) = \frac{qH^*(\omega)}{q|H(\omega)|^2 + r} \tag{22-38}$$

This IIR digital Wiener deconvolution filter (i.e., two-sided least-squares inverse filter) was, to the best of our knowledge, first derived by Berkhout (1977).

Theorem 22-4 (Chi and Mendel, 1984). *The steady-state MVD filter, whose IR is given by* $h_{MV}(k)$, *is exactly the same as Berkhout's IIR digital Wiener deconvolution filter.* □

The steady-state MVD filter is a recursive implementation of Berkhout's infinite-length filter. Of course, the MVD filter is also applicable to time-varying and nonstationary systems, whereas his filter is not.

MAXIMUM-LIKELIHOOD DECONVOLUTION

In Example 14-2 we began with the deconvolution linear model $\mathcal{Z}(N) = \mathcal{H}(N-1)\boldsymbol{\mu} + \mathcal{V}(N)$, used the product model for $\boldsymbol{\mu}$ (i.e., $\boldsymbol{\mu} = \mathbf{Q}_q\mathbf{r}$), and showed that a separation principle exists for the determination of $\hat{\mathbf{r}}_{MAP}$ and $\hat{\mathbf{q}}_{MAP}$. We showed that first one must determine $\hat{\mathbf{q}}_{MAP}$, after which $\hat{\mathbf{r}}_{MAP}$ can be computed using (14-57). We repeat (14-57) here for convenience,

$$\hat{\mathbf{r}}_{MAP}(N|\hat{\mathbf{q}}) = \sigma_r^2\mathbf{Q}_{\hat{q}}\mathcal{H}'(N-1) [\sigma_r^2\mathcal{H}(N-1)\mathbf{Q}_{\hat{q}}\mathcal{H}'(N-1) + \rho\mathbf{I}]^{-1}\mathcal{Z}(N) \tag{22-39}$$

where $\hat{\mathbf{q}}$ is short for $\hat{\mathbf{q}}_{MAP}$. In terms of our conditioning notation used in state estimation, the elements of $\hat{\mathbf{r}}_{MAP}(N|\hat{\mathbf{q}})$ are $\hat{r}_{MAP}(k|N;\hat{\mathbf{q}})$, $k = 1, 2, \ldots, N$. Equation (22-39) is terribly unwieldy because of the $N \times N$ matrix,

$\sigma_r^2\mathcal{H}(N-1)\mathbf{Q}_{\hat{\mathbf{q}}}\mathcal{H}'(N-1)+\rho\mathbf{I}$, that must be inverted. The following theorem provides a more practical way to compute $\hat{r}_{\text{MAP}}(k|N;\hat{\mathbf{q}})$.

Theorem 22-5 (Mendel, 1983b). *Unconditional maximum-likelihood (i.e., MAP) estimates of* **r** *can be obtained by applying MVD formulas to the state-variable model*

$$\mathbf{x}(k+1)=\mathbf{\Phi x}(k)+\boldsymbol{\gamma}\hat{q}_{\text{MAP}}(k)r(k) \tag{22-40}$$

$$z(k)=\mathbf{h}'\mathbf{x}(k)+v(k) \tag{22-41}$$

where $\hat{q}_{\text{MAP}}(k)$ *is a MAP estimate of* $q(k)$.

Proof. Example 14-2 showed that a MAP estimate of **q** can be obtained prior to finding a MAP estimate of **r**. By using the product model for $\mu(k)$, and $\hat{\mathbf{q}}_{\text{MAP}}$, our state-variable model in (22-2) and (22-3) can be expressed as in (22-40) and (22-41). Applying (14-41) to this system, we see that

$$\hat{r}_{\text{MAP}}(k|N)=\hat{r}_{\text{MS}}(k|N) \tag{22-42}$$

but, by comparing (22-40) and (22-2), and (22-41) and (22-3), we see that $\hat{r}_{\text{MS}}(k|N)$ can be found from the MVD algorithm in Theorem 22-2 in which we replace $\mu(k)$ by $r(k)$ and set $q(k)=\sigma_r^2\hat{q}_{\text{MAP}}(k)$. □

RECURSIVE WAVESHAPING

In Lesson 19 we described the design of a FIR waveshaping filter (e.g., see Figure 19-4). In this section we shall develop a recursive waveshaping filter in the framework of state-variable models and mean-squared estimation theory. Other approaches to the design of recursive waveshaping filters have been given by Shanks (1967) and Aguilara, et al. (1970).

We direct our attention at the situation depicted in Figure 22-6. To

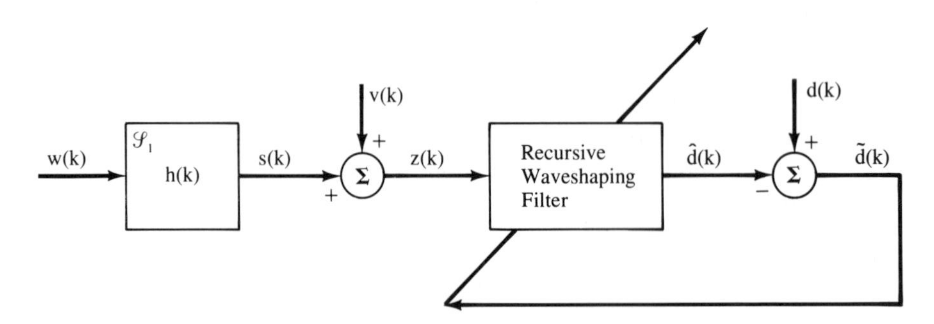

Figure 22-6 Waveshaping problem studied in this section. Information about $\tilde{d}(k)$, not necessarily $\tilde{d}(k)$ itself, is used to drive the (time-varying) recursive waveshaping filter, (Mendel, 1983a, © 1983, IEEE).

begin, we must obtain the following state-variable models for $h(k)$ and $d(k)$ [i.e., we must use an approximate realization procedure (Kung, 1978) or any other viable technique to map $\{h(i), i = 0, 1, \ldots, I_1\}$ into $\{\mathbf{\Phi}_1, \boldsymbol{\gamma}_1, \mathbf{h}_1\}$, and $\{d(i), i = 0, 1, \ldots, I_2\}$ into $\{\mathbf{\Phi}_2, \boldsymbol{\gamma}_2, \mathbf{h}_2\}$]:

$$\mathbf{x}_1(k + 1) = \mathbf{\Phi}_1\mathbf{x}_1(k) + \boldsymbol{\gamma}_1\delta(k) \quad (22\text{-}43)$$

$$h(k) = \mathbf{h}_1'\mathbf{x}_1(k) \quad (22\text{-}44)$$

and

$$\mathbf{x}_2(k + 1) = \mathbf{\Phi}_2\mathbf{x}_2(k) + \boldsymbol{\gamma}_2\delta(k) \quad (22\text{-}45)$$

$$d(k) = \mathbf{h}_2'\mathbf{x}_2(k) \quad (22\text{-}46)$$

State vectors $\mathbf{x}_1$ and $\mathbf{x}_2$ are $n_1 \times 1$ and $n_2 \times 1$, respectively. Signal $\delta(k)$ is the unit spike.

In the stochastic situation depicted in Figure 22-6, where $h(k)$ is excited by the white sequence $w(k)$ and noise, $v(k)$, corrupts $s(k)$, the best we can possibly hope to achieve by waveshaping is to make $z(k) = w(k)*h(k) + v(k)$ look like $w(k)*d(k)$ (Figure 22-7). This is because both $h(k)$ and $d(k)$ must be excited by the same random input, $w(k)$, for the waveshaping problem in this situation to be well posed. The state-variable model for this situation is

$$\mathcal{S}_1: \begin{cases} \mathbf{x}_1(k + 1) = \mathbf{\Phi}_1\mathbf{x}_1(k) + \boldsymbol{\gamma}_1 w(k) & (22\text{-}47) \\ z(k) = \mathbf{h}_1'\mathbf{x}_1(k) + v(k) & (22\text{-}48) \end{cases}$$

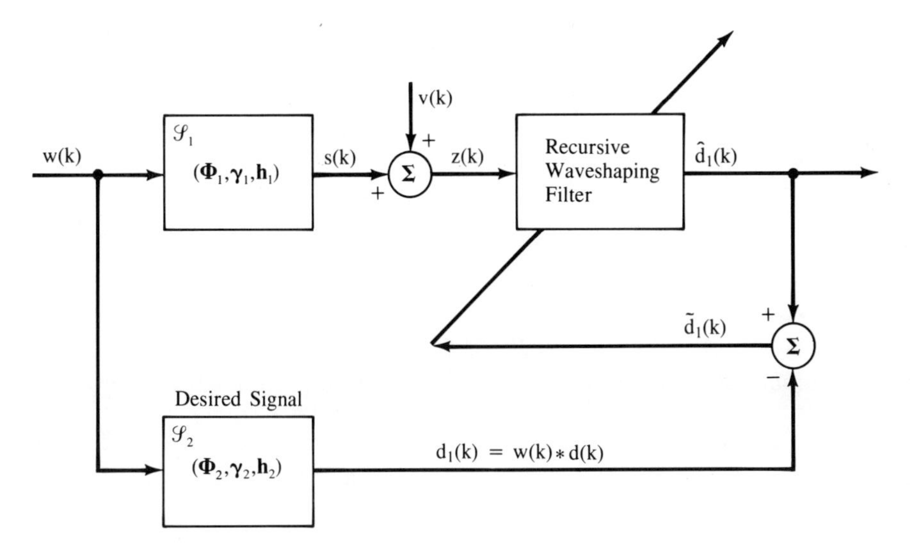

Figure 22-7 State-variable formulation of recursive waveshaping problem (Mendel, 1983a, © 1983, IEEE).

and

$$\mathcal{S}_2:\begin{cases}\mathbf{x}_2(k+1) = \mathbf{\Phi}_2\mathbf{x}_2(k) + \boldsymbol{\gamma}_2 w(k) & \text{(22-49)}\\ d_1(k) = \mathbf{h}_2'\mathbf{x}_2(k) & \text{(22-50)}\end{cases}$$

Observe that both $\mathcal{S}_1$ and $\mathcal{S}_2$ are excited by the same input, $w(k)$. Additionally, $w(k)$ and $v(k)$ are zero-mean mutually uncorrelated white noise sequences, for which

$$\mathbf{E}\{w^2(k)\} = q \tag{22-51}$$

and

$$\mathbf{E}\{v^2(k)\} = r \tag{22-52}$$

We now proceed to formulate the recursive waveshaping filter design problem in the context of mean-squared estimation theory. Our *filter design problem* is: Given the measurements $z(1), z(2), \ldots, z(j)$ determine an estimator $\hat{d}_1(k|j)$ such that the mean-squared error

$$\mathbf{J}[\tilde{d}_1(k|j)] = \mathbf{E}\{[\tilde{d}_1(k|j)]^2\} \tag{22-53}$$

is minimized. The solution to this problem is given next.

Theorem 22-6 (Structure of Minimum-Variance Waveshaper, Dai and Mendel, 1986). *The Minimum-Variance Waveshaping filter consists of two components 1. stochastic inversion, and 2. waveshaping.*

Proof. According to the Fundamental Theorem of Estimation Theory, Theorem 13-1, the unbiased, minimum-variance estimator of $d_1(k)$ based on the measurements $\{z(1), z(2), \ldots, z(j)\}$ is

$$\hat{d}_1(k|j) = \mathbf{E}\{d_1(k)|\mathfrak{Z}(j)\} \tag{22-54}$$

where $\mathfrak{Z}(j) = \text{col}\,(z(1), z(2), \ldots, z(j))$. Observe, from Figure 22-7, that

$$d_1(k) = w(k) * d(k) \tag{22-55}$$

hence,

$$\hat{d}_1(k|j) = \mathbf{E}\{w(k) * d(k)|\mathfrak{Z}(j)\}$$
$$= \mathbf{E}\{w(k)|\mathfrak{Z}(j)\} * d(k) = \hat{w}(k|j) * d(k) \tag{22-56}$$

Equation (22-56) tells us that there are two steps to obtain $\hat{d}_1(k|j)$:

1. first obtain $\hat{w}(k|j)$, and
2. then convolve the desired signal with $\hat{w}(k|j)$.

Step 1 removes the effects of the original wavelet and noise from the measurements. It is the problem of stochastic inversion and can be performed by means of minimum-variance deconvolution. □

In this book we have only discussed fixed-interval MVD, from which we obtain $\hat{w}(k|N)$. Fixed point algorithms are also available (e.g., Mendel, 1983).

Theorem 22-7 (Recursive waveshaping, Mendel, 1983a, pg. 600). *Let* $\hat{w}(k|N)$ *denote the fixed-interval estimate of* w(k), *which can be obtained from* $\mathbf{S}_1$ *via minimum-variance deconvolution (MVD), (Theorem 22-2). Then* $\hat{d}_1(k|N)$ *is obtained from the waveshaping filter*

$$\hat{d}_1(k|N) = \mathbf{h}_2'\hat{\mathbf{x}}_2(k|N) \tag{22-57}$$

$$\hat{\mathbf{x}}_2(k+1|N) = \mathbf{\Phi}_2\hat{\mathbf{x}}_2(k|N) + \boldsymbol{\gamma}_2\hat{w}(k|N) \tag{22-58}$$

where $k = 0, 1, \ldots, N-1$, *and* $\hat{\mathbf{x}}_2(0|N) = \mathbf{0}$. □

We leave the proof of this theorem as an exercise for the reader.

Some observations about Theorems 22-6 and 22-7 are in order. First, MVD is analogous to solving a stochastic inverse problem; hence, an MVD filter can be thought of as an optimal inverse filter. Second, if $w(k)$ was deterministic and $v(k) = 0$, then our intuition tells us that the recursive waveshaping filter should consist of the following two distinct components: an inverse filter, to remove the effects of $H(z)$, followed by a waveshaping filter, whose transfer function is $D(z)$. The transfer function of the recursive waveshaping filter would then be $\frac{1}{H(z)}D(z)$, so that $D(z) = H(z)\left[\frac{D(z)}{H(z)}\right] = D(z)$. Finally, the results in these theorems support our intuition even in the stochastic case; for, $\hat{d}_1(k|N)$, for example, is also obtained in two steps. As shown in Figure 22-8, first $\hat{w}(k|N)$ is obtained via MVD. Then this signal is reshaped to give $\hat{d}_1(k|N)$.

Note, also, that in order to compute $\hat{d}_1(k|N)$ it is not really necessary to use the state-variable model for $\hat{d}_1(k|N)$. Observe, from (22-56), that

$$\hat{d}_1(k|N) = d(k)*\hat{w}(k|N) \tag{22-59}$$

Example 22-3

In this example we describe a simulation study for the Bernoulli-Gaussian input sequence [i.e., $w(k)$] depicted in Figure 22-9. When this sequence is convolved with the fourth-order IR depicted in Figure 22-1a and noise is added to the result we obtain measurements $z(k)$, $k = 1, 2, \ldots, 1000$, depicted in Figure 22-10.

In Figure 22-11 we see $\hat{w}(k|N)$. Observe that the large spikes in $w(k)$ have been estimated quite well. When $\hat{w}(k|N)$ is convolved with a first-order decaying exponential [$d_1(t) = e^{-300t}$], we obtain the shaped signal depicted in Figure 22-12. Some "smoothing" of the data occurs when $\hat{w}(k|N)$ is convolved with the exponential $d_1(k)$.

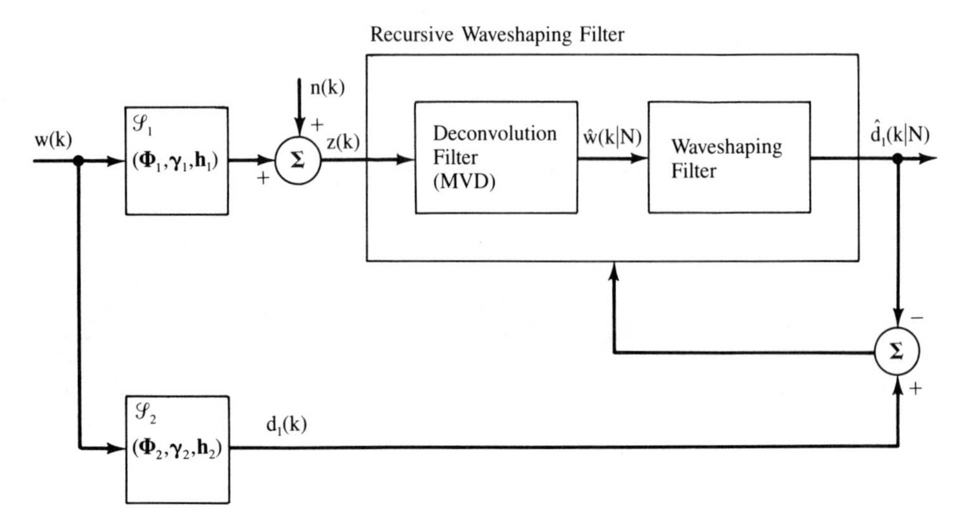

Figure 22-8 Details of recursive linear waveshaping filter. (Mendel, 1983a, © 1983 IEEE.)

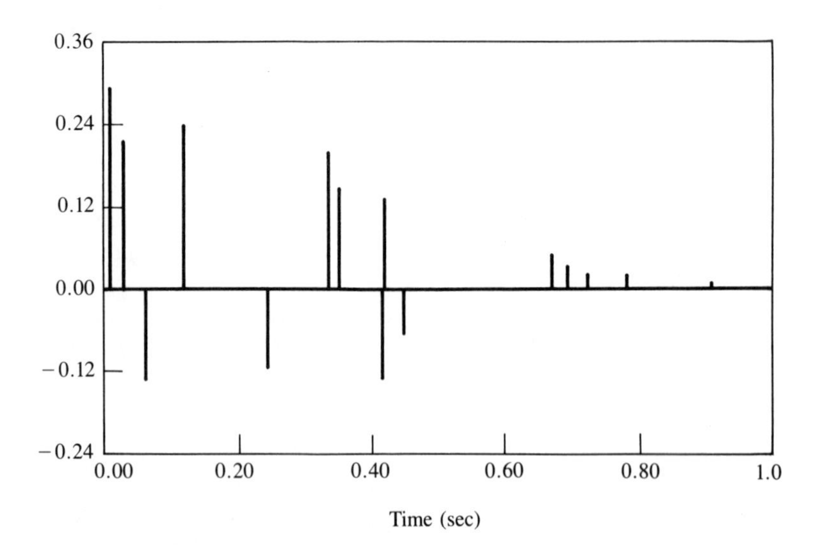

Figure 22-9 Bernoulli-Gaussian input sequence (Mendel, 1983a, © 1983, IEEE).

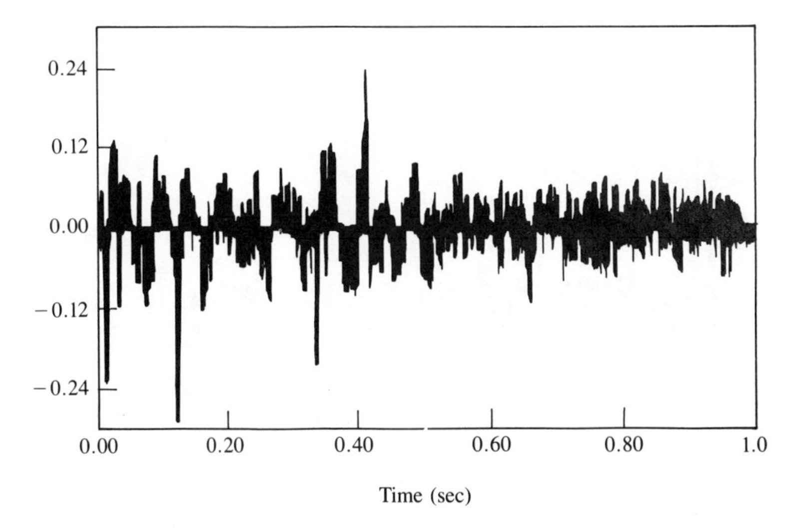

Figure 22-10 Noise-corrupted signal $z(k)$. Signal-to-noise ratio chosen equal to ten (Mendel, 1983a, © 1983, IEEE).

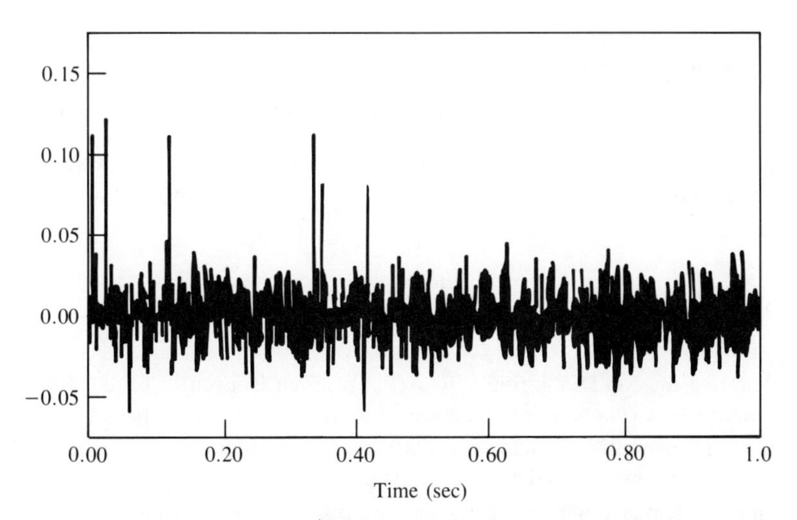

Figure 22-11 $\hat{w}(k|N)$, (Mendel, 1983a, © 1983, IEEE).

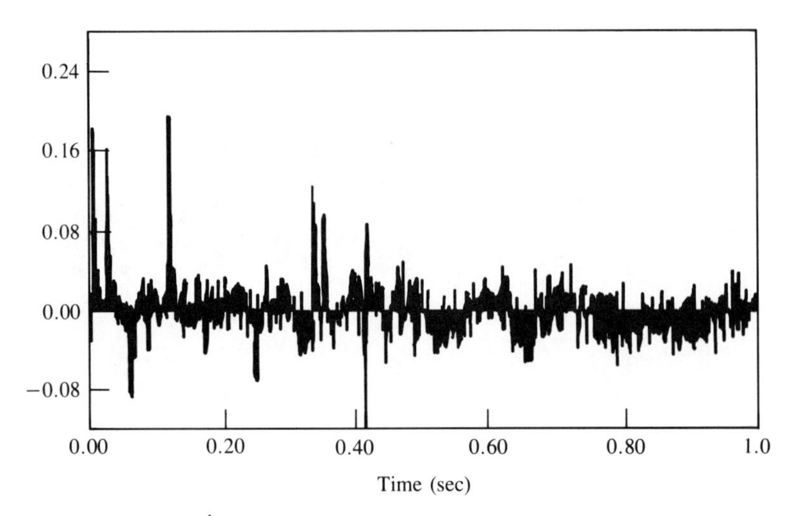

Figure 22-12 $\hat{d}_1(k|N)$ when $d_1(t) = e^{-300t}$ (Mendel 1983a, © 1983, IEEE).

More smoothing is achieved when $\hat{w}(k|N)$ is convolved with a zero-phase waveform. □

Finally, note that, because of Theorem 22-3, we know that perfect waveshaping is not possible. For example, the signal component of $\hat{d}_1(k|N)$, $\hat{d}_{1s}(k|N)$, is given by the expression

$$\hat{d}_{1s}(k|N) = d(k)*R(k)*w(k) \tag{22-60}$$

How much the auto-correlation function $R(k)$ will distort $\hat{d}_{1s}(k|N)$ from $d(k)*w(k)$, depends, of course, on bandwidth and signal-to-noise ratio considerations.

PROBLEMS

22-1. Rederive the MVD algorithm for $\hat{\mu}(k|N)$, which is given in (22-5), from the Fundamental Theorem of Estimation Theory, i.e., $\hat{\mu}(k|N) = \mathbf{E}\{\mu(k)|\mathcal{Z}(N)\}$.

22-2. Prove Theorem 22-3. Explain why part (b) of the theorem means that $\hat{\mu}_s(k|N)$ is a zero-phase waveshaped version of $\mu(k)$.

22-3. This problem is a memory refresher. You probably have either seen or carried out the calculations asked for in a course on random processes.
(a) Derive Equation (22-34);
(b) Derive Equation (22-37).

22-4. Prove the recursive waveshaping Theorem 22-7.

Lesson 23

State Estimation for the Not-So-Basic State-Variable Model

INTRODUCTION

In deriving all of our state estimators we assumed that our dynamical system could be modeled as in Lesson 15, i.e., as our *basic state-variable model*. The results so obtained are applicable only for systems that satisfy all the conditions of that model: the noise processes $\mathbf{w}(k)$ and $\mathbf{v}(k)$ are both zero mean, white, and mutually uncorrelated, no known bias functions appear in the state or measurement equations, and no measurements are noise-free (i.e., perfect). The following cases frequently occur in practice:

1. either nonzero–mean noise processes or known bias functions or both in the state or measurement equations
2. correlated noise processes,
3. colored noise processes, and
4. some perfect measurements.

In this lesson we show how to modify some of our earlier results in order to treat these important special cases. In order to see the forest from the trees, we consider each of these four cases separately. In practice, some or all of them may occur together.

BIASES

Here we assume that our basic state-variable model, given in (15-17) and (15-18), has been modified to

$$\mathbf{x}(k+1) = \mathbf{\Phi}(k+1,k)\mathbf{x}(k) + \mathbf{\Gamma}(k+1,k)\mathbf{w}_1(k) + \mathbf{\Psi}(k+1,k)\mathbf{u}(k) \tag{23-1}$$

$$\mathbf{z}(k+1) = \mathbf{H}(k+1)\mathbf{x}(k+1) + \mathbf{G}(k+1)\mathbf{u}(k+1) + \mathbf{v}_1(k+1) \tag{23-2}$$

where $\mathbf{w}_1(k)$ and $\mathbf{v}_1(k)$ are nonzero mean, individually and mutually uncorrelated Gaussian noise sequences, i.e.,

$$\mathbf{E}\{\mathbf{w}_1(k)\} = \mathbf{m}_{\mathbf{w}_1}(k) \neq \mathbf{0} \qquad \mathbf{m}_{\mathbf{w}_1}(k) \text{ known} \tag{23-3}$$

$$\mathbf{E}\{\mathbf{v}_1(k)\} = \mathbf{m}_{\mathbf{v}_1}(k) \neq \mathbf{0} \qquad \mathbf{m}_{\mathbf{v}_1}(k) \text{ known} \tag{23-4}$$

$\mathbf{E}\{[\mathbf{w}_1(i) - \mathbf{m}_{\mathbf{w}_1}(i)][\mathbf{w}_1(j) - \mathbf{m}_{\mathbf{w}_1}(j)]'\} = \mathbf{Q}(i)\delta_{ij}$, $\mathbf{E}\{[\mathbf{v}_1(i) - \mathbf{m}_{\mathbf{v}_1}(i)][\mathbf{v}_1(j) - \mathbf{m}_{\mathbf{v}_1}(j)]'\} = \mathbf{R}(i)\delta_{ij}$, and $\mathbf{E}\{[\mathbf{w}_1(i) - \mathbf{m}_{\mathbf{w}_1}(i)][\mathbf{v}_1(j) - \mathbf{m}_{\mathbf{v}_1}(j)]'\} = \mathbf{0}$.

This case is handled by reducing (23-1) and (23-2) to our previous basic state-variable model, using the following simple transformations. Let

$$\mathbf{w}(k) \triangleq \mathbf{w}_1(k) - \mathbf{m}_{\mathbf{w}_1}(k) \tag{23-5}$$

and

$$\mathbf{v}(k) \triangleq \mathbf{v}_1(k) - \mathbf{m}_{\mathbf{v}_1}(k) \tag{23-6}$$

Observe that both $\mathbf{w}(k)$ and $\mathbf{v}(k)$ are zero-mean white noise processes, with covariances $\mathbf{Q}(k)$ and $\mathbf{R}(k)$, respectively. Adding and subtracting $\mathbf{\Gamma}(k+1,k)\mathbf{m}_{\mathbf{w}_1}(k)$ in state equation (23-1) and $\mathbf{m}_{\mathbf{v}_1}(k+1)$ in measurement equation (23-2), these equations can be expressed as

$$\mathbf{x}(k+1) = \mathbf{\Phi}(k+1,k)\mathbf{x}(k) + \mathbf{\Gamma}(k+1,k)\mathbf{w}(k) + \mathbf{u}_1(k) \tag{23-7}$$

and

$$\mathbf{z}_1(k+1) = \mathbf{H}(k+1)\mathbf{x}(k+1) + \mathbf{v}(k+1) \tag{23-8}$$

where

$$\mathbf{u}_1(k) = \mathbf{\Psi}(k+1,k)\mathbf{u}(k) + \mathbf{\Gamma}(k+1,k)\mathbf{m}_{\mathbf{w}_1}(k) \tag{23-9}$$

and

$$\mathbf{z}_1(k+1) = \mathbf{z}(k+1) - \mathbf{G}(k+1)\mathbf{u}(k+1) - \mathbf{m}_{\mathbf{v}_1}(k+1) \tag{23-10}$$

Clearly, (23-7) and (23-8) is once again a basic state-variable model, one in which $\mathbf{u}_1(k)$ plays the role of $\mathbf{\Psi}(k+1,k)\mathbf{u}(k)$ and $\mathbf{z}_1(k+1)$ plays the role of $\mathbf{z}(k+1)$.

Theorem 23-1. *When biases are present in a state-variable model, then that model can always be reduced to a basic state-variable model [e.g., (23-7) to (23-10)]. All of our previous state estimators can be applied to this basic state-variable model by replacing* $\mathbf{z}(k)$ *by* $\mathbf{z}_1(k)$ *and* $\mathbf{\Psi}(k+1,k)\mathbf{u}(k)$ *by* $\mathbf{u}_1(k)$. □

CORRELATED NOISES

Here we assume that our basic state-variable model is given by (15-17) and (15-18), except that now $\mathbf{w}(k)$ and $\mathbf{v}(k)$ are correlated, i.e.,

$$\mathbf{E}\{\mathbf{w}(k)\mathbf{v}'(k)\} = \mathbf{S}(k) \neq \mathbf{0} \tag{23-11}$$

There are many approaches for treating correlated process and measurement noises, some leading to a recursive predictor, some to a recursive filter, and others to a filter in predictor-corrector form, as in the following:

Theorem 23-2. *When $\mathbf{w}(k)$ and $\mathbf{v}(k)$ are correlated, then a predictor-corrector form of the Kalman filter is*

$$\hat{\mathbf{x}}(k+1|k) = \mathbf{\Phi}(k+1,k)\hat{\mathbf{x}}(k|k) + \mathbf{\Psi}(k+1,k)\mathbf{u}(k) + \mathbf{\Gamma}(k+1,k)\mathbf{S}(k)[\mathbf{H}(k)\mathbf{P}(k|k-1)\mathbf{H}'(k) + \mathbf{R}(k)]^{-1}\tilde{\mathbf{z}}(k|k-1) \tag{23-12}$$

and

$$\hat{\mathbf{x}}(k+1|k+1) = \hat{\mathbf{x}}(k+1|k) + \mathbf{K}(k+1)\tilde{\mathbf{z}}(k+1|k) \tag{23-13}$$

where Kalman gain matrix, $\mathbf{K}(k+1)$, is given by (17-12), filtering-error covariance matrix, $\mathbf{P}(k+1|k+1)$, is given by (17-14), and, prediction-error covariance matrix, $\mathbf{P}(k+1|k)$, is given by

$$\mathbf{P}(k+1|k) = \mathbf{\Phi}_1(k+1,k)\mathbf{P}(k|k)\mathbf{\Phi}_1'(k+1,k) + \mathbf{Q}_1(k) \tag{23-14}$$

in which

$$\mathbf{\Phi}_1(k+1,k) = \mathbf{\Phi}(k+1,k) - \mathbf{\Gamma}(k+1,k)\mathbf{S}(k)\mathbf{R}^{-1}(k)\mathbf{H}(k) \tag{23-15}$$

and

$$\mathbf{Q}_1(k) = \mathbf{\Gamma}(k+1,k)\mathbf{Q}(k)\mathbf{\Gamma}'(k+1,k) - \mathbf{\Gamma}(k+1,k)\mathbf{S}(k)\mathbf{R}^{-1}(k)\mathbf{S}'(k)\mathbf{\Gamma}'(k+1,k) \tag{23-16}$$

Observe that, if $\mathbf{S}(k) = \mathbf{0}$, then (23-12) reduces to the more familiar predictor equation (16-4), and (23-14) reduces to the more familiar (17-13).

Proof. The derivation of correction equation (23-13) is exactly the same, when $\mathbf{w}(k)$ and $\mathbf{v}(k)$ are correlated, as it was when $\mathbf{w}(k)$ and $\mathbf{v}(k)$ were assumed uncorrelated. See the proof of part (a) of Theorem 17-1 for the details.

In order to derive predictor equation (23-12), we begin with the Fundamental Theorem of Estimation Theory, i.e ,

$$\hat{\mathbf{x}}(k+1|k) = \mathbf{E}\{\mathbf{x}(k+1)|\mathfrak{Z}(k)\} \tag{23-17}$$

Substitute state equation (15-17) into (23-17), to show that

$$\hat{\mathbf{x}}(k+1|k) = \mathbf{\Phi}(k+1,k)\hat{\mathbf{x}}(k|k) + \mathbf{\Psi}(k+1,k)\mathbf{u}(k) + \mathbf{\Gamma}(k+1,k)\mathbf{E}\{\mathbf{w}(k)|\mathfrak{Z}(k)\} \tag{23-18}$$

Next, we develop an expression for $\mathbf{E}\{\mathbf{w}(k)|\mathfrak{Z}(k)\}$.

Let $\mathfrak{Z}(k) = \text{col}\,(\mathfrak{Z}(k-1), \mathbf{z}(k))$; then,

$$\begin{aligned}\mathbf{E}\{\mathbf{w}(k)|\mathfrak{Z}(k)\} &= \mathbf{E}\{\mathbf{w}(k)|\mathfrak{Z}(k-1),\mathbf{z}(k)\} \\ &= \mathbf{E}\{\mathbf{w}(k)|\mathfrak{Z}(k-1),\tilde{\mathbf{z}}(k|k-1)\} \\ &= \mathbf{E}\{\mathbf{w}(k)|\mathfrak{Z}(k-1)\} + \mathbf{E}\{\mathbf{w}(k)|\tilde{\mathbf{z}}(k|k-1)\} \\ &\quad - \mathbf{E}\{\mathbf{w}(k)\} \\ &= \mathbf{E}\{\mathbf{w}(k)|\tilde{\mathbf{z}}(k|k-1)\}\end{aligned} \tag{23-19}$$

In deriving (23-19) we used the facts that $\mathbf{w}(k)$ is zero mean, and $\mathbf{w}(k)$ and $\mathfrak{Z}(k-1)$ are statistically independent. Because $\mathbf{w}(k)$ and $\tilde{\mathbf{z}}(k|k-1)$ are jointly Gaussian,

$$\mathbf{E}\{\mathbf{w}(k)|\tilde{\mathbf{z}}(k|k-1)\} = \mathbf{P}_{\mathbf{w}\tilde{\mathbf{z}}}(k,k|k-1)\mathbf{P}_{\tilde{\mathbf{z}}\tilde{\mathbf{z}}}^{-1}(k|k-1)\tilde{\mathbf{z}}(k|k-1) \tag{23-20}$$

where $\mathbf{P}_{\tilde{\mathbf{z}}\tilde{\mathbf{z}}}$ is given by (16-33), and

$$\begin{aligned}\mathbf{P}_{\mathbf{w}\tilde{\mathbf{z}}}(k,k|k-1) &= \mathbf{E}\{\mathbf{w}(k)\tilde{\mathbf{z}}'(k|k-1)\} \\ &= \mathbf{E}\{\mathbf{w}(k)[\mathbf{H}(k)\tilde{\mathbf{x}}(k|k-1) + \mathbf{v}(k)]'\} \\ &= \mathbf{S}(k)\end{aligned} \tag{23-21}$$

In deriving (23-21) we used the facts that $\tilde{\mathbf{x}}(k|k-1)$ and $\mathbf{w}(k)$ are statistically independent, and $\mathbf{w}(k)$ is zero mean. Substituting (23-21) and (16-33) into (23-20), we find that

$$\mathbf{E}\{\mathbf{w}(k)|\tilde{\mathbf{z}}(k|k-1)\} = \mathbf{S}(k)[\mathbf{H}(k)\mathbf{P}(k|k-1)\mathbf{H}'(k) + \mathbf{R}(k)]^{-1} \tag{23-22}$$

Substituting (23-22) into (23-19), and the resulting equation into (23-18) completes our derivation of the recursive predictor equation (23-12).

We leave the derivation of (23-14) as an exercise. It is straightforward but algebraically tedious. □

Recall that the recursive predictor plays the predominant role in smoothing; hence, we present

Corollary 23-1. *When* $\mathbf{w}(k)$ *and* $\mathbf{v}(k)$ *are correlated, then a recursive predictor for* $\mathbf{x}(k+1)$, *is*

$$\hat{\mathbf{x}}(k+1|k) = \mathbf{\Phi}(k+1,k)\hat{\mathbf{x}}(k|k-1) + \mathbf{\Psi}(k+1,k)\mathbf{u}(k) + \mathbf{L}(k)\tilde{\mathbf{z}}(k|k-1) \tag{23-23}$$

where

$$\mathbf{L}(k) = [\mathbf{\Phi}(k+1,k)\mathbf{P}(k|k-1)\mathbf{H}'(k) + \mathbf{\Gamma}(k+1,k)\mathbf{S}(k)][\mathbf{H}(k)\mathbf{P}(k|k-1)\mathbf{H}'(k) + \mathbf{R}(k)]^{-1} \tag{23-24}$$

and

$$\begin{aligned}\mathbf{P}(k+1|k) = [&\mathbf{\Phi}(k+1,k) - \mathbf{L}(k)\mathbf{H}(k)]\mathbf{P}(k|k-1)\\ &[\mathbf{\Phi}(k+1,k) - \mathbf{L}(k)\mathbf{H}(k)]'\\ &+\mathbf{\Gamma}(k+1,k)\mathbf{Q}(k)\mathbf{\Gamma}'(k+1,k)\\ &-\mathbf{\Gamma}(k+1,k)\mathbf{S}(k)\mathbf{L}'(k)\\ &-\mathbf{L}(k)\mathbf{S}(k)\mathbf{\Gamma}'(k+1,k)\\ &+\mathbf{L}(k)\mathbf{R}(k)\mathbf{L}'(k)\end{aligned} \tag{23-25}$$

Proof. These results follow directly from Theorem 23-2; or, they can be derived in an independent manner, as explained in Problem 23-2. □

Corollary 23-2. *When* $\mathbf{w}(k)$ *and* $\mathbf{v}(k)$ *are correlated, then a recursive filter for* $\mathbf{x}(k+1)$ *is*

$$\hat{\mathbf{x}}(k+1|k+1) = \mathbf{\Phi}_1(k+1,k)\hat{\mathbf{x}}(k|k) + \mathbf{\Psi}(k+1,k)\mathbf{u}(k) + \mathbf{D}(k)\mathbf{z}(k) + \mathbf{K}(k+1)\tilde{\mathbf{z}}(k+1|k) \tag{23-26}$$

where

$$\mathbf{D}(k) = \mathbf{\Gamma}(k+1,k)\mathbf{S}(k)\mathbf{R}^{-1}(k) \tag{23-27}$$

and all other quantities have been defined above.

Proof. Again, these results follow directly from Theorem 23-2; however, they can also be derived, in a much more elegant and independent manner, as described in Problem 23-3. □

COLORED NOISES

Quite often, some or all of the elements of either $\mathbf{v}(k)$ or $\mathbf{w}(k)$ or both are colored (i.e., have finite bandwidth). The following three-step procedure is used in these cases:

1. model each colored noise by a low-order difference equation that is excited by white Gaussian noise;
2. augment the states associated with the step 1 colored noise models to the original state-variable model;
3. apply the recursive filter or predictor to the augmented system.

We try to model colored noise processes by low-order Markov processes, i.e., low-order difference equations. Usually, first- or second-order models are quite adequate. Consider the following first-order model for colored noise process $\omega(k)$,

$$\omega(k+1) = \alpha\omega(k) + n(k) \tag{23-28}$$

In this model $n(k)$ is white noise with variance σ_n^2; thus, this model contains two parameters, α and σ_n^2, which must be determined from a priori knowledge about $\omega(k)$. We may know the amplitude spectrum of $\omega(k)$, correlation information about $\omega(k)$, steady-state variance of $\omega(k)$, etc. Two independent pieces of information are needed in order to uniquely identify α and σ_n^2.

Example 23-1

We are given the facts that scalar noise $w(k)$ is stationary with the properties $\mathbf{E}\{w(k)\} = 0$ and $\mathbf{E}\{w(i)w(j)\} = e^{-2|j-i|}$. A first-order Markov model for $w(k)$ can easily be obtained as

$$\xi(k+1) = e^{-2}\xi(k) + \sqrt{1-e^{-4}}\,n(k) \tag{23-29}$$

$$w(k) = \xi(k) \tag{23-30}$$

where $\mathbf{E}\{\xi(0)\} = 0$, $\mathbf{E}\{\xi^2(0)\} = 1$, $\mathbf{E}\{n(k)\} = 0$ and $\mathbf{E}\{n(i)n(j)\} = \delta_{ij}$. □

Example 23-2

Here we illustrate the state augmentation procedure for the first-order system

$$\chi(k+1) = a_1\chi(k) + \omega(k) \tag{23-31}$$

$$z(k+1) = h\chi(k+1) + v(k+1) \tag{23-32}$$

where $\omega(k)$ is a first-order Markov process, i.e.,

$$\omega(k+1) = a_2\omega(k) + n(k) \tag{23-33}$$

and $v(k)$ and $n(k)$ are white noise processes. We "augment" (23-33) to (23-31), as follows. Let

$$\mathbf{x}(k) = \text{col}\,(\chi(k), \omega(k)) \tag{23-34}$$

then (23-31) and (23-33) can be combined, to give

$$\underbrace{\begin{pmatrix} \chi(k+1) \\ \omega(k+1) \end{pmatrix}}_{\mathbf{x}(k+1)} = \underbrace{\begin{pmatrix} a_1 & 1 \\ 0 & a_2 \end{pmatrix}}_{\mathbf{\Phi}} \underbrace{\begin{pmatrix} \chi(k) \\ \omega(k) \end{pmatrix}}_{\mathbf{x}(k)} + \underbrace{\begin{pmatrix} 0 \\ 1 \end{pmatrix}}_{\boldsymbol{\gamma}} n(k) \tag{23-35}$$

Equation (23-25) is our *augmented state equation*. Observe that it is once again excited by a white noise process, just as our basic state equation (15-17) is.

In order to complete the description of the augmented state-variable model, we must express measurement $z(k+1)$ in terms of the augmented state vector, $\mathbf{x}(k+1)$, i.e.,

$$z(k+1) = \underbrace{(h \quad 0)}_{\mathbf{H}} \underbrace{\begin{pmatrix} \chi(k+1) \\ \omega(k+1) \end{pmatrix}}_{\mathbf{x}(k+1)} + v(k+1) \tag{23-36}$$

Equations (23-35) and (23-36) constitute the augmented state-variable model. We observe that, when the original process noise is colored and the measurement noise is white, the state augmentation procedure leads us once again to a basic (augmented) state-variable model, one that is of higher dimension than the original model because of the modeled colored process noise. Hence, in this case we can apply all of our state estimation algorithms to the augmented state-variable model. □

Example 23-3

Here we consider the situation where the process noise is white but the measurement noise is colored, again for a first-order system,

$$\mathcal{X}(k+1) = a_1\mathcal{X}(k) + \omega(k) \tag{23-37}$$

$$z(k+1) = h\mathcal{X}(k+1) + \nu(k+1) \tag{23-38}$$

As in the preceding example, we model $\nu(k)$ by the following first-order Markov process

$$\nu(k+1) = a_2\nu(k) + n(k) \tag{23-39}$$

where $n(k)$ is white noise. Augmenting (23-39) to (23-37) and reexpressing (23-38) in terms of the augmented state vector $\mathbf{x}(k)$, where

$$\mathbf{x}(k) = \text{col}\,(\mathcal{X}(k), \nu(k)) \tag{23-40}$$

we obtain the following augmented state-variable model,

$$\underbrace{\begin{pmatrix} \mathcal{X}(k+1) \\ \nu(k+1) \end{pmatrix}}_{\mathbf{x}(k+1)} = \underbrace{\begin{pmatrix} a_1 & 0 \\ 0 & a_2 \end{pmatrix}}_{\mathbf{\Phi}} \underbrace{\begin{pmatrix} \mathcal{X}(k) \\ \nu(k) \end{pmatrix}}_{\mathbf{x}(k)} + \underbrace{\begin{pmatrix} 1 & 0 \\ 0 & 1 \end{pmatrix}}_{\mathbf{\Gamma}} \underbrace{\begin{pmatrix} \omega(k) \\ n(k) \end{pmatrix}}_{\mathbf{w}(k)} \tag{23-41}$$

and

$$z(k+1) = \underbrace{(h \quad 1)}_{\mathbf{H}} \underbrace{\begin{pmatrix} \mathcal{X}(k+1) \\ \nu(k+1) \end{pmatrix}}_{\mathbf{x}(k+1)} \tag{23-42}$$

Observe that a vector process noise now excites the augmented state equation and that there is no measurement noise in the measurement equation. This second observation can lead to serious numerical problems in our state estimators, because it means that we must set $\mathbf{R} = \mathbf{0}$ in those estimators, and, when we do this, covariance matrices become and remain singular. □

Let us examine what happens to $\mathbf{P}(k+1|k+1)$ when covariance matrix $\mathbf{R}$ is set equal to zero. From (17-14) and (17-12) (in which we set $\mathbf{R} = \mathbf{0}$), we find that

$$\begin{aligned} \mathbf{P}(k+1|k+1) &= \mathbf{P}(k+1|k) - \mathbf{P}(k+1|k)\mathbf{H}'(k+1) \\ &\quad [\mathbf{H}(k+1)\mathbf{P}(k+1|k)\mathbf{H}'(k+1)]^{-1}\mathbf{H}(k+1)\mathbf{P}(k+1|k) \end{aligned} \tag{23-43}$$

Multiplying both sides of (23-43) on the right by $\mathbf{H}'(k+1)$, we find that

$$\mathbf{P}(k+1|k+1)\mathbf{H}'(k+1) = \mathbf{0} \tag{23-44}$$

Because $\mathbf{H}'(k+1)$ is a nonzero matrix, (23-44) implies that $\mathbf{P}(k+1|k+1)$ must be a singular matrix. We leave it to the reader to show that once $\mathbf{P}(k+1|k+1)$ becomes singular it remains singular for all other values of k.

PERFECT MEASUREMENTS: REDUCED-ORDER ESTIMATORS

We have just seen that when $\mathbf{R} = \mathbf{0}$ (or, in fact, even if some, but not all, measurements are perfect) numerical problems can occur in the Kalman filter. One way to circumvent these problems is ad hoc, and that is to use small values for the elements of covariance matrix $\mathbf{R}$, even through measurements are thought to be perfect. Doing this has a stabilizing effect on the numerics of the Kalman filter.

A second way to circumvent these problems is to recognize that a set of "perfect" measurements reduces the number of states that have to be estimated. Suppose, for example, that there are l perfect measurements and that state vector $\mathbf{x}(k)$ is $n \times 1$. Then, *we conjecture that we ought to be able to estimate* $\mathbf{x}(\mathrm{k})$ *by a Kalman filter whose dimension is no greater than* $\mathrm{n} - \ell$. Such an estimator will be referred to as a *reduced-order estimator*. The payoff for using a reduced-order estimator is fewer computations and less storage.

In order to illustrate an approach to designing a reduced-order estimator, we limit our discussions in this section to the following time-invariant and stationary basic state-variable model in which $\mathbf{u}(k) \triangleq \mathbf{0}$ and *all* measurements are perfect,

$$\mathbf{x}(k+1) = \mathbf{\Phi}\mathbf{x}(k) + \mathbf{\Gamma}\mathbf{w}(k) \tag{23-45}$$

$$\mathbf{y}(k+1) = \mathbf{H}\mathbf{x}(k+1) \tag{23-46}$$

In this model $\mathbf{y}$ is $l \times 1$. What makes the design of a reduced-order estimator challenging is the fact that the l perfect measurements are linearly related to the n states, i.e., $\mathbf{H}$ is rectangular.

To begin we introduce a *reduced-order state vector*, $\mathbf{p}(k)$, whose dimension is $(n - l) \times 1$; $\mathbf{p}(k)$ is assumed to be a linear transformation of $\mathbf{x}(k)$, i.e.,

$$\mathbf{p}(k) \triangleq \mathbf{C}\mathbf{x}(k) \tag{23-47}$$

Augmenting (23-47) to (23-46), we obtain

$$\begin{pmatrix} \mathbf{y}(k) \\ \mathbf{p}(k) \end{pmatrix} = \begin{pmatrix} \mathbf{H} \\ \hline \mathbf{C} \end{pmatrix} \mathbf{x}(k) \tag{23-48}$$

Design matrix $\mathbf{C}$ is chosen so that $\begin{pmatrix} \mathbf{H} \\ \hline \mathbf{C} \end{pmatrix}$ is invertible. Of course, many different choices of $\mathbf{C}$ are possible; thus, this first step of our reduced-order estimator design procedure is nonunique. Let

$$\begin{pmatrix} \mathbf{H} \\ \hline \mathbf{C} \end{pmatrix}^{-1} = (\mathbf{L}_1 | \mathbf{L}_2) \tag{23-49}$$

where $\mathbf{L}_1$ is $n \times l$ and $\mathbf{L}_2$ is $n \times (n - l)$; thus,

$$\mathbf{x}(k) = \mathbf{L}_1\mathbf{y}(k) + \mathbf{L}_2\mathbf{p}(k) \tag{23-50}$$

In order to obtain a filtered estimate of $\mathbf{x}(k)$, we operate on both sides of (23-50) with $\mathbf{E}\{\cdot|\mathcal{Y}(k)\}$, where

$$\mathcal{Y}(k) = \text{col}\,(\mathbf{y}(1), \mathbf{y}(2), \ldots, \mathbf{y}(k)) \tag{23-51}$$

Doing this, we find that

$$\boxed{\hat{\mathbf{x}}(k|k) = \mathbf{L}_1\mathbf{y}(k) + \mathbf{L}_2\hat{\mathbf{p}}(k|k)} \tag{23-52}$$

which is a *reduced-order estimator for* $\mathbf{x}(k)$. Of course, in order to evaluate $\hat{\mathbf{x}}(k|k)$ we must develop a reduced-order Kalman filter to estimate $\mathbf{p}(k)$. Knowing $\hat{\mathbf{p}}(k|k)$ and $\mathbf{y}(k)$, it is then a simple matter to compute $\hat{\mathbf{x}}(k|k)$, using (23-52).

In order to obtain $\hat{\mathbf{p}}(k|k)$, using our previously-derived Kalman filter algorithm, we first must establish a state-variable model for $\mathbf{p}(k)$. A state equation for $\mathbf{p}$ is easily obtained, as follows:

$$\begin{aligned}\mathbf{p}(k+1) &= \mathbf{Cx}(k+1) = \mathbf{C}[\mathbf{\Phi x}(k) + \mathbf{\Gamma w}(k)]\\ &= \mathbf{C\Phi}[\mathbf{L}_1\mathbf{y}(k) + \mathbf{L}_2\mathbf{p}(k)] + \mathbf{C\Gamma w}(k)\\ &= \mathbf{C\Phi L}_2\mathbf{p}(k) + \mathbf{C\Phi L}_1\mathbf{y}(k) + \mathbf{C\Gamma w}(k)\end{aligned} \tag{23-53}$$

Observe that this state equation is driven by white noise $\mathbf{w}(k)$ and the known *forcing function*, $\mathbf{y}(k)$.

A measurement equation is obtained from (23-46), as

$$\begin{aligned}\mathbf{y}(k+1) &= \mathbf{Hx}(k+1) = \mathbf{H}[\mathbf{\Phi x}(k) + \mathbf{\Gamma w}(k)]\\ &= \mathbf{H\Phi}[\mathbf{L}_1\mathbf{y}(k) + \mathbf{L}_2\mathbf{p}(k)] + \mathbf{H\Gamma w}(k)\\ &= \mathbf{H\Phi L}_2\mathbf{p}(k) + \mathbf{H\Phi L}_1\mathbf{y}(k) + \mathbf{H\Gamma w}(k)\end{aligned} \tag{23-54}$$

At time $k + 1$ we know $\mathbf{y}(k)$; hence, we can reexpress (23-54) as

$$\mathbf{y}_1(k+1) = \mathbf{H\Phi L}_2\mathbf{p}(k) + \mathbf{H\Gamma w}(k) \tag{23-55}$$

where

$$\mathbf{y}_1(k+1) \triangleq \mathbf{y}(k+1) - \mathbf{H\Phi L}_1\mathbf{y}(k) \tag{23-56}$$

Before proceeding any farther, we make some important observations about our state-variable model in (23-53) and (23-55). First, the new measurement $\mathbf{y}_1(k+1)$ represents a weighted difference between measurements $\mathbf{y}(k+1)$ and $\mathbf{y}(k)$. The technique for obtaining our reduced-order state-variable model is, therefore, sometimes referred to as a *measurement-differencing technique* (e.g., Bryson and Johansen, 1965). Because we have already used $\mathbf{y}(k)$ to reduce the dimension of $\mathbf{x}(k)$ from n to $n - l$, we cannot again use $\mathbf{y}(k)$ alone as the measurements in our reduced-order state-variable model. As we have just seen, we must use both $\mathbf{y}(k)$ and $\mathbf{y}(k+1)$.

Second, measurement equation (23-55) appears to be a combination of signal and noise. Unless $\mathbf{H\Gamma} = \mathbf{0}$, the term $\mathbf{H\Gamma w}(k)$ will act as the measurement noise in our reduced-order state-variable model. Its covariance matrix is $\mathbf{H\Gamma Q\Gamma' H'}$. Unfortunately, it is possible for $\mathbf{H\Gamma}$ to equal the zero matrix. From linear system theory, we know that $\mathbf{H\Gamma}$ is the matrix of first Markov parameters for our original system in (23-45) and (23-46), and $\mathbf{H\Gamma}$ may equal zero. If this occurs, then we must repeat all of the above until we obtain a reduced-order state vector whose measurement equation is excited by white noise. We see, therefore, that depending upon system dynamics, it is possible to obtain a reduced-order estimator of $\mathbf{x}(k)$ that uses a reduced-order Kalman filter of dimension less than $n - l$.

Third, the noises, which appear in state equation (23-53) and measurement equation (23-55) are the same, namely $\mathbf{w}(k)$; hence, the reduced-order state-variable model involves the correlated noise case that we described before in this chapter in the section entitled Correlated Noises.

Finally, and most important, measurement equation (23-55) is nonstandard, in that it expresses $\mathbf{y}_1$ at $k + 1$ in terms of $\mathbf{p}$ at k rather than $\mathbf{p}$ at $k + 1$. Recall that the measurement equation in our basic state-variable model is $\mathbf{z}(k + 1) = \mathbf{Hx}(k + 1) + \mathbf{v}(k + 1)$. We cannot immediately apply our Kalman filter equations to (23-53) and (23-55) until we express (23-55) in the standard way.

To proceed, we let

$$\zeta(k) \triangleq \mathbf{y}_1(k + 1) \tag{23-57}$$

so that

$$\zeta(k) = \mathbf{H\Phi L}_2\mathbf{p}(k) + \mathbf{H\Gamma w}(k) \tag{23-58}$$

Measurement equation (23-58) is now in the standard form; however, because $\zeta(k)$ equals a future value of $\mathbf{y}_1$, namely $\mathbf{y}_1(k + 1)$, we must be very careful in applying our estimator formulas to our reduced-order model (23-53) and (23-58).

In order to see this more clearly, we define the following two data sets,

$$\mathfrak{D}_{\mathbf{y}_1}(k + 1) = \{\mathbf{y}_1(1), \mathbf{y}_1(2), \ldots, \mathbf{y}_1(k + 1), \ldots\} \tag{23-59}$$

and

$$\mathfrak{D}_{\zeta}(k) = \{\zeta(0), \zeta(1), \ldots, \zeta(k), \ldots\} \tag{23-60}$$

Obviously,

$$\mathfrak{D}_{\mathbf{y}_1}(k + 1) = \mathfrak{D}_{\zeta}(k) \tag{23-61}$$

Letting

$$\hat{\mathbf{p}}_{\mathbf{y}_1}(k + 1|k + 1) = \mathbf{E}\{\mathbf{p}(k + 1)|\mathfrak{D}_{\mathbf{y}_1}(k + 1)\} \tag{23-62}$$

and

$$\hat{\mathbf{p}}_{\zeta}(k + 1|k) = \mathbf{E}\{\mathbf{p}(k + 1)|\mathfrak{D}_{\zeta}(k)\} \tag{23-63}$$

we see that

$$\hat{\mathbf{p}}_{\mathbf{y}_1}(k+1|k+1) = \hat{\mathbf{p}}_{\zeta}(k+1|k) \tag{23-64}$$

Equation (23-64) tells us to obtain a recursive filter for our reduced-order model, that is in terms of data set $\mathfrak{D}_{\mathbf{y}_1}(k+1)$, we must first obtain a recursive predictor for that model, which is in terms of data set $\mathfrak{D}_{\zeta}(k)$. Then, wherever $\zeta(k)$ appears in the recursive predictor, it can be replaced by $\mathbf{y}_1(k+1)$.

Using Corollary 23-1, applied to the reduced-order model in (23-53) and (23-58), we find that

$$\hat{\mathbf{p}}_{\zeta}(k+1|k) = \mathbf{C\Phi L}_2\hat{\mathbf{p}}_{\zeta}(k|k-1) + \mathbf{C\Phi L}_1\mathbf{y}(k) + \mathbf{L}(k)[\zeta(k) - \mathbf{H\Phi L}_2\hat{\mathbf{p}}_{\zeta}(k|k-1)] \tag{23-65}$$

thus,

$$\hat{\mathbf{p}}_{\mathbf{y}_1}(k+1|k+1) = \mathbf{C\Phi L}_2\hat{\mathbf{p}}_{\mathbf{y}_1}(k|k) + \mathbf{C\Phi L}_1\mathbf{y}(k) + \mathbf{L}(k)[\mathbf{y}_1(k+1) - \mathbf{H\Phi L}_2\hat{\mathbf{p}}_{\mathbf{y}_1}(k|k)] \tag{23-66}$$

Equation (23-66) is our reduced-order Kalman filter. It provides filtered estimates of $\mathbf{p}(k+1)$ and is only of dimension $(n-l) \times 1$. Of course, when $\mathbf{L}(k)$ and $\mathbf{P}_{\mathbf{y}_1}(k+1|k+1)$ are computed using (23-13) and (23-14), respectively, we must make the following substitutions: $\mathbf{\Phi}(k+1,k) \rightarrow \mathbf{C\Phi L}_2$, $\mathbf{H}(k) \rightarrow \mathbf{H\Phi L}_2$, $\mathbf{\Gamma}(k+1,k) \rightarrow \mathbf{C\Gamma}$, $\mathbf{Q}(k) \rightarrow \mathbf{Q}$, $\mathbf{S}(k) \rightarrow \mathbf{Q\Gamma}'\mathbf{H}'$, and $\mathbf{R}(k) \rightarrow \mathbf{H\Gamma Q\Gamma}'\mathbf{H}'$.

FINAL REMARK

In order to see the forest from the trees, we have considered each of our special cases separately. In actual practice, some or all of them may occur simultaneously. The exercises at the end of this lesson will permit the reader to gain experience with such cases.

PROBLEMS

23-1. Derive the prediction-error covariance equation (23-14).

23-2. Derive the recursive predictor, given in (23-23), by expressing $\hat{\mathbf{x}}(k+1|k)$ as $\mathbf{E}\{\mathbf{x}(k+1)|\mathfrak{Z}(k)\} = \mathbf{E}\{\mathbf{x}(k+1)|\mathfrak{Z}(k-1),\tilde{\mathbf{z}}(k|k-1)\}$.

23-3. Here we derive the recursive filter, given in (23-26), by first adding a convenient form of zero to state equation (15-17), in order to decorrelate the process noise in this modified basic state-variable model from the measurement noise $\mathbf{v}(k)$. Add $\mathbf{D}(k)[\mathbf{z}(k) - \mathbf{H}(k)\mathbf{x}(k) - \mathbf{v}(k)]$ to (15-17). The process noise, $\mathbf{w}_1(k)$, in the modified basic state-variable model, is equal to $\mathbf{\Gamma}(k+1,k)\mathbf{w}(k) - \mathbf{D}(k)\mathbf{v}(k)$. Choose "decorrelation" matrix $\mathbf{D}(k)$ so that $\mathbf{E}\{\mathbf{w}_1(k)\mathbf{v}'(k)\} = \mathbf{0}$. Then complete the derivation of (23-26). Observe that (23-14) can be obtained by inspection, via this derivation.

23-4. In solving Problem 23-3, one arrives at the following predictor equation,

$$\hat{\mathbf{x}}(k+1|k) = \mathbf{\Phi}_1(k+1,k)\hat{\mathbf{x}}(k|k) + \mathbf{\Psi}(k+1,k)\mathbf{u}(k) + \mathbf{D}(k)\mathbf{z}(k)$$

Beginning with this predictor equation, and corrector equation (23-13), derive the recursive predictor given in (23-23).

23-5. Show that once $\mathbf{P}(k+1|k+1)$ becomes singular it remains singular for all other values of k.

23-6. Assume that $\mathbf{R} = \mathbf{0}$, $\mathbf{H\Gamma} = \mathbf{0}$, and $\mathbf{H\Phi\Gamma} \neq \mathbf{0}$. Obtain the reduced-order estimator and its associated reduced-order Kalman filter for this situation. Contrast this situation with the case given in the text, for which $\mathbf{H\Gamma} \neq \mathbf{0}$.

23-7. Develop a reduced-order estimator and its associated reduced-order Kalman filter for the case when l measurements are perfect and $m - l$ measurements are noisy.

23-8. Consider the first-order system $x(k+1) = \frac{3}{4}x(k) + w_1(k)$ and $z(k+1) = x(k+1) + v(k+1)$, where $\mathbf{E}\{w_1(k)\} = 3$, $\mathbf{E}\{v(k)\} = 0$, $w_1(k)$ and $v(k)$ are both white and Gaussian, $\mathbf{E}\{w_1^2(k)\} = 10$, $\mathbf{E}\{v^2(k)\} = 2$, and, $w_1(k)$ and $v(k)$ are correlated, i.e., $\mathbf{E}\{w_1(k)v(k)\} = 1$.

(a) Obtain the steady-state recursive Kalman filter for this system.

(b) What is the steady-state filter error variance, and how does it compare with the steady-state predictor error variance?

23-9. Consider the first-order system $x(k+1) = \frac{1}{2}x(k) + w(k)$ and $z(k+1) = x(k+1) + v(k+1)$, where $w(k)$ is a first-order Markov process and $v(k)$ is Gaussian white noise with $\mathbf{E}\{v(k)\} = 4$ and $r = 1$.

(a) Let the model for $w(k)$ be $w(k+1) = \alpha w(k) + u(k)$, where $u(k)$ is a zero-mean white Gaussian noise sequence for which $\mathbf{E}\{u^2(k)\} = \sigma_u^2$. Additionally, $\mathbf{E}\{w(k)\} = 0$. What value must α have if $\mathbf{E}\{w^2(k)\} = W$ for all k?

(b) Suppose $W^2 = 2$ and $\sigma_u^2 = 1$. What are the Kalman filter equations for estimation of $x(k)$ and $w(k)$?

23-10. Consider the first-order system $x(k+1) = -\frac{1}{8}x(k) + w(k)$ and $z(k+1) = x(k+1) + v(k+1)$, where $w(k)$ is white and Gaussian $[w(k) \sim N(w(k); 0, 1)]$ and $v(k)$ is also a noise process. The model for $v(k)$ is summarized in Figure P23-10.

(a) Verify that a correct state-variable model for $v(k)$ is,

$$x_1(k+1) = -\tfrac{1}{2}x_1(k) + \tfrac{3}{2}n(k)$$
$$v(k) = x_1(k) + n(k)$$

(b) Show that $v(k)$ is also a white process.

(c) Noise $n(k)$ is white and Gaussian $[n(k) \sim N(n(k); 0, 1/4)]$. What are the Kalman filter equations for finding $\hat{x}(k+1|k+1)$?

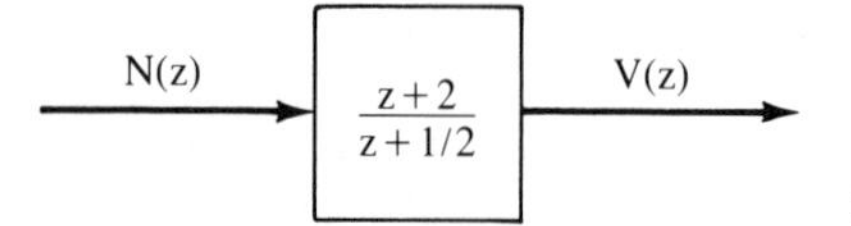

Figure P23-10

23-11. Obtain the equations from which we can find $\hat{x}_1(k+1|k+1)$, $\hat{x}_2(k+1|k+1)$ and $\hat{\nu}(k+1|k+1)$ for the following system:

$$x_1(k+1) = -x_1(k) + x_2(k)$$
$$x_2(k+1) = x_2(k) + w(k)$$
$$z(k+1) = x_1(k+1) + \nu(k+1)$$

where $\nu(k)$ is a colored noise process, i.e.,

$$\nu(k+1) = -\tfrac{1}{2}\nu(k) + n(k)$$

Assume that $w(k)$ and $n(k)$ are white processes and are mutually uncorrelated, and, $\sigma_w^2(k) = 4$ and $\sigma_n^2(k) = 2$. Include a block diagram of the interconnected system and reduced-order KF.

23-12. Consider the system $\mathbf{x}(k+1) = \mathbf{\Phi}\mathbf{x}(k) + \boldsymbol{\gamma}\mu(k)$ and $z(k+1) = \mathbf{h}'\mathbf{x}(k+1) + v(k+1)$, where $\mu(k)$ is a colored noise sequence and $v(k)$ is zero-mean white noise. What are the formulas for computing $\hat{\mu}(k|k+1)$? Filter $\hat{\mu}(k|k+1)$ is a deconvolution filter.

23-13. Consider the scalar moving average (MA) time-series model

$$z(k) = r(k) + r(k-1)$$

where $r(k)$ is a unit variance, white Gaussian sequence. Show that the optimal one-step predictor for this model is [assume $P(0|0) = 1$]

$$\hat{z}(k+1|k) = \frac{k}{k+1}[z(k) - \hat{z}(k|k-1)]$$

(*Hint*: Express the MA model in state-space form.)

23-14. Consider the basic state variable model for the stationary time-invariant case. assume also that $\mathbf{w}(k)$ and $\mathbf{v}(k)$ are *correlated*, i.e., $\mathbf{E}\{\mathbf{w}(k)\mathbf{v}'(k)\} = \mathbf{S}$.

(a) Show, from first principles, that the single-stage smoother of $\mathbf{x}(k)$, i.e., $\hat{\mathbf{x}}(k|k+1)$ is given by

$$\hat{\mathbf{x}}(k|k+1) = \hat{\mathbf{x}}(k|k) + M(k|k+1)\tilde{\mathbf{z}}(k+1|k)$$

where $\mathbf{M}(k|k+1)$ is an appropriate smoother gain matrix.

(b) Derive a closed form solution for $\mathbf{M}(k|k+1)$ as a function of the correlation matrix $\mathbf{S}$ and other quantities of the basic state-variable model.

Lesson 24

Linearization and Discretization of Nonlinear Systems

INTRODUCTION

Many real-world systems are continuous-time in nature and quite a few are also nonlinear. For example, the state equations associated with the motion of a satellite of mass m about a spherical planet of mass M, in a planet-centered coordinate system, are nonlinear, because the planet's force field obeys an inverse square law. Figure 24-1 depicts a situation where the measurement equation is nonlinear. The measurement is angle ϕ_i and is expressed in a rectangular coordinate system, i.e., $\phi_i = \tan^{-1}[y/(x - l_i)]$. Sometimes the state equation may be nonlinear and the measurement equation linear, or vice-versa, or they may both be nonlinear. Occasionally, the coordinate system in which one chooses to work causes the two former situations. For example, equations of motion in a polar coordinate system are nonlinear, whereas the measurement equations are linear. In a polar coordinate system, where ϕ is a state-variable, the measurement equation for the situation depicted in Figure 24-1 is $z_i = \phi_i$, which is linear. In a rectangular coordinate system, on the other hand, equations of motion are linear, but the measurement equations are nonlinear.

Finally, we may begin with a linear system that contains some unknown parameters. When these parameters are modeled as first-order Markov processes, and these models are augmented to the original system, the augmented model is nonlinear, because the parameters that appeared in the original "linear" model are treated as states. We shall describe this situation in much more detail in Lesson 25.

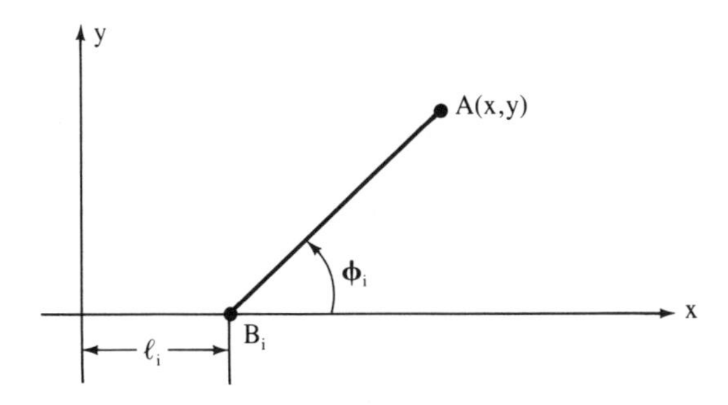

Figure 24-1 Coordinate system for an angular measurement between two objects *A* and *B*.

The purpose of this lesson is to explain how to linearize and discretize a nonlinear differential equation model. We do this so that we will be able to apply our digital estimators to the resulting discrete-time system.

A DYNAMICAL MODEL

The starting point for this lesson is the nonlinear state-variable model

$$\dot{\mathbf{x}}(t) = \mathbf{f}[\mathbf{x}(t),\mathbf{u}(t),t] + \mathbf{G}(t)\mathbf{w}(t) \tag{24-1}$$

$$\mathbf{z}(t) = \mathbf{h}[\mathbf{x}(t),\mathbf{u}(t),t] + \mathbf{v}(t) \tag{24-2}$$

We shall assume that measurements are only available at specific values of time, namely at $t = t_i$, $i = 1, 2, \ldots$; thus, our measurement equation will be treated as a discrete-time equation, whereas our state equation will be treated as a continuous-time equation. State vector $\mathbf{x}(t)$ is $n \times 1$; $\mathbf{u}(t)$ is an $l \times 1$ vector of known inputs; measurement vector $\mathbf{z}(t)$ is $m \times 1$; $\dot{\mathbf{x}}(t)$ is short for $d\mathbf{x}(t)/dt$; nonlinear functions $\mathbf{f}$ and $\mathbf{h}$ may depend both implicitly and explicitly on t, and we assume that both $\mathbf{f}$ and $\mathbf{h}$ are continuous and continuously differentiable with respect to all the elements of $\mathbf{x}$ and $\mathbf{u}$; $\mathbf{w}(t)$ is a continuous-time white noise process, i.e., $\mathbf{E}\{\mathbf{w}(t)\} = \mathbf{0}$ and

$$\mathbf{E}\{\mathbf{w}(t)\mathbf{w}'(\tau)\} = \mathbf{Q}(t)\delta(t - \tau); \tag{24-3}$$

$\mathbf{v}(t_i)$ is a discrete-time white noise process, i.e., $\mathbf{E}\{\mathbf{v}(t_i)\} = \mathbf{0}$ for $t = t_i$, $i = 1, 2, \ldots$, and

$$\mathbf{E}\{\mathbf{v}(t_i)\mathbf{v}'(t_j)\} = \mathbf{R}(t_i)\delta_{ij}; \tag{24-4}$$

and, $\mathbf{w}(t)$ and $\mathbf{v}(t_i)$ are mutually uncorrelated at all $t = t_i$, i.e.,

$$\mathbf{E}\{\mathbf{w}(t)\mathbf{v}'(t_i)\} = \mathbf{0} \qquad \text{for } t = t_i \qquad i = 1, 2, \ldots \tag{24-5}$$

Example 24-1

Here we expand upon the previously mentioned satellite-planet example. Our example is taken from Meditch (1969, pp. 60–61), who states . . . "Assuming that the planet's force field obeys an inverse square law, and that the only other forces present are the satellite's two thrust forces $u_r(t)$ and $u_\theta(t)$ (see Figure 24-2), and that the satellite's initial position and velocity vectors lie in the plane, we know from elementary particle mechanics that the satellite's motion is confined to the plane and is governed by the two equations

$$\ddot{r} = r\dot{\theta}^2 - \frac{\gamma}{r^2} + \frac{1}{m}u_r(t) \tag{24-6}$$

and

$$\ddot{\theta} = -\frac{2\dot{r}\dot{\theta}}{r} + \frac{1}{m}u_\theta(t) \tag{24-7}$$

where $\gamma = GM$ and G is the universal gravitational constant.

"Defining $x_1 = r$, $x_2 = \dot{r}$, $x_3 = \theta$, $x_4 = \dot{\theta}$, $u_1 = u_r$, and $u_2 = u_\theta$, we have

$$\left.\begin{aligned} \dot{x}_1 &= x_2 \\ \dot{x}_2 &= x_1 x_4^2 - \frac{\gamma}{x_1^2} + \frac{1}{m}u_1(t) \\ \dot{x}_3 &= x_4 \\ \dot{x}_4 &= -\frac{2x_2x_4}{x_1} + \frac{1}{m}u_2(t) \end{aligned}\right\} \tag{24-8}$$

which is of the form in (24-1). . . . Assuming . . . that the measurement made on the satellite during its motion is simply its distance from the surface of the planet, we have the scalar measurement equation

$$z(t) = r(t) - r_0 + v(t) = x_1(t) - r_0 + v(t) \tag{24-9}$$

where r_0 is the planet's radius."

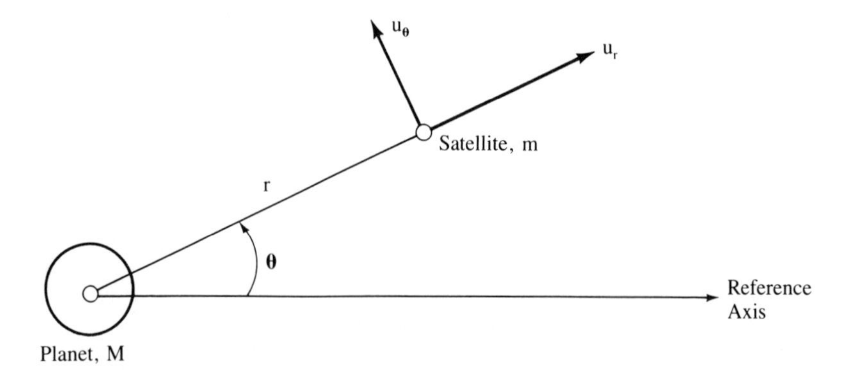

Figure 24-2 Schematic for satellite-planet system (Copyright 1969, McGraw-Hill).

Comparing (24-8) and (24-1), and (24-9) and (24-2), we conclude that

$$\mathbf{f}[\mathbf{x}(t),\mathbf{u}(t),t] = \text{col}\left[x_2, x_1x_4^2 - \frac{\gamma}{x_1^2} + \frac{1}{m}u_1, x_4, -\frac{2x_2x_4}{x_1} + \frac{1}{m}u_2\right] \tag{24-10}$$

and

$$\mathbf{h}[\mathbf{x}(t),\mathbf{u}(t),t] = x_1 - r_0 \tag{24-11}$$

Observe that, in this example, only the state equation is nonlinear. □

LINEAR PERTURBATION EQUATIONS

In this section we shall linearize our nonlinear dynamical model in (24-1) and (24-2) about nominal values of $\mathbf{x}(t)$ and $\mathbf{u}(t)$, $\mathbf{x}^*(t)$ and $\mathbf{u}^*(t)$, respectively. If we are given a nominal input, $\mathbf{u}^*(t)$, then $\mathbf{x}^*(t)$ satisfies the following nonlinear differential equation,

$$\dot{\mathbf{x}}^*(t) = \mathbf{f}[\mathbf{x}^*(t),\mathbf{u}^*(t),t] \tag{24-12}$$

and associated with $\mathbf{x}^*(t)$ and $\mathbf{u}^*(t)$ is the following nominal measurement, $\mathbf{z}^*(t)$, where

$$\mathbf{z}^*(t) = \mathbf{h}[\mathbf{x}^*(t),\mathbf{u}^*(t)] \qquad t = t_i \qquad i = 1, 2, \ldots \tag{24-13}$$

Throughout this lesson, we shall assume that $\mathbf{x}^*(t)$ exists. We discuss two methods for choosing $\mathbf{x}^*(t)$ in Lesson 25. Obviously, one is just to solve (24-12) for $\mathbf{x}^*(t)$.

Note that $\mathbf{x}^*(t)$ must provide a good approximation to the actual behavior of the system. The approximation is considered good if the difference between the nominal and actual solutions can be described by a system of linear differential equations, called *linear perturbation equations*. We derive these equations next.

Let

$$\delta\mathbf{x}(t) = \mathbf{x}(t) - \mathbf{x}^*(t) \tag{24-14}$$

and

$$\delta\mathbf{u}(t) = \mathbf{u}(t) - \mathbf{u}^*(t) \tag{24-15}$$

then,

$$\frac{d}{dt}\delta\mathbf{x}(t) = \delta\dot{\mathbf{x}}(t) = \dot{\mathbf{x}}(t) - \dot{\mathbf{x}}^*(t) = \mathbf{f}[\mathbf{x}(t),\mathbf{u}(t),t] + \mathbf{G}(t)\mathbf{w}(t) - \mathbf{f}[\mathbf{x}^*(t),\mathbf{u}^*(t),t] \tag{24-16}$$

Fact 1. *When* $\mathbf{f}[\mathbf{x}(t),\mathbf{u}(t),t]$ *is expanded in a Taylor series about* $\mathbf{x}^*(t)$ *and* $\mathbf{u}^*(t)$, *we obtain*

$$\mathbf{f}[\mathbf{x}(t),\mathbf{u}(t),t] = \mathbf{f}[\mathbf{x}^*(t),\mathbf{u}^*(t),t] + \mathbf{F}_\mathbf{x}[\mathbf{x}^*(t),\mathbf{u}^*(t),t]\delta\mathbf{x}(t) + \mathbf{F}_\mathbf{u}[\mathbf{x}^*(t),\mathbf{u}^*(t),t]\delta\mathbf{u}(t) + \text{higher-order terms} \tag{24-17}$$

where $\mathbf{F}_\mathbf{x}$ *and* $\mathbf{F}_\mathbf{u}$ *are* $n \times n$ *and* $n \times \ell$ *Jacobian matrices, i.e.,*

$$\mathbf{F}_\mathbf{x}[\mathbf{x}^*(t),\mathbf{u}^*(t),t] = \begin{pmatrix} \partial f_1/\partial x_1^* & \cdots & \partial f_1/\partial x_n^* \\ \vdots & \ddots & \vdots \\ \partial f_n/\partial x_1^* & \cdots & \partial f_n/\partial x_n^* \end{pmatrix} \tag{24-18}$$

and

$$\mathbf{F}_\mathbf{u}[\mathbf{x}^*(t),\mathbf{u}^*(t),t] = \begin{pmatrix} \partial f_1/\partial u_1^* & \cdots & \partial f_1/\partial u_l^* \\ \vdots & \ddots & \vdots \\ \partial f_n/\partial u_1^* & \cdots & \partial f_n/\partial u_l^* \end{pmatrix} \tag{24-19}$$

In these expressions $\partial f_i/\partial x_j^*$ *and* $\partial f_i/\partial u_j^*$ *are short for*

$$\frac{\partial f_i}{\partial x_j^*} = \left.\frac{\partial f_i[\mathbf{x}(t),\mathbf{u}(t),t]}{\partial x_j(t)}\right|_{\mathbf{x}(t) = \mathbf{x}^*(t),\mathbf{u}(t) = \mathbf{u}^*(t)} \tag{24-20}$$

and

$$\frac{\partial f_i}{\partial u_j^*} = \left.\frac{\partial f_i[\mathbf{x}(t),\mathbf{u}(t),t]}{\partial u_j(t)}\right|_{\mathbf{x}(t) = \mathbf{x}^*(t),\mathbf{u}(t) = \mathbf{u}^*(t)} \tag{24-21}$$

Proof. The Taylor series expansion of the ith component of $\mathbf{f}[\mathbf{x}(t),\mathbf{u}(t),t]$ is

$$\begin{aligned} f_i[\mathbf{x}(t),\mathbf{u}(t),t] = f_i[\mathbf{x}^*(t),\mathbf{u}^*(t),t] &+ \frac{\partial f_i}{\partial x_1^*}[x_1(t) - x_1^*(t)] + \cdots \\ &+ \frac{\partial f_i}{\partial x_n^*}[x_n(t) - x_n^*(t)] + \frac{\partial f_i}{\partial u_1^*}[u_1(t) - u_1^*(t)] + \cdots \\ &+ \frac{\partial f_i}{\partial u_l^*}[u_l(t) - u_l^*(t)] + \text{higher-order terms} \end{aligned} \tag{24-22}$$

where $i = 1, 2, \ldots, n$. Collecting these n equations together in vector-matrix format, we obtain (24-17), in which $\mathbf{F}_\mathbf{x}$ and $\mathbf{F}_\mathbf{u}$ are defined in (24-18) and (24-19), respectively. □

Substituting (24-17) into (24-16) and neglecting the "higher-order terms," we obtain the following *perturbation state equation*

$$\boxed{\delta\dot{\mathbf{x}}(t) = \mathbf{F}_\mathbf{x}[\mathbf{x}^*(t),\mathbf{u}^*(t),t]\delta\mathbf{x}(t) + \mathbf{F}_\mathbf{u}[\mathbf{x}^*(t),\mathbf{u}^*(t),t]\delta\mathbf{u}(t) + \mathbf{G}(t)\mathbf{w}(t)} \tag{24-23}$$

Observe that, even if our original nonlinear differential equation is not an explicit function of time {i.e., $\mathbf{f}[\mathbf{x}(t),\mathbf{u}(t),t] = \mathbf{f}[\mathbf{x}(t),\mathbf{u}(t)]$}, our perturbation state equation is always time-varying because Jacobian matrices $\mathbf{F_x}$ and $\mathbf{F_u}$ vary with time, because $\mathbf{x}^*$ and $\mathbf{u}^*$ vary with time.

Next, let

$$\delta\mathbf{z}(t) = \mathbf{z}(t) - \mathbf{z}^*(t) \tag{24-24}$$

Fact 2. *When* $\mathbf{h}[\mathbf{x}(\mathrm{t}),\mathbf{u}(\mathrm{t}),\mathrm{t}]$ *is expanded in a Taylor series about* $\mathbf{x}^*(\mathrm{t})$ *and* $\mathbf{u}^*(\mathrm{t})$, *we obtain*

$$\begin{aligned}\mathbf{h}[\mathbf{x}(t),\mathbf{u}(t),t] &= \mathbf{h}[\mathbf{x}^*(t),\mathbf{u}^*(t),t] + \mathbf{H_x}[\mathbf{x}^*(t),\mathbf{u}^*(t),t]\delta\mathbf{x}(t) \\ &\quad + \mathbf{H_u}[\mathbf{x}^*(t),\mathbf{u}^*(t),t]\delta\mathbf{u}(t) + \text{higher-order terms}\end{aligned} \tag{24-25}$$

where $\mathbf{H_x}$ *and* $\mathbf{H_u}$ *are* $\mathrm{m} \times \mathrm{n}$ *and* $\mathrm{m} \times \ell$ *Jacobian matrices, i.e.,*

$$\mathbf{H_x}[\mathbf{x}^*(t),\mathbf{u}^*(t),t] = \begin{pmatrix} \partial h_1/\partial x_1^* & \cdots & \partial h_1/\partial x_n^* \\ \vdots & \ddots & \vdots \\ \partial h_m/\partial x_1^* & \cdots & \partial h_m/\partial x_n^* \end{pmatrix} \tag{24-26}$$

and

$$\mathbf{H_u}[\mathbf{x}^*(t),\mathbf{u}^*(t),t] = \begin{pmatrix} \partial h_1/\partial u_1^* & \cdots & \partial h_1/\partial u_l^* \\ \vdots & \ddots & \vdots \\ \partial h_m/\partial u_1^* & \cdots & \partial h_m/\partial u_l^* \end{pmatrix} \tag{24-27}$$

In these expressions $\partial \mathrm{h_i}/\partial \mathrm{x_j^*}$ *and* $\partial \mathrm{h_i}/\partial \mathrm{u_j^*}$ *are short for*

$$\frac{\partial h_i}{\partial x_j^*} = \left.\frac{\partial h_i[\mathbf{x}(t),\mathbf{u}(t),t]}{\partial x_j(t)}\right|_{\mathbf{x}(t) = \mathbf{x}^*(t),\mathbf{u}(t) = \mathbf{u}^*(t)} \tag{24-28}$$

and

$$\frac{\partial h_i}{\partial u_j^*} = \left.\frac{\partial h_i[\mathbf{x}(t),\mathbf{u}(t),t]}{\partial u_j(t)}\right|_{\mathbf{x}(t) = \mathbf{x}^*(t),\mathbf{u}(t) = \mathbf{u}^*(t)} \quad \square \tag{24-29}$$

We leave the derivation of this fact to the reader, because it is analogous to the derivation of the Taylor series expansion of $\mathbf{f}[\mathbf{x}(t),\mathbf{u}(t),t]$.

Substituting (24-25) into (24-24) and neglecting the "higher-order terms," we obtain the following *perturbation measurement equation*

$$\boxed{\begin{aligned}\delta\mathbf{z}(t) &= \mathbf{H_x}[\mathbf{x}^*(t),\mathbf{u}^*(t),t]\delta\mathbf{x}(t) \\ &\quad + \mathbf{H_u}[\mathbf{x}^*(t),\mathbf{u}^*(t),t]\delta\mathbf{u}(t) + \mathbf{v}(t) \qquad t = t_i, \qquad i = 1, 2, \ldots\end{aligned}} \tag{24-30}$$

Equations (24-23) and (24-30) constitute our linear perturbation equations, or our linear *perturbation state-variable model.*

Example 24-2

Returning to our satellite-planet Example 24-1, we find that

$$\mathbf{F}_{\mathbf{x}}[\mathbf{x}^*(t),\mathbf{u}^*(t),t] = \mathbf{F}_{\mathbf{x}}[\mathbf{x}^*(t)] = \begin{pmatrix} 0 & 1 & 0 & 0 \\ x_4^2(t) + \dfrac{2\gamma}{x_1^3(t)} & 0 & 0 & 2x_1(t)x_4(t) \\ 0 & 0 & 0 & 1 \\ \dfrac{2x_2(t)x_4(t)}{x_1^2(t)} & \dfrac{-2x_4(t)}{x_1(t)} & 0 & \dfrac{-2x_2(t)}{x_1(t)} \end{pmatrix}_*$$

$$\mathbf{F}_{\mathbf{u}}[\mathbf{x}^*(t),\mathbf{u}^*(t),t] = \mathbf{F}_{\mathbf{u}} = \begin{pmatrix} 0 & 0 \\ \dfrac{1}{m} & 0 \\ 0 & 0 \\ 0 & \dfrac{1}{m} \end{pmatrix}$$

$$\mathbf{H}_{\mathbf{x}}[\mathbf{x}^*(t),\mathbf{u}^*(t),t] = \mathbf{H}_{\mathbf{x}} = (1 \quad 0 \quad 0 \quad 0)$$

and

$$\mathbf{H}_{\mathbf{u}}[\mathbf{x}^*(t),\mathbf{u}^*(t),t] = \mathbf{0}$$

In the equation for $\mathbf{F}_{\mathbf{x}}[\mathbf{x}^*(t)]$, the notation $(\quad)_*$ means that all $x_i(t)$ within the matrix are nominal values, i.e., $x_i(t) = x_i^*(t)$.

Observe that the linearized satellite-planet system is time-varying, because its linearized plant matrix, $\mathbf{F}_{\mathbf{x}}[\mathbf{x}^*(t)]$, depends upon the nominal trajectory $\mathbf{x}^*(t)$. □

DISCRETIZATION OF A LINEAR TIME-VARYING STATE-VARIABLE MODEL

In this section we describe how one discretizes the general linear, time-varying state-variable model

$$\dot{\mathbf{x}}(t) = \mathbf{F}(t)\mathbf{x}(t) + \mathbf{C}(t)\mathbf{u}(t) + \mathbf{G}(t)\mathbf{w}(t) \tag{24-31}$$

$$\mathbf{z}(t) = \mathbf{H}(t)\mathbf{x}(t) + \mathbf{v}(t) \qquad t = t_i, \qquad i = 1, 2, \ldots \tag{24-32}$$

The application of this section's results to the perturbation state-variable model is given in the following section.

In (24-31) and (24-32), $\mathbf{x}(t)$ is $n \times 1$, control input $\mathbf{u}(t)$ is $l \times 1$, process noise $\mathbf{w}(t)$ is $p \times 1$, and $\mathbf{z}(t)$ and $\mathbf{v}(t)$ are each $m \times 1$. Additionally, $\mathbf{w}(t)$ is a continuous-time white noise process, $\mathbf{v}(t_i)$ is a discrete-time white noise process, and, $\mathbf{w}(t)$ and $\mathbf{v}(t_i)$ are mutually uncorrelated at all $t = t_i$, $i = 1, 2, \ldots,$ i.e., $\mathbf{E}\{\mathbf{w}(t)\} = \mathbf{0}$ for all t, $\mathbf{E}\{\mathbf{v}(t_i)\} = \mathbf{0}$ for all t_i, and

$$\mathbf{E}\{\mathbf{w}(t)\mathbf{w}'(\tau)\} = \mathbf{Q}(t)\delta(t - \tau) \tag{24-33}$$

$$\mathbf{E}\{\mathbf{v}(t_i)\mathbf{v}'(t_j)\} = \mathbf{R}(t_i)\delta_{ij} \tag{24-34}$$

and

$$\mathbf{E}\{\mathbf{w}(t)\mathbf{v}'(t_i)\} = \mathbf{0} \qquad \text{for } t = t_i \qquad i = 1, 2, \ldots \tag{24-35}$$

Our approach to discretizing state equation (24-30) begins with the solution of that equation.

Theorem 24-1. *The solution to state equation (24-31) can be expressed as*

$$\mathbf{x}(t) = \mathbf{\Phi}(t,t_0)\mathbf{x}(t_0) + \int_{t_0}^{t} \mathbf{\Phi}(t,\tau)[\mathbf{C}(\tau)\mathbf{u}(\tau) + \mathbf{G}(\tau)\mathbf{w}(\tau)]d\tau \tag{24-36}$$

where state transition matrix $\mathbf{\Phi}(t,\tau)$ *is the solution to the following matrix homogeneous differential equation,*

$$\left.\begin{array}{l}\dot{\mathbf{\Phi}}(t,\tau) = \mathbf{F}(t)\mathbf{\Phi}(t,\tau)\\ \mathbf{\Phi}(t,t) = \mathbf{I}\end{array}\right\} \quad \square \tag{24-37}$$

This result should be a familiar one to the readers of this book; hence, we omit its proof.

Next, we assume that $\mathbf{u}(t)$ is a piecewise constant function of time for $t \in [t_k, t_{k+1}]$, and set $t_0 = t_k$ and $t = t_{k+1}$ in (24-36), to obtain

$$\mathbf{x}(t_{k+1}) = \mathbf{\Phi}(t_{k+1},t_k)\mathbf{x}(t_k) + \left[\int_{t_k}^{t_{k+1}} \mathbf{\Phi}(t_{k+1},\tau)\mathbf{C}(\tau)d\tau\right]\mathbf{u}(t_k) + \int_{t_k}^{t_{k+1}} \mathbf{\Phi}(t_{k+1},\tau)\mathbf{G}(\tau)\mathbf{w}(\tau)d\tau \tag{24-38}$$

which can also be written as

$$\mathbf{x}(k+1) = \mathbf{\Phi}(k+1,k)\mathbf{x}(k) + \mathbf{\Psi}(k+1,k)\mathbf{u}(k) + \mathbf{w}_d(k) \tag{24-39}$$

where

$$\mathbf{\Phi}(k+1,k) = \mathbf{\Phi}(t_{k+1},t_k) \tag{24-40}$$

$$\mathbf{\Psi}(k+1,k) = \int_{t_k}^{t_{k+1}} \mathbf{\Phi}(t_{k+1},\tau)\mathbf{C}(\tau)d\tau \tag{24-41}$$

and $\mathbf{w}_d(k)$ is a discrete-time white Gaussian sequence that is *statistically equivalent* to

$$\int_{t_k}^{t_{k+1}} \mathbf{\Phi}(t_{k+1},\tau)\mathbf{G}(\tau)\mathbf{w}(\tau)d\tau$$

The mean and covariance matrices of $\mathbf{w}_d(k)$ are

$$\mathbf{E}\{\mathbf{w}_d(k)\} = \mathbf{E}\left\{\int_{t_k}^{t_{k+1}} \mathbf{\Phi}(t_{k+1},\tau)\mathbf{G}(\tau)\mathbf{w}(\tau)d\tau\right\} = \mathbf{0} \tag{24-42}$$

and

$$\begin{aligned}\mathbf{E}\{\mathbf{w}_d(k)\mathbf{w}_d'(k)\} &\triangleq \mathbf{Q}_d(k)\\ &= \mathbf{E}\left\{\int_{t_k}^{t_{k+1}} \mathbf{\Phi}(t_{k+1},\tau)\mathbf{G}(\tau)\mathbf{w}(\tau)d\tau \int_{t_k}^{t_{k+1}} \mathbf{w}'(\xi)\mathbf{G}'(\xi)\mathbf{\Phi}'(t_{k+1},\xi)d\xi\right\}\\ &= \int_{t_k}^{t_{k+1}} \mathbf{\Phi}(t_{k+1},\tau)\mathbf{G}(\tau)\mathbf{Q}(\tau)\mathbf{G}'(\tau)\mathbf{\Phi}'(t_{k+1},\tau)d\tau \end{aligned} \tag{24-43}$$

respectively.

Observe, from the right-hand side of Equations (24-40), (24-41), and (24-43), that these quantities can be computed from knowledge about $\mathbf{F}(t)$, $\mathbf{C}(t)$, $\mathbf{G}(t)$, and $\mathbf{Q}(t)$. In general, we must compute $\mathbf{\Phi}(k+1,k)$, $\mathbf{\Psi}(k+1,k)$, and $\mathbf{Q}_d(k)$ using numerical integration, and, these matrices change from one time interval to the next because $\mathbf{F}(t)$, $\mathbf{C}(t)$, $\mathbf{G}(t)$, and $\mathbf{Q}(t)$ usually change from one time interval to the next.

Because our measurements have been assumed to be available only at sampled values of t, namely at $t = t_i$, $i = 1, 2, \ldots$, we can express (24-32) as

$$\mathbf{z}(k+1) = \mathbf{H}(k+1)\mathbf{x}(k+1) + \mathbf{v}(k+1) \tag{24-44}$$

Equations (24-39) and (24-44) constitute our discretized state-variable model.

Example 24-3

Great simplifications of the calculations in (24-40), (24-41) and (24-43) occur if $\mathbf{F}(t)$, $\mathbf{C}(t)$, $\mathbf{G}(t)$, and $\mathbf{Q}(t)$ are approximately constant during the time interval $[t_k, t_{k+1}]$, i.e., if

$$\left.\begin{array}{lll}\mathbf{F}(t) \simeq \mathbf{F}_k, & \mathbf{C}(t) \simeq \mathbf{C}_k, & \mathbf{G}(t) \simeq \mathbf{G}_k, \text{ and}\\ \mathbf{Q}(t) \simeq \mathbf{Q}_k & \text{for } t \in [t_k, t_{k+1}] & \end{array}\right\} \tag{24-45}$$

To begin, (24-37) is easily integrated to yield

$$\mathbf{\Phi}(t,\tau) = e^{\mathbf{F}_k(t-\tau)} \tag{24-46}$$

hence,

$$\mathbf{\Phi}(k+1,k) = e^{\mathbf{F}_k T} \tag{24-47}$$

where we have assumed that $t_{k+1} - t_k = T$. The matrix exponential is given by the infinite series

$$e^{\mathbf{F}_k T} = \mathbf{I} + \mathbf{F}_k T + \mathbf{F}_k^2 \frac{T^2}{2} + \mathbf{F}_k^3 \frac{T^3}{3!} + \cdots \tag{24-48}$$

and, for sufficiently small values of T

$$e^{\mathbf{F}_k T} \simeq \mathbf{I} + \mathbf{F}_k T \tag{24-49}$$

We use this approximation for $e^{\mathbf{F}_k T}$ in deriving simpler expressions for $\mathbf{\Psi}(k+1,k)$ and $\mathbf{Q}_d(k)$. Comparable results can be obtained for higher-order truncations of $e^{\mathbf{F}_k T}$.

Substituting (24-46) into (24-41), we find that

$$\begin{aligned}\boldsymbol{\Psi}(k+1,k) &= \int_{t_k}^{t_{k+1}} \boldsymbol{\Phi}(t_{k+1},\tau)\mathbf{C}_k d\tau = \int_{t_k}^{t_{k+1}} e^{\mathbf{F}_k(t_{k+1}-\tau)}\mathbf{C}_k d\tau \\ &\simeq \int_{t_k}^{t_{k+1}} [\mathbf{I} + \mathbf{F}_k(t_{k+1}-\tau)]\mathbf{C}_k d\tau \\ &\simeq \mathbf{C}_k T + \mathbf{F}_k\mathbf{C}_k t_{k+1} T - \mathbf{F}_k\mathbf{C}_k \int_{t_k}^{t_{k+1}} \tau\, d\tau \\ &\simeq \mathbf{C}_k T + \mathbf{F}_k\mathbf{C}_k \frac{T^2}{2} \simeq \mathbf{C}_k T\end{aligned} \tag{24-50}$$

where we have truncated $\boldsymbol{\Psi}(k+1,k)$ to its first-order term in T. Proceeding in a similar manner for $\mathbf{Q}_d(k)$, it is straightforward to show that

$$\mathbf{Q}_d(k) \simeq \mathbf{G}_k\mathbf{Q}_k\mathbf{G}_k' T \tag{24-51}$$

Note that (24-47), (24-49), (24-50), and (24-51), while much simpler than their original expressions, can change in values from one time-interval to another, because of their dependence upon k. □

DISCRETIZED PERTURBATION STATE-VARIABLE MODEL

Applying the results of the preceding section to the perturbation state-variable model in (24-23) and (24-30), we obtain the following *discretized perturbation state-variable model*

$$\delta\mathbf{x}(k+1) = \boldsymbol{\Phi}(k+1,k;^*)\delta\mathbf{x}(k) + \boldsymbol{\Psi}(k+1,k;^*)\delta\mathbf{u}(k) + \mathbf{w}_d(k) \tag{24-52}$$

$$\begin{aligned}\delta\mathbf{z}(k+1) = \mathbf{H}_\mathbf{x}(k+1;^*)\delta\mathbf{x}(k+1) \\ + \mathbf{H}_\mathbf{u}(k+1;^*)\delta\mathbf{u}(k+1) + \mathbf{v}(k+1)\end{aligned} \tag{24-53}$$

The notation $\boldsymbol{\Phi}(k+1,k;^*)$, for example, denotes the fact that this matrix depends on $\mathbf{x}^*(t)$ and $\mathbf{u}^*(t)$. More specifically,

$$\boldsymbol{\Phi}(k+1,k;^*) = \boldsymbol{\Phi}(t_{k+1},t_k;^*) \tag{24-54}$$

where

$$\begin{aligned}\dot{\boldsymbol{\Phi}}(t,\tau;^*) &= \mathbf{F}_\mathbf{x}[\mathbf{x}^*(t),\mathbf{u}^*(t),t]\boldsymbol{\Phi}(t,\tau;^*) \\ \boldsymbol{\Phi}(t,t;^*) &= \mathbf{I}\end{aligned} \tag{24-55}$$

Additionally,

$$\boldsymbol{\Psi}(k+1,k;^*) = \int_{t_k}^{t_{k+1}} \boldsymbol{\Phi}(t_{k+1},\tau;^*)\mathbf{F}_\mathbf{u}[\mathbf{x}^*(\tau),\mathbf{u}^*(\tau),\tau]d\tau \tag{24-56}$$

and

$$\mathbf{Q}_d(k;^*) = \int_{t_k}^{t_{k+1}} \boldsymbol{\Phi}(t_{k+1},\tau;^*)\mathbf{G}(\tau)\mathbf{Q}(\tau)\mathbf{G}'(\tau)\boldsymbol{\Phi}'(t_{k+1},\tau;^*)d\tau \tag{24-57}$$

PROBLEMS

24-1. Derive the Taylor series expansion of $\mathbf{h}[\mathbf{x}(t), \mathbf{u}(t), t]$ given in (24-25).

24-2. Derive the formula for $\mathbf{Q}_d(k)$ given in (24-51).

24-3. Derive formulas for $\mathbf{\Psi}(k+1,k)$ and $\mathbf{Q}_d(k)$ that include first- and second-order effects of T, using the first three terms in the expansion of $e^{\mathbf{F}_k T}$.

24-4. Let a zero-mean stationary Gaussian random process $v(t)$ have the autocorrelation function $\phi_v(\tau)$ given by $\phi_v(\tau) = e^{-|\tau|} + e^{-2|\tau|}$.

(a) Show that this colored-noise process can be generated by passing white noise $\mu(t)$ through the linear system whose transfer function is

$$\sqrt{6}\frac{s+\sqrt{2}}{(s+1)(s+2)}$$

(b) Obtain a discrete-time state-variable model for this colored noise process (assume $T = 1$ msec).

24-5. This problem presents a model for estimation of the altitude, velocity, and constant ballistic coefficient of a vertically falling body (Athans, et al., 1968). The measurements are made at discrete instants of time by a radar that measures range in the presence of discrete-time white Gaussian noise. The state equations for the falling body are

$$\dot{x}_1 = -x_2$$
$$\dot{x}_2 = -e^{-\gamma x_1}x_2^2 x_3$$
$$\dot{x}_3 = 0$$

where $\gamma = 5 \times 10^{-5}$, $x_1(t)$ is altitude, $x_2(t)$ is downward velocity, and x_3 is a constant ballistic parameter. Measured range is given by

$$z(k) = \sqrt{M^2 + [x_1(k) - H]^2} + v(k) \qquad k = 1, 2, \ldots,$$

where M is horizontal distance and H is radar altitude. Obtain the discretized perturbation state-variable model for this system.

24-6. Normalized equations of a stirred reactor are

$$\dot{x}_1 = -(c_1 + c_4)x_1 + c_3(1 + x_4)^2 \exp[K_1 x_1/(1 + x_1)] + c_4 x_2 + c_1 u_1$$
$$\dot{x}_2 = -(c_5 + c_6)x_2 + c_5 x_1 + c_6 x_3$$
$$\dot{x}_3 = -(c_7 + c_8)x_3 + c_8 x_2 + c_7 u_2$$
$$\dot{x}_4 = -c_1 x_4 - c_2(1 + x_4)^2 \exp[K_1 x_1/(1 + x_1)] + c_1 u_3$$

in which u_1, u_2, and u_3 are control inputs. Measurements are

$$z_i(k) = x_i(k) + v_i(k) \qquad i = 1, 2, 3,$$

where $v_i(k)$ are zero-mean white Gaussian noises with variances r_i. Obtain the discretized perturbation state-variable model for this system.

24-7. Obtain the discretized perturbation-state equation for each of the following nonlinear systems:

(a) Equation for the unsteady operation of a synchronous motor:

$$\ddot{x}(t) + C\dot{x}(t) + p \sin x(t) = L(t)$$

(b) Duffing's equation:

$$\ddot{x}(t) + C\dot{x}(t) + \alpha x(t) + \beta x^3(t) = F \cos \omega t$$

(c) Van der Pol's equation:

$$\ddot{x}(t) - \epsilon \dot{x}(t)\left[1 - \frac{1}{3}\dot{x}^2(t)\right] + x(t) = m(t)$$

(d) Hill's equation:

$$\ddot{x}(t) - ax(t) + bp(t)x(t) = m(t)$$

where $p(t)$ is a known periodic function.

Lesson 25

Iterated Least Squares and Extended Kalman Filtering

INTRODUCTION

This lesson is primarily devoted to the extended Kalman filter (EKF), which is a form of the Kalman filter "extended" to nonlinear dynamical systems of the type described in Lesson 24. We shall show that the EKF is related to the method of iterated least squares (ILS), the major difference being that the EKF is for dynamical systems whereas ILS is not.

ITERATED LEAST SQUARES

We shall illustrate the method of ILS for the nonlinear model described in Example 2-5 of Lesson 2, i.e., for the model

$$z(k) = f(\theta,k) + v(k) \tag{25-1}$$

where $k = 1, 2, \ldots, N$.

Iterated least squares is basically a four step procedure.

1. Linearize $f(\theta,k)$ about a nominal value of θ, θ^*. Doing this, we obtain the *perturbation measurement equation*

$$\delta z(k) = F_\theta(k;\theta^*)\delta\theta + v(k) \qquad k = 1, 2, \ldots, N \tag{25-2}$$

where

$$\delta z(k) = z(k) - z^*(k) = z(k) - f(\theta^*,k) \tag{25-3}$$

$$\delta\theta = \theta - \theta^* \tag{25-4}$$

and

$$F_\theta(k;\theta^*) = \left.\frac{\partial f(\theta,k)}{\partial \theta}\right|_{\theta = \theta^*} \tag{25-5}$$

2. Concatenate (25-2) and compute $\hat{\delta\theta}_{\text{WLS}}(N)$ [or $\hat{\delta\theta}_{\text{LS}}(N)$] using our Lesson 3 formulas.
3. Solve the equation

$$\hat{\delta\theta}_{\text{WLS}}(N) = \hat{\theta}_{\text{WLS}}(N) - \theta^* \tag{25-6}$$

for $\hat{\theta}_{\text{WLS}}(N)$, i.e.,

$$\hat{\theta}_{\text{WLS}}(N) = \theta^* + \hat{\delta\theta}_{\text{WLS}}(N) \tag{25-7}$$

4. Replace θ^* with $\hat{\theta}_{\text{WLS}}(N)$ and return to Step 1. Iterate through these steps until convergence occurs. Let $\hat{\theta}^i_{\text{WLS}}(N)$ and $\hat{\theta}^{i+1}_{\text{WLS}}(N)$ denote estimates of θ obtained at the ith and $(i + 1)$st iterations, respectively. Convergence of the ILS method occurs when

$$|\hat{\theta}^{i+1}_{\text{WLS}}(N) - \hat{\theta}^i_{\text{WLS}}(N)| < \epsilon \tag{25-8}$$

where ϵ is a prespecified small positive number.

We observe, from this four-step procedure, that ILS uses the estimate obtained from the linearized model to generate the nominal value of θ about which the nonlinear model is *relinearized.* Additionally, in each complete cycle of this procedure, *we use both the nonlinear and linearized models.* The nonlinear model is used to compute $z^*(k)$, and subsequently $\delta z(k)$, using (25-3).

The notions of relinearizing about a filter output and using both the nonlinear and linearized models are also at the very heart of the EKF.

EXTENDED KALMAN FILTER

The nonlinear dynamical system of interest to us is the one described in Lesson 24. For convenience to the reader, we summarize aspects of that system next. The nonlinear state-variable model is

$$\dot{\mathbf{x}}(t) = \mathbf{f}[\mathbf{x}(t),\mathbf{u}(t),t] + \mathbf{G}(t)\mathbf{w}(t) \tag{25-9}$$

$$\mathbf{z}(t) = \mathbf{h}[\mathbf{x}(t),\mathbf{u}(t),t] + \mathbf{v}(t) \qquad t = t_i \qquad i = 1, 2, \ldots \tag{25-10}$$

Given a nominal input, $\mathbf{u}^*(t)$, and assuming that a nominal trajectory, $\mathbf{x}^*(t)$, exists, $\mathbf{x}^*(t)$ and its associated nominal measurement satisfy the following nominal system model,

$$\dot{\mathbf{x}}^*(t) = \mathbf{f}[\mathbf{x}^*(t),\mathbf{u}^*(t),t] \tag{25-11}$$

$$\mathbf{z}^*(t) = \mathbf{h}[\mathbf{x}^*(t),\mathbf{u}^*(t),t] \qquad t = t_i \qquad i = 1, 2, \ldots \tag{25-12}$$

Letting $\delta\mathbf{x}(t) = \mathbf{x}(t) - \mathbf{x}^*(t)$, $\delta\mathbf{u}(t) = \mathbf{u}(t) - \mathbf{u}^*(t)$, and $\delta\mathbf{z}(t) = \mathbf{z}(t) - \mathbf{z}^*(t)$, we also have the following *discretized perturbation state-variable model* that is associated with a linearized version of the original nonlinear state-variable model,

$$\delta\mathbf{x}(k+1) = \mathbf{\Phi}(k+1,k;^*)\delta\mathbf{x}(k) + \mathbf{\Psi}(k+1,k;^*)\delta\mathbf{u}(k) + \mathbf{w}_d(k) \qquad (25\text{-}13)$$

$$\delta\mathbf{z}(k+1) = \mathbf{H}_{\mathbf{x}}(k+1;^*)\delta\mathbf{x}(k+1) + \mathbf{H}_{\mathbf{u}}(k+1;^*)\delta\mathbf{u}(k+1) + \mathbf{v}(k+1) \qquad (25\text{-}14)$$

In deriving (25-13) and (25-14), we made the important assumption that higher-order terms in the Taylor series expansions of $\mathbf{f}[\mathbf{x}(t),\mathbf{u}(t),t]$ and $\mathbf{h}[\mathbf{x}(t),\mathbf{u}(t),t]$ could be neglected. Of course, this is only correct as long as $\mathbf{x}(t)$ is "close" to $\mathbf{x}^*(t)$ and $\mathbf{u}(t)$ is "close" to $\mathbf{u}^*(t)$.

If $\mathbf{u}(t)$ is an input derived from a feedback control law, so that $\mathbf{u}(t) = \mathbf{u}[\mathbf{x}(t),t]$, then $\mathbf{u}(t)$ can differ from $\mathbf{u}^*(t)$, because $\mathbf{x}(t)$ will differ from $\mathbf{x}^*(t)$. On the other hand, if $\mathbf{u}(t)$ does not depend on $\mathbf{x}(t)$ then usually $\mathbf{u}(t)$ is the same as $\mathbf{u}^*(t)$, in which case $\delta\mathbf{u}(t) = \mathbf{0}$. We see, therefore, that $\mathbf{x}^*(t)$ is the critical quantity in the calculation of our discretized perturbation state-variable model.

Suppose $\mathbf{x}^*(t)$ is given a priori; then we can compute predicted, filtered, or smoothed estimates of $\delta\mathbf{x}(k)$ by applying all of our previously derived estimators to the discretized perturbation state-variable model in (25-13) and (25-14). We can precompute $\mathbf{x}^*(t)$ by solving the nominal differential equation (25-11). The Kalman filter associated with using a precomputed $\mathbf{x}^*(t)$ is known as a *relinearized KF.*

A relinearized KF usually gives poor results, because it relies on an open-loop strategy for choosing $\mathbf{x}^*(t)$. When $\mathbf{x}^*(t)$ is precomputed there is no way of forcing $\mathbf{x}^*(t)$ to remain close to $\mathbf{x}(t)$, and this must be done or else the perturbation state-variable model is invalid. Divergence of the relinearized KF often occurs; hence, we do not recommend the relinearized KF.

The relinearized KF is based only on the discretized perturbation state-variable model. It does not use the nonlinear nature of the original system in an active manner. The extended Kalman filter relinearizes the nonlinear system about each new estimate as it becomes available; i.e., at $k = 0$, the system is linearized about $\hat{\mathbf{x}}(0|0)$. Once $\mathbf{z}(1)$ is processed by the EKF, so that $\hat{\mathbf{x}}(1|1)$ is obtained, the system is linearized about $\hat{\mathbf{x}}(1|1)$. By "linearize about $\hat{\mathbf{x}}(1|1)$," we mean $\hat{\mathbf{x}}(1|1)$ is used to calculate all the quantities needed to make the transition from $\hat{\mathbf{x}}(1|1)$ to $\hat{\mathbf{x}}(2|1)$, and subsequently $\hat{\mathbf{x}}(2|2)$. This phrase will become clear below. The purpose of relinearizing about the filter's output is to use a better reference trajectory for $\mathbf{x}^*(t)$. Doing this, $\delta\mathbf{x} = \mathbf{x} - \hat{\mathbf{x}}$ will be held as small as possible, so that our linearization assumptions are less likely to be violated than in the case of the relinearized KF.

The EKF is developed below in predictor-corrector format (Jazwinski, 1970). Its prediction equation is obtained by integrating the nominal differential equation for $\mathbf{x}^*(t)$, from t_k to t_{k+1}. In order to do this, we need to know how to choose $\mathbf{x}^*(t)$ for the entire interval of time $t \in [t_k, t_{k+1}]$. Thus far, we have only mentioned how $\mathbf{x}^*(t)$ is chosen at t_k, i.e., as $\hat{\mathbf{x}}(k|k)$.

Theorem 25-1. *As a consequence of relinearizing about* $\hat{\mathbf{x}}(k|k)$ (k = 0, 1, . . .),

$$\delta\hat{\mathbf{x}}(t|t_k) = \mathbf{0} \qquad \text{for } all\ t \in [t_k, t_{k+1}] \tag{25-15}$$

This means that

$$\mathbf{x}^*(t) = \hat{\mathbf{x}}(t|t_k) \text{ for } all\ t \in [t_k, t_{k+1}] \tag{25-16}$$

Before proving this important result, we observe that it provides us with a choice of $\mathbf{x}^*(t)$ over the entire interval of time $t \in [t_k, t_{k+1}]$, and, it states that at the left-hand side of this time interval $\mathbf{x}^*(t_k) = \hat{\mathbf{x}}(k|k)$, whereas at the right-hand side of this time interval $\mathbf{x}^*(t_{k+1}) = \hat{\mathbf{x}}(k+1|k)$. The transition from $\hat{\mathbf{x}}(k+1|k)$ to $\hat{\mathbf{x}}(k+1|k+1)$ will be made using the EKF's correction equation.

Proof. Let t_l be an arbitrary value of t lying in the interval between t_k and t_{k+1} (see Figure 25-1). For the purposes of this derivation, we can assume that $\delta\mathbf{u}(k) = \mathbf{0}$ [i.e., perturbation input $\delta\mathbf{u}(k)$ takes on no new values in the interval from t_k to t_{k+1}; recall the piecewise-constant assumption made about $\mathbf{u}(t)$ in the derivation of (24-37)], i.e.,

$$\delta\mathbf{x}(k+1) = \mathbf{\Phi}(k+1, k; {}^*)\delta\mathbf{x}(k) + \mathbf{w}_d(k) \tag{25-17}$$

Using our general state-predictor results given in (16-14), we see that (remember that k is short for t_k, and that $t_{k+1} - t_k$ does not have to be a constant; this is true in all of our predictor, filter, and smoother formulas)

$$\begin{aligned}\delta\hat{\mathbf{x}}(t_l|t_k) &= \mathbf{\Phi}(t_l, t_k; {}^*)\delta\hat{\mathbf{x}}(t_k|t_k)\\ &= \mathbf{\Phi}(t_l, t_k; {}^*)[\hat{\mathbf{x}}(k|k) - \mathbf{x}^*(k)]\end{aligned} \tag{25-18}$$

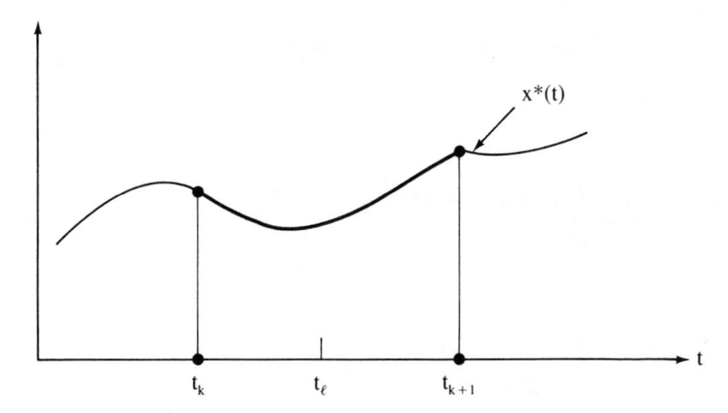

Figure 25-1 Nominal state trajectory $x^*(t)$.

In the EKF we set $\mathbf{x}^*(k) = \hat{\mathbf{x}}(k|k)$; thus, when this is done,

$$\delta\hat{\mathbf{x}}(t_l|t_k) = \mathbf{0} \tag{25-19}$$

and, because $t_l \in [t_k, t_{k+1}]$,

$$\delta\hat{\mathbf{x}}(t_l|t_k) = \mathbf{0} \qquad \text{for } all\ t_l \in [t_k, t_{k+1}] \tag{25-20}$$

which is (25-15). Equation (25-16) follows from (25-20) and the fact that $\delta\hat{\mathbf{x}}(t_l|t_k) = \hat{\mathbf{x}}(t_l|t_k) - \mathbf{x}^*(t_l)$. □

We are now able to derive the EKF. As mentioned above, the EKF must be obtained in predictor-corrector format. We begin the derivation by obtaining the predictor equation for $\hat{\mathbf{x}}(k+1|k)$.

Recall that $\mathbf{x}^*(t)$ is the solution of the nominal state equation (25-11). Using (25-16) in (25-11), we find that

$$\frac{d}{dt}\hat{\mathbf{x}}(t|t_k) = \mathbf{f}[\hat{\mathbf{x}}(t|t_k), \mathbf{u}^*(t), t] \tag{25-21}$$

Integrating this equation from $t = t_k$ to $t = t_{k+1}$, we obtain

$$\boxed{\hat{\mathbf{x}}(k+1|k) = \hat{\mathbf{x}}(k|k) + \int_{t_k}^{t_{k+1}} \mathbf{f}[\hat{\mathbf{x}}(t|t_k), \mathbf{u}^*(t), t]\,dt} \tag{25-22}$$

which is the *EKF prediction equation*. Observe that the nonlinear nature of the system's state equation is used to determine $\hat{\mathbf{x}}(k+1|k)$. The integral in (25-22) is evaluated by means of numerical integration formulas that are initialized by $\mathbf{f}[\hat{\mathbf{x}}(t_k|t_k), \mathbf{u}^*(t_k), t_k]$.

The corrector equation for $\hat{\mathbf{x}}(k+1|k+1)$ is obtained from the Kalman filter associated with the discretized perturbation state-variable model in (25-13) and (25-14), and is

$$\begin{aligned}\delta\hat{\mathbf{x}}(k+1|k+1) = \delta\hat{\mathbf{x}}(k+1|k) + \mathbf{K}(k+1;*)[\delta\mathbf{z}(k+1) \\ - \mathbf{H}_\mathbf{x}(k+1;*)\delta\hat{\mathbf{x}}(k+1|k) - \mathbf{H}_\mathbf{u}(k+1;*)\delta\mathbf{u}(k+1)]\end{aligned} \tag{25-23}$$

As a consequence of relinearizing about $\hat{\mathbf{x}}(k|k)$, we know that

$$\delta\hat{\mathbf{x}}(k+1|k) = \mathbf{0} \tag{25-24}$$

$$\begin{aligned}\delta\hat{\mathbf{x}}(k+1|k+1) &= \hat{\mathbf{x}}(k+1|k+1) - \mathbf{x}^*(k+1) \\ &= \hat{\mathbf{x}}(k+1|k+1) - \hat{\mathbf{x}}(k+1|k)\end{aligned} \tag{25-25}$$

and

$$\begin{aligned}\delta\mathbf{z}(k+1) &= \mathbf{z}(k+1) - \mathbf{z}^*(k+1) \\ &= \mathbf{z}(k+1) - \mathbf{h}[\mathbf{x}^*(k+1), \mathbf{u}^*(k+1), k+1] \\ &= \mathbf{z}(k+1) - \mathbf{h}[\hat{\mathbf{x}}(k+1|k), \mathbf{u}^*(k+1), k+1]\end{aligned} \tag{25-26}$$

Substituting these three equations into (25-23) we obtain

$$\hat{\mathbf{x}}(k+1|k+1) = \hat{\mathbf{x}}(k+1|k) + \mathbf{K}(k+1;^*)\{\mathbf{z}(k+1) - \mathbf{h}[\hat{\mathbf{x}}(k+1|k),\mathbf{u}^*(k+1),k+1] - \mathbf{H}_{\mathbf{u}}(k+1;^*)\delta\mathbf{u}(k+1)\} \tag{25-27}$$

which is the *EKF correction equation*. Observe that the nonlinear nature of the system's measurement equation is used to determine $\hat{\mathbf{x}}(k+1|k+1)$. One usually sees this equation for the case when $\delta\mathbf{u} = \mathbf{0}$, in which case the last term on the right-hand side of (25-27) is not present.

In order to compute $\hat{\mathbf{x}}(k+1|k+1)$, we must compute the EKF gain matrix $\mathbf{K}(k+1;^*)$. This matrix, as well as its associated $\mathbf{P}(k+1|k;^*)$ and $\mathbf{P}(k+1|k+1;^*)$ matrices, depends upon the nominal $\mathbf{x}^*(t)$ that results from prediction, namely $\hat{\mathbf{x}}(k+1|k)$. Observe, from (25-16), that $\mathbf{x}^*(k+1) = \hat{\mathbf{x}}(k+1|k)$, and that the argument of $\mathbf{K}$ in the correction equation is $k+1$; hence, we are indeed justified to use $\hat{\mathbf{x}}(k+1|k)$ as the nominal value of $\mathbf{x}^*$ during the calculations of $\mathbf{K}(k+1;^*)$, $\mathbf{P}(k+1|k;^*)$, and $\mathbf{P}(k+1|k+1;^*)$. These three quantities are computed from

$$\mathbf{K}(k+1;^*) = \mathbf{P}(k+1|k;^*)\mathbf{H}_{\mathbf{x}}'(k+1;^*)[\mathbf{H}_{\mathbf{x}}(k+1;^*) \mathbf{P}(k+1|k;^*)\mathbf{H}_{\mathbf{x}}'(k+1;^*) + \mathbf{R}(k+1)]^{-1} \tag{25-28}$$

$$\mathbf{P}(k+1|k;^*) = \mathbf{\Phi}(k+1,k;^*)\mathbf{P}(k|k;^*)\mathbf{\Phi}'(k+1,k;^*) + \mathbf{Q}_d(k;^*) \tag{25-29}$$

$$\mathbf{P}(k+1|k+1;^*) = [\mathbf{I} - \mathbf{K}(k+1;^*)\mathbf{H}_{\mathbf{x}}(k+1;^*)]\mathbf{P}(k+1|k;^*) \tag{25-30}$$

Remember that in these three equations * denotes the use of $\hat{\mathbf{x}}(k+1|k)$.

The EKF is very widely used, especially in the aerospace industry; however, it does not provide an optimal estimate of $\mathbf{x}(k)$. The optimal estimate of $\mathbf{x}(k)$ is still $\mathbf{E}\{\mathbf{x}(k)|\mathfrak{Z}(k)\}$, regardless of the linear or nonlinear nature of the system's model. The EKF is a first-order approximation of $\mathbf{E}\{\mathbf{x}(k)|\mathfrak{Z}(k)\}$ that sometimes works quite well, but cannot be guaranteed always to work well. No convergence results are known for the EKF; hence, the EKF must be viewed as an ad hoc filter. Alternatives to the EKF, which are based on nonlinear filtering, are quite complicated and are rarely used.

The EKF is designed to work well as long as $\delta\mathbf{x}(k)$ is "small." The *iterated EKF* (Jazwinski, 1970), depicted in Figure 25-2, is designed to keep $\delta\mathbf{x}(k)$ as small as possible. The iterated EKF differs from the EKF in that it iterates the correction equation L times until $\|\hat{\mathbf{x}}_L(k+1|k+1) - \hat{\mathbf{x}}_{L-1}(k+1|k+1)\| \le \epsilon$. Corrector #1 computes $\mathbf{K}(k+1;^*)$, $\mathbf{P}(k+1|k;^*)$, and $\mathbf{P}(k+1|k+1;^*)$ using $\mathbf{x}^* = \hat{\mathbf{x}}(k+1|k)$; corrector #2 computes these quantities using $\mathbf{x}^* = \hat{\mathbf{x}}_1(k+1|k+1)$; corrector #3 computes these quantities using $\mathbf{x}^* = \hat{\mathbf{x}}_2(k+1|k+1)$; etc.

Often, just adding one additional corrector (i.e., $L = 2$) leads to substantially better results for $\hat{\mathbf{x}}(k+1|k+1)$ than are obtained using the EKF.

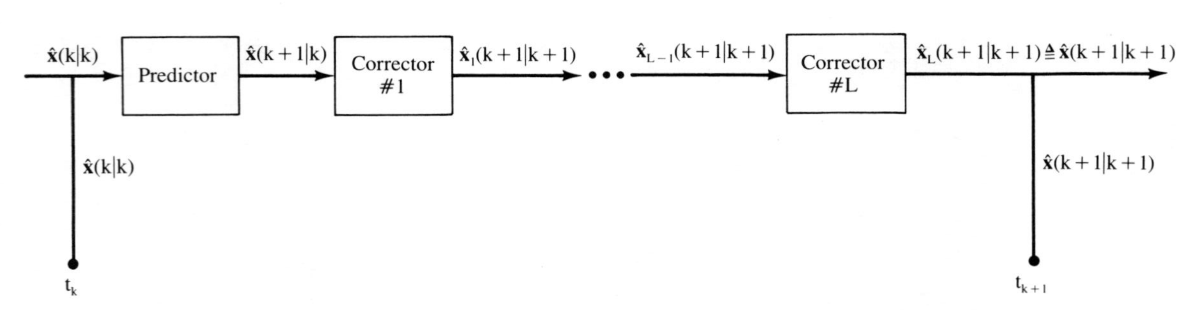

Figure 25-2 Iterated EKF. All of the calculations provide us with a refined estimate of $\mathbf{x}(k+1)$, $\hat{\mathbf{x}}(k+1|k+1)$, starting with $\hat{\mathbf{x}}(k|k)$.

APPLICATION TO PARAMETER ESTIMATION

One of the earliest applications of the extended Kalman filter was to parameter estimation (Kopp and Orford, 1963). Consider the continuous-time linear system

$$\dot{\mathbf{x}}(t) = \mathbf{A}\mathbf{x}(t) + \mathbf{w}(t) \tag{25-31a}$$

$$\mathbf{z}(t) = \mathbf{H}\mathbf{x}(t) + \mathbf{v}(t) \qquad t = t_i \qquad i = 1, 2, \ldots \tag{25-31b}$$

Matrices **A** and **H** contain some unknown parameters, and our objective is to estimate these parameters from the measurements $\mathbf{z}(t_i)$ as they become available.

To begin, we assume differential equation models for the unknown parameters, i.e., either

$$\dot{a}_l(t) = 0 \qquad l = 1, 2, \ldots, l^* \tag{25-32a}$$

$$\dot{h}_j(t) = 0 \qquad j = 1, 2, \ldots, j^* \tag{25-32b}$$

or

$$\dot{a}_l(t) = c_l a_l(t) + n_l(t) \qquad l = 1, 2, \ldots, l^* \tag{25-33a}$$

$$\dot{h}_j(t) = d_j h_j(t) + \eta_j(t) \qquad j = 1, 2, \ldots, j^* \tag{25-33b}$$

In the latter models $n_l(t)$ and $\eta_j(t)$ are white noise processes, and one often chooses $c_l = 0$ and $d_j = 0$. The noises $n_l(t)$ and $\eta_j(t)$ introduce uncertainty about the "constancy" of the a_l and h_j parameters.

Next, we augment the parameter differential equations to (25-31a) and (25-31b). The resulting system is nonlinear, because it contains products of states [e.g., $a_l(t)x_i(t)$]. The augmented system can be expressed as in (25-9) and (25-10), which means we have reduced the problem of parameter estimation in a linear system to state estimation in a nonlinear system.

Finally, we apply the EKF to the augmented state-variable model to obtain $\hat{a}_l(k|k)$ and $\hat{h}_j(k|k)$.

Ljung (1979) has studied the convergence properties of the EKF applied to parameter estimation, and has shown that parameter estimates do not converge to their true values. He shows that another term must be added to the EKF corrector equation in order to guarantee convergence. For details, see his paper.

Example 25-1

Consider the satellite and planet Example 24-1, in which the satellite's motion is governed by the two equations

$$\ddot{r} = r\dot{\theta}^2 - \frac{\gamma}{r^2} + \frac{1}{m}u_r(t) \tag{25-34}$$

and

$$\ddot{\theta} = -\frac{2\dot{r}\dot{\theta}}{r} + \frac{1}{m}u_\theta(t) \tag{25-35}$$

We shall assume that m and γ are unknown constants, and shall model them as

$$\dot{m}(t) = 0 \tag{25-36}$$

$$\dot{\gamma}(t) = 0 \tag{25-37}$$

Defining $x_1 = r$, $x_2 = \dot{r}$, $x_3 = \theta$, $x_4 = \dot{\theta}$, $x_5 = m$, $x_6 = \gamma$, $u_1 = u_r$, and $u_2 = u_\theta$, we have

$$\left.\begin{aligned} \dot{x}_1 &= x_2 \\ \dot{x}_2 &= x_1 x_4^2 - \frac{x_6}{x_1^2} + \frac{1}{x_5}u_1(t) \\ \dot{x}_3 &= x_4 \\ \dot{x}_4 &= -\frac{2x_2x_4}{x_1} + \frac{1}{x_5}u_2(t) \\ \dot{x}_5 &= 0 \\ \dot{x}_6 &= 0 \end{aligned}\right\} \tag{25-38}$$

Now,

$$\mathbf{f}[\mathbf{x}(t),\mathbf{u}(t),t] = \text{col}\left[x_2,\, x_1x_4^2 - \frac{x_6}{x_1^2} + \frac{1}{x_5}u_1,\, x_4,\, -\frac{2x_2x_4}{x_1} + \frac{1}{x_5}u_2,\, 0,\, 0\right] \quad \square \tag{25-39}$$

We note, finally, that the modeling and augmentation approach to parameter estimation, described above, is not restricted to continuous-time linear systems. Additional situations are described in the exercises.

PROBLEMS

25-1. In the first-order system, $x(k+1) = ax(k) + w(k)$, and $z(k+1) = x(k+1) + v(k+1)$, $k = 1, 2, \ldots, N$, a is an unknown parameter that is to be estimated. Sequences $w(k)$ and $v(k)$ are, as usual, mutually uncorrelated and white, and, $w(k) \sim N(w(k);0,1)$ and $v(k) \sim N(v(k);0,1/2)$. Explain, using equations and a flowchart, how parameter a can be estimated using an EKF.

25-2. Repeat the preceding problem where all conditions are the same except that now $w(k)$ and $v(k)$ are correlated, and $\mathbf{E}\{w(k)v(k)\} = 1/4$.

25-3. The system of differential equations describing the motion of an aerospace vehicle about its pitch axis can be written as (Kopp and Orford, 1963)

$$\dot{x}_1(t) = x_2(t)$$
$$\dot{x}_2(t) = a_1(t)x_2(t) + a_2(t)x_1(t) + a_3(t)u(t)$$

where $x_1 = \dot{\theta}(t)$, which is the actual pitch rate. Sampled measurements are made of the pitch rate, i.e.,

$$z(t) = x_1(t) + v(t) \qquad t = t_i \qquad i = 1, 2, \ldots, N$$

Noise $v(t_i)$ is white and Gaussian, and $\sigma_v^2(t_i)$ is given. The control signal $u(t)$ is the sum of a desired control signal $u^*(t)$ and additive noise, i.e.,

$$u(t) = u^*(t) + \delta u(t)$$

The additive noise $\delta u(t)$ is a normally distributed random variable modulated by a function of the desired control signal, i.e.,

$$\delta u(t) = S[u^*(t)]w_0(t)$$

where $w_0(t)$ is zero-mean white noise with intensity σ_{w0}^2. Parameters $a_1(t)$, $a_2(t)$ and $a_3(t)$ may be unknown and are modeled as

$$\dot{a}_i(t) = \alpha_i(t)[a_i(t) - \bar{a}_i(t)] + w_i(t) \qquad i = 1, 2, 3$$

In this model the parameters $\alpha_i(t)$ are assumed given, as are the a priori values of $a_i(t)$ and $\bar{a}_i(t)$, and, $w_i(t)$ are zero-mean white noises with intensities $\sigma_{w_i}^2$.

(a) What are the EKF formulas for estimation of x_1, x_2, a_1, and a_2, assuming that a_3 is known?

(b) Repeat (a) but now assume that a_3 is unknown.

25-4. Suppose we begin with the nonlinear discrete-time system,

$$\mathbf{x}(k+1) = \mathbf{f}[\mathbf{x}(k),k] + \mathbf{w}(k)$$
$$\mathbf{z}(k) = \mathbf{h}[\mathbf{x}(k),k] + \mathbf{v}(k) \qquad k = 1, 2, \ldots$$

Develop the EKF for this system [*Hint*: expand $\mathbf{f}[\mathbf{x}(k),k]$ and $\mathbf{h}[\mathbf{x}(k),k]$ in Taylor series about $\hat{\mathbf{x}}(k|k)$ and $\hat{\mathbf{x}}(k|k-1)$, respectively].

25-5. Refer to Problem 24-7. Obtain the EKF for

(a) Equation for the unsteady operation of a synchronous motor, in which C and p are unknown;

(b) Duffing's equation, in which C, α, and β are unknown;

(c) Van der Pol's equation, in which ϵ is unknown; and

(d) Hill's equation, in which a and b are unknown.

Lesson 26

Maximum-Likelihood State and Parameter Estimation

INTRODUCTION

In Lesson 11 we studied the problem of obtaining maximum-likelihood estimates of a collection of parameters, $\boldsymbol{\theta} = \text{col}$ (elements of $\boldsymbol{\Phi}$, $\boldsymbol{\Psi}$, $\mathbf{H}$, and $\mathbf{R}$), that appear in the state-variable model

$$\mathbf{x}(k+1) = \boldsymbol{\Phi}\mathbf{x}(k) + \boldsymbol{\Psi}\mathbf{u}(k) \tag{26-1}$$

$$\mathbf{z}(k+1) = \mathbf{H}\mathbf{x}(k+1) + \mathbf{v}(k+1) \qquad k = 0, 1, \ldots, N-1 \tag{26-2}$$

We determined the log-likelihood function to be

$$L(\boldsymbol{\theta}|\mathcal{Z}) = -\frac{1}{2}\sum_{i=1}^{N}[\mathbf{z}(i) - \mathbf{H}_{\theta}\mathbf{x}_{\theta}(i)]'\mathbf{R}_{\theta}^{-1}[\mathbf{z}(i) - \mathbf{H}_{\theta}\mathbf{x}_{\theta}(i)] - \frac{N}{2}\ln|\mathbf{R}_{\theta}| \tag{26-3}$$

where quantities that are subscripted $\boldsymbol{\theta}$ denote a dependence on $\boldsymbol{\theta}$. Finally, we pointed out that the state equation (26-1), written as

$$\mathbf{x}_{\theta}(k+1) = \boldsymbol{\Phi}_{\theta}\mathbf{x}_{\theta}(k) + \boldsymbol{\Psi}_{\theta}\mathbf{u}(k) \qquad \mathbf{x}_{\theta}(0) \text{ known} \tag{26-4}$$

acts as a constraint that is associated with the computation of the log-likelihood function. Parameter vector $\boldsymbol{\theta}$ must be determined by maximizing $L(\boldsymbol{\theta}|\mathcal{Z})$ subject to the constraint (26-4). This can only be done using mathematical programming techniques (i.e., an optimization algorithm such as steepest descent or Marquardt-Levenberg).

In this lesson we study the problem of obtaining maximum-likelihood estimates of a collection of parameters, also denoted $\boldsymbol{\theta}$, that appear in our basic state-variable model,

$$\mathbf{x}(k+1) = \boldsymbol{\Phi}\mathbf{x}(k) + \boldsymbol{\Gamma}\mathbf{w}(k) + \boldsymbol{\Psi}\mathbf{u}(k) \tag{26-5}$$

$$\mathbf{z}(k+1) = \mathbf{H}\mathbf{x}(k+1) + \mathbf{v}(k+1) \qquad k = 0, 1, \ldots, N-1 \tag{26-6}$$

Now, however,

$$\boldsymbol{\theta} = \text{col (elements of } \boldsymbol{\Phi}, \boldsymbol{\Gamma}, \boldsymbol{\Psi}, \mathbf{H}, \mathbf{Q}, \text{ and } \mathbf{R}) \tag{26-7}$$

and, we assume that $\boldsymbol{\theta}$ is $d \times 1$. As in Lesson 11, *we shall assume that* $\boldsymbol{\theta}$ *is identifiable.*

Before we can determine $\hat{\boldsymbol{\theta}}_{ML}$ we must establish the log-likelihood function for our basic state-variable model.

A LOG-LIKELIHOOD FUNCTION FOR THE BASIC STATE-VARIABLE MODEL

As always, we must compute $p(\mathfrak{Z}|\boldsymbol{\theta}) = p(\mathbf{z}(1), \mathbf{z}(2), \ldots, \mathbf{z}(N)|\boldsymbol{\theta})$. This is difficult to do for the basic state-variable model, because

$$p(\mathbf{z}(1), \mathbf{z}(2), \ldots, \mathbf{z}(N)|\boldsymbol{\theta}) \neq p(\mathbf{z}(1)|\boldsymbol{\theta})p(\mathbf{z}(2)|\boldsymbol{\theta}) \cdots p(\mathbf{z}(N)|\boldsymbol{\theta}) \tag{26-8}$$

The measurements are all correlated due to the presence of either the process noise, $\mathbf{w}(k)$, or random initial conditions or both. This represents the major difference between our basic state-variable model, (26-5) and (26-6), and the state-variable model studied earlier, in (26-1) and (26-2). Fortunately, the measurements and innovations are causally invertible, and the innovations are all uncorrelated, so that it is still relatively easy to determine the log-likelihood function for the basic state-variable model.

Theorem 26-1. *The log-likelihood function for our basic state-variable model in (26-5) and (26-6) is*

$$L(\boldsymbol{\theta}|\mathfrak{Z}) = -\frac{1}{2}\sum_{j=1}^{N}[\tilde{\mathbf{z}}_{\boldsymbol{\theta}}'(j|j-1)\mathcal{N}_{\boldsymbol{\theta}}^{-1}(j|j-1)\tilde{\mathbf{z}}_{\boldsymbol{\theta}}(j|j-1) + \ln|\mathcal{N}_{\boldsymbol{\theta}}(j|j-1)|] \tag{26-9}$$

where $\tilde{\mathbf{z}}_{\boldsymbol{\theta}}(j|j-1)$ *is the innovations process, and* $\mathcal{N}_{\boldsymbol{\theta}}(j|j-1)$ *is the covariance of that process* [in Lesson 16 we used the symbol $\mathbf{P}_{\tilde{z}\tilde{z}}(j|j-1)$ for this covariance],

$$\mathcal{N}_{\boldsymbol{\theta}}(j|j-1) = \mathbf{H}_{\boldsymbol{\theta}}\mathbf{P}_{\boldsymbol{\theta}}(j|j-1)\mathbf{H}_{\boldsymbol{\theta}}' + \mathbf{R}_{\boldsymbol{\theta}} \tag{26-10}$$

This theorem is also applicable to either time-varying or nonstationary systems or both. Within the structure of these more complicated systems there must still be a collection of unknown but constant parameters. It is these parameters that are estimated by maximizing $L(\boldsymbol{\theta}|\mathfrak{Z})$.

Proof (Mendel, 1983b, pp. 101–103). We must first obtain the joint density function $p(\mathfrak{Z}|\boldsymbol{\theta}) = p(\mathbf{z}(1), \ldots, \mathbf{z}(N)|\boldsymbol{\theta})$. In Lesson 17 we saw that the innovations process $\tilde{\mathbf{z}}(i|i-1)$ and measurement $\mathbf{z}(i)$ are causally invertible; thus, the density function

$$\tilde{p}(\tilde{\mathbf{z}}(1|0), \tilde{\mathbf{z}}(2|1), \ldots, \tilde{\mathbf{z}}(N|N-1)|\boldsymbol{\theta})$$

contains the same data information as $p(\mathbf{z}(1), \ldots, \mathbf{z}(N)|\boldsymbol{\theta})$ does. Consequently, $L(\boldsymbol{\theta}|\mathfrak{Z})$ can be replaced by $\tilde{L}(\boldsymbol{\theta}|\tilde{\mathfrak{Z}})$, where

$$\tilde{\mathfrak{Z}} = \text{col}\,(\tilde{\mathbf{z}}(1|0), \ldots, \tilde{\mathbf{z}}(N|N-1)) \tag{26-11}$$

and

$$\tilde{L}(\boldsymbol{\theta}|\tilde{\mathfrak{Z}}) = \ln \tilde{p}(\tilde{\mathbf{z}}(1|0), \ldots, \tilde{\mathbf{z}}(N|N-1)|\boldsymbol{\theta}) \tag{26-12}$$

Now, however, we use the fact that the innovations process is Gaussian white noise to express $\tilde{L}(\boldsymbol{\theta}|\tilde{\mathfrak{Z}})$ as

$$\tilde{L}(\boldsymbol{\theta}|\tilde{\mathfrak{Z}}) = \ln \prod_{j=1}^{N} \tilde{p}_j\,(\tilde{\mathbf{z}}(j|j-1)|\boldsymbol{\theta}) \tag{26-13}$$

For our basic state-variable model, the innovations are Gaussian distributed, which means that $\tilde{p}_j\,(\tilde{\mathbf{z}}(j|j-1)|\boldsymbol{\theta}) = \tilde{p}(\tilde{\mathbf{z}}(j|j-1)|\boldsymbol{\theta})$ for $j = 1, \ldots, N$; hence,

$$\tilde{L}(\boldsymbol{\theta}|\tilde{\mathfrak{Z}}) = \ln \prod_{j=1}^{N} \tilde{p}(\tilde{\mathbf{z}}(j|j-1)|\boldsymbol{\theta}) \tag{26-14}$$

From part (b) of Theorem 16-2 in Lesson 16 we know that

$$\tilde{p}(\tilde{\mathbf{z}}(j|j-1)|\boldsymbol{\theta}) = [(2\pi)^m|\mathcal{N}(j|j-1)|]^{-1/2} \exp\left[-\frac{1}{2}\tilde{\mathbf{z}}'(j|j-1)\mathcal{N}^{-1}(j|j-1)\tilde{\mathbf{z}}(j|j-1)\right] \tag{26-15}$$

Substitute (26-15) into (26-14) to show that

$$\tilde{L}(\boldsymbol{\theta}|\tilde{\mathfrak{Z}}) = -\frac{1}{2}\sum_{j=1}^{N} [\tilde{\mathbf{z}}'(j|j-1)\mathcal{N}^{-1}(j|j-1)\tilde{\mathbf{z}}(j|j-1) + \ln|\mathcal{N}(j|j-1)|] \tag{26-16}$$

where by convention we have neglected the constant term $-\ln(2\pi)^{m/2}$ because it does not depend on $\boldsymbol{\theta}$.

Because $\tilde{p}(\,\#\,|\boldsymbol{\theta})$ and $p(\,\#\,|\boldsymbol{\theta})$ contain the same information about the data, $\tilde{L}(\boldsymbol{\theta}|\tilde{\mathfrak{Z}})$ and $L(\boldsymbol{\theta}|\mathfrak{Z})$ must also contain the same information about the data; hence, we can use $L(\boldsymbol{\theta}|\mathfrak{Z})$ to denote the right-hand side of (26-16), as in (26-9). To indicate which quantities on the right-hand side of (26-9) may depend on $\boldsymbol{\theta}$, we have subscripted all such quantities with $\boldsymbol{\theta}$. □

The innovations process $\tilde{\mathbf{z}}_{\boldsymbol{\theta}}(j|j-1)$ can be generated by a Kalman filter; hence, *the Kalman filter acts as a constraint that is associated with the computation of the log-likelihood function for the basic state-variable model.*

In the present situation, where the true values of θ, θ_T, are not known but are being estimated, the estimate of $\mathbf{x}(j)$ obtained from a Kalman filter will be suboptimal due to wrong values of θ being used by that filter. In fact, we must use $\hat{\theta}_{ML}$ in the implementation of the Kalman filter, because $\hat{\theta}_{ML}$ will be the best information available about θ_T at t_j. If $\hat{\theta}_{ML} \rightarrow \theta_T$ as $N \rightarrow \infty$, then $\tilde{\mathbf{z}}_{\hat{\theta}_{ML}}(j|j-1) \rightarrow \tilde{\mathbf{z}}_{\theta_T}(j|j-1)$ as $N \rightarrow \infty$, and the suboptimal Kalman filter will approach the optimal Kalman filter. This result is about the best that one can hope for in maximum-likelihood estimation of parameters in our basic state-variable model.

Note also, that although we began with a parameter estimation problem, we wound up with a simultaneous state and parameter estimation problem. This is due to the uncertainties present in our state equation, which necessitated state estimation using a Kalman filter.

ON COMPUTING $\hat{\theta}_{ML}$

How do we determine $\hat{\theta}_{ML}$ for $L(\theta|\mathfrak{Z})$ given in (26-9) (subject to the constraint of the Kalman filter)? No simple closed-form solution is possible, because θ enters into $L(\theta|\mathfrak{Z})$ in a complicated *nonlinear* manner. The only way presently known to obtain $\hat{\theta}_{ML}$ is by means of mathematical programming.

The most effective optimization methods to determine $\hat{\theta}_{ML}$ require the computation of the gradient of $L(\theta|\mathfrak{Z})$ as well as the Hessian matrix, or a pseudo-Hessian matrix of $L(\theta|\mathfrak{Z})$. The Marquardt-Levenberg algorithm (also known as the Levenberg-Marquardt algorithm [Bard, 1970; Marquardt, 1963]), for example, has the form

$$\hat{\theta}_{ML}^{i+1} = \hat{\theta}_{ML}^{i} - (\mathbf{H}_i + \mathbf{D}_i)^{-1}\mathbf{g}_i \qquad i = 0, 1, \ldots \tag{26-17}$$

where $\mathbf{g}_i$ denotes the gradient

$$\mathbf{g}_i = \left.\frac{\partial L(\theta|\mathfrak{Z})}{\partial \theta}\right|_{\theta = \hat{\theta}_{ML}^{i}} \tag{26-18}$$

$\mathbf{H}_i$ denotes a pseudo-Hessian

$$\mathbf{H}_i \simeq \left.\frac{\partial^2 L(\theta|\mathfrak{Z})}{\partial \theta^2}\right|_{\theta = \hat{\theta}_{ML}^{i}} \tag{26-19}$$

and $\mathbf{D}_i$ is a diagonal matrix chosen to force $\mathbf{H}_i + \mathbf{D}_i$ to be positive definite, so that $(\mathbf{H}_i + \mathbf{D}_i)^{-1}$ will always be computable.

We do not propose to discuss details of the Marquardt-Levenberg algorithm. The interested reader should consult the preceding references for general discussions, and Mendel (1983) or Gupta and Mehra (1974) for discussions directly related to the application of this algorithm to the present problem, maximization of $L(\theta|\mathfrak{Z})$.

We direct our attention to the calculations of $\mathbf{g}_i$ and $\mathbf{H}_i$. The gradient of $L(\boldsymbol{\theta}|\mathfrak{Z})$ will require the calculations of

$$\frac{\partial \tilde{\mathbf{z}}_{\boldsymbol{\theta}}(j|j-1)}{\partial \boldsymbol{\theta}} \quad \text{and} \quad \frac{\partial \mathcal{N}_{\boldsymbol{\theta}}(j|j-1)}{\partial \boldsymbol{\theta}}$$

The innovations depend upon $\hat{\mathbf{x}}_{\boldsymbol{\theta}}(j|j-1)$; hence, in order to compute $\partial \tilde{\mathbf{z}}_{\boldsymbol{\theta}}(j|j-1)/\partial \boldsymbol{\theta}$, we must compute $\partial \hat{\mathbf{x}}_{\boldsymbol{\theta}}(j|j-1)/\partial \boldsymbol{\theta}$. A Kalman filter must be used to compute $\hat{\mathbf{x}}_{\boldsymbol{\theta}}(j|j-1)$; but this filter requires the following sequence of calculations: $\mathbf{P}_{\boldsymbol{\theta}}(k|k) \rightarrow \mathbf{P}_{\boldsymbol{\theta}}(k+1|k) \rightarrow \mathbf{K}_{\boldsymbol{\theta}}(k+1) \rightarrow \hat{\mathbf{x}}_{\boldsymbol{\theta}}(k+1|k) \rightarrow \hat{\mathbf{x}}_{\boldsymbol{\theta}}(k+1|k+1)$. Taking the partial derivative of the prediction equation with respect to θ_i we find that

$$\frac{\partial \hat{\mathbf{x}}_{\boldsymbol{\theta}}(k+1|k)}{\partial \theta_i} = \boldsymbol{\Phi}_{\boldsymbol{\theta}} \frac{\partial \hat{\mathbf{x}}_{\boldsymbol{\theta}}(k|k)}{\partial \theta_i} + \frac{\partial \boldsymbol{\Phi}_{\boldsymbol{\theta}}}{\partial \theta_i} \hat{\mathbf{x}}_{\boldsymbol{\theta}}(k|k) + \frac{\partial \boldsymbol{\Psi}_{\boldsymbol{\theta}}}{\partial \theta_i} \mathbf{u}(k) \qquad i = 1, 2, \ldots, d \tag{26-20}$$

We see that to compute $\partial \hat{\mathbf{x}}_{\boldsymbol{\theta}}(k+1|k)/\partial \theta_i$, we must also compute $\partial \hat{\mathbf{x}}_{\boldsymbol{\theta}}(k|k)/\partial \theta_i$. Taking the partial derivative of the correction equation with respect to θ_i, we find that

$$\begin{aligned}\frac{\partial \hat{\mathbf{x}}_{\boldsymbol{\theta}}(k+1|k+1)}{\partial \theta_i} = {} & \frac{\partial \hat{\mathbf{x}}_{\boldsymbol{\theta}}(k+1|k)}{\partial \theta_i} + \frac{\partial \mathbf{K}_{\boldsymbol{\theta}}(k+1)}{\partial \theta_i}[\mathbf{z}(k+1) - \mathbf{H}_{\boldsymbol{\theta}}\hat{\mathbf{x}}_{\boldsymbol{\theta}}(k+1|k)] \\ & - \mathbf{K}_{\boldsymbol{\theta}}(k+1)\left[\frac{\partial \mathbf{H}_{\boldsymbol{\theta}}}{\partial \theta_i}\hat{\mathbf{x}}_{\boldsymbol{\theta}}(k+1|k)\right. \\ & \left. + \mathbf{H}_{\boldsymbol{\theta}} \frac{\partial \hat{\mathbf{x}}_{\boldsymbol{\theta}}(k+1|k)}{\partial \theta_i}\right] \qquad i = 1, 2, \ldots, d\end{aligned} \tag{26-21}$$

Observe that to compute $\partial \hat{\mathbf{x}}_{\boldsymbol{\theta}}(k+1|k+1)/\partial \theta_i$, we must also compute $\partial \mathbf{K}_{\boldsymbol{\theta}}(k+1)/\partial \theta_i$. We leave it to the reader to show that the calculation of $\partial \mathbf{K}_{\boldsymbol{\theta}}(k+1)/\partial \theta_i$ requires the calculation of $\partial \mathbf{P}_{\boldsymbol{\theta}}(k+1|k)/\partial \theta_i$, which in turn requires the calculation of $\partial \mathbf{P}_{\boldsymbol{\theta}}(k+1|k+1)/\partial \theta_i$.

The system of equations

$$\begin{array}{lll} \partial \hat{\mathbf{x}}_{\boldsymbol{\theta}}(k+1|k)/\partial \theta_i & \partial \hat{\mathbf{x}}_{\boldsymbol{\theta}}(k+1|k+1)/\partial \theta_i & \\ \partial \mathbf{K}_{\boldsymbol{\theta}}(k+1)/\partial \theta_i & \partial \mathbf{P}_{\boldsymbol{\theta}}(k+1|k)/\partial \theta_i & \partial \mathbf{P}_{\boldsymbol{\theta}}(k+1|k+1)/\partial \theta_i \end{array}$$

is called a Kalman filter sensitivity system. It is a linear system of equations, just as the Kalman filter, which is not only driven by measurements $\mathbf{z}(k+1)$ [e.g., see (26-21)] but is also driven by the Kalman filter [e.g., see (26-20) and (26-21)]. We need a total of d such sensitivity systems, one for each of the d unknown parameters in $\boldsymbol{\theta}$.

Each system of sensitivity equations requires about as much computation as a Kalman filter. Observe, however, that Kalman filter quantities are used by the sensitivity equations; hence, the Kalman filter must be run together with the d sets of sensitivity equations. This procedure for recursively calculating the gradient $\partial L(\boldsymbol{\theta}|\mathfrak{Z})/\partial\boldsymbol{\theta}$ therefore requires about as much computation as $d + 1$ Kalman filters. The sensitivity systems are totally uncoupled and lend themselves quite naturally to parallel processing (see Figure 26-1).

The Hessian matrix of $L(\boldsymbol{\theta}|\mathfrak{Z})$ is quite complicated, involving not only first derivatives of $\tilde{\mathbf{z}}_{\boldsymbol{\theta}}(j|j-1)$ and $\mathcal{N}_{\boldsymbol{\theta}}(j|j-1)$, but also their second derivatives. The pseudo-Hessian matrix of $L(\boldsymbol{\theta}|\mathfrak{Z})$ ignores all the second derivative terms; hence, it is relatively easy to compute because all the first derivative terms have already been computed in order to calculate the gradient of $L(\boldsymbol{\theta}|\mathfrak{Z})$. Justification for neglecting the second derivative terms is given by Gupta and Mehra (1974), who show that as $\hat{\boldsymbol{\theta}}_{ML}$ approaches $\boldsymbol{\theta}_T$, the expected value of the dropped terms goes to zero.

The estimation literature is filled with many applications of maximum-likelihood state and parameter estimation. For example, Mendel (1983b) applies it to seismic data processing, Mehra and Tyler (1973) apply it to aircraft parameter identification and McLaughlin (1980) applies it to groundwater flow.

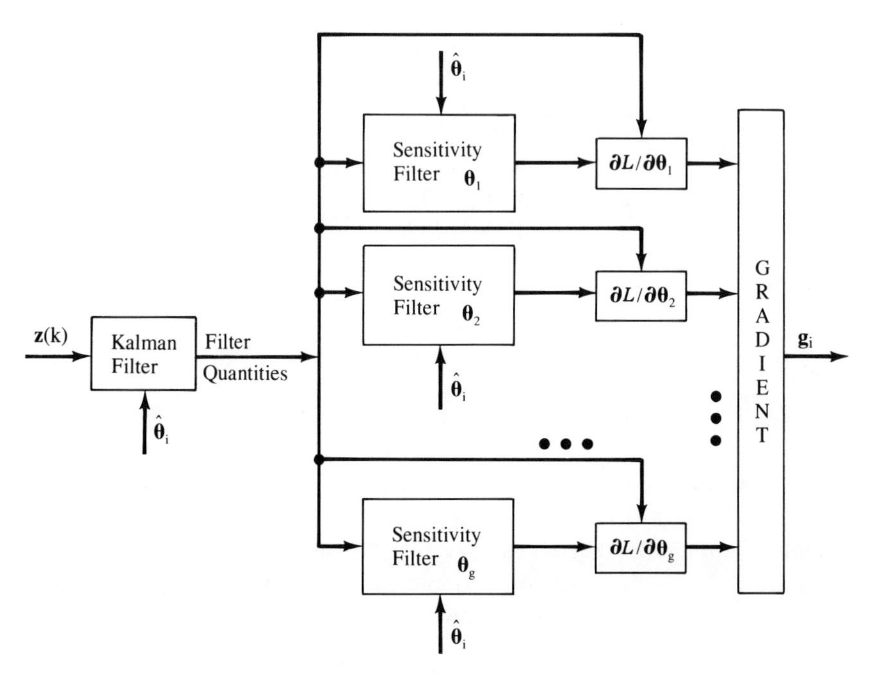

Figure 26-1 Calculations needed to compute gradient vector $\mathbf{g}_i$. Note that $\boldsymbol{\theta}_j$ denotes the jth component of $\boldsymbol{\theta}$ (Mendel, 1983b, © 1983, Academic Press, Inc.).

A STEADY-STATE APPROXIMATION

Suppose our basic state variable model is time-invariant and stationary so that $\bar{\mathbf{P}} = \lim_{j\to\infty} \mathbf{P}(j|j-1)$ exists. Let

$$\bar{\mathcal{N}} = \mathbf{H}\bar{\mathbf{P}}\mathbf{H}' + \mathbf{R} \tag{26-22}$$

and

$$\bar{L}(\boldsymbol{\theta}|\mathfrak{X}) = -\frac{1}{2}\sum_{j=1}^{N} \tilde{\mathbf{z}}'(j|j-1)\bar{\mathcal{N}}^{-1}\tilde{\mathbf{z}}(j|j-1) - \frac{1}{2}N\ln|\bar{\mathcal{N}}| \tag{26-23}$$

Log-likelihood function $\bar{L}(\boldsymbol{\theta}|\mathfrak{X})$ is a steady-state approximation of $L(\boldsymbol{\theta}|\mathfrak{X})$. The steady-state Kalman filter used to compute $\tilde{\mathbf{z}}(j|j-1)$ and $\bar{\mathcal{N}}$ is

$$\hat{\mathbf{x}}(k+1|k) = \boldsymbol{\Phi}\hat{\mathbf{x}}(k|k) + \boldsymbol{\Psi}\mathbf{u}(k) \tag{26-24}$$

$$\begin{aligned}\hat{\mathbf{x}}(k+1|k+1) = \hat{\mathbf{x}}(k+1|k) + \bar{\mathbf{K}}[\mathbf{z}(k+1) \\ - \mathbf{H}\hat{\mathbf{x}}(k+1|k)]\end{aligned} \tag{26-25}$$

in which $\bar{\mathbf{K}}$ is the steady-state Kalman gain matrix.

Recall that

$$\boldsymbol{\theta} = \text{col (elements of } \boldsymbol{\Phi}, \boldsymbol{\Gamma}, \boldsymbol{\Psi}, \mathbf{H}, \mathbf{Q}, \text{ and } \mathbf{R}) \tag{26-26}$$

We now make the following transformations of variables:

$$\left.\begin{aligned} &\boldsymbol{\Phi}\to\boldsymbol{\Phi} \\ &\boldsymbol{\Psi}\to\boldsymbol{\Psi} \\ &\mathbf{H}\to\mathbf{H} \\ &(\boldsymbol{\Phi},\boldsymbol{\Gamma},\boldsymbol{\Psi},\mathbf{H},\mathbf{Q},\mathbf{R})\to\bar{\mathcal{N}} \\ &(\boldsymbol{\Phi},\boldsymbol{\Gamma},\boldsymbol{\Psi},\mathbf{H},\mathbf{Q},\mathbf{R})\to\bar{\mathbf{K}} \end{aligned}\right\} \tag{26-27}$$

We ignore $\boldsymbol{\Gamma}$ (initially) for reasons that are explained below Equation (26-32). Let

$$\boldsymbol{\phi} = \text{col (elements of } \boldsymbol{\Phi}, \boldsymbol{\Psi}, \mathbf{H}, \bar{\mathcal{N}}, \text{ and } \bar{\mathbf{K}}) \tag{26-28}$$

where $\boldsymbol{\phi}$ is $p \times 1$, and view $\bar{L}$ as a function of $\boldsymbol{\phi}$, i.e.,

$$\bar{L}(\boldsymbol{\phi}|\mathfrak{X}) = -\frac{1}{2}\sum_{j=1}^{N} \tilde{\mathbf{z}}'_{\phi}(j|j-1)\bar{\mathcal{N}}_{\phi}^{-1}\tilde{\mathbf{z}}_{\phi}(j|j-1) - \frac{1}{2}N\ln|\bar{\mathcal{N}}_{\phi}| \tag{26-29}$$

Instead of finding $\hat{\boldsymbol{\theta}}_{ML}$ that maximizes $\bar{L}(\boldsymbol{\theta}|\mathfrak{X})$, subject to the constraints of a full-blown Kalman filter, we now propose to find $\hat{\boldsymbol{\phi}}_{ML}$ that maximizes $\bar{L}(\boldsymbol{\phi}|\mathfrak{X})$, subject to the constraints of the following filter:

$$\hat{\mathbf{x}}_{\phi}(k+1|k) = \boldsymbol{\Phi}_{\phi}\hat{\mathbf{x}}_{\phi}(k|k) + \boldsymbol{\Psi}_{\phi}\mathbf{u}(k) \tag{26-30}$$

$$\hat{\mathbf{x}}_{\phi}(k+1|k+1) = \hat{\mathbf{x}}_{\phi}(k+1|k) + \bar{\mathbf{K}}_{\phi}\tilde{\mathbf{z}}_{\phi}(k+1|k) \tag{26-31}$$

and

$$\tilde{\mathbf{z}}_{\phi}(k+1|k) = \mathbf{z}(k+1) - \mathbf{H}_{\phi}\hat{\mathbf{x}}_{\phi}(k+1|k) \qquad (26\text{-}32)$$

Once we have computed $\hat{\boldsymbol{\phi}}_{\mathrm{ML}}$ we can compute $\hat{\boldsymbol{\theta}}_{\mathrm{ML}}$ by inverting the transformations in (26-27). Of course, when we do this, we are also using the invariance property of maximum-likelihood estimates.

Observe that $\overline{L}(\boldsymbol{\phi}|\mathfrak{Z})$ in (26-29) and the filter in (26-30)–(26-32) do not depend on $\boldsymbol{\Gamma}$; hence, we have not included $\boldsymbol{\Gamma}$ in any definition of $\boldsymbol{\phi}$. We explain how to reconstruct $\boldsymbol{\Gamma}$ from $\hat{\boldsymbol{\phi}}_{\mathrm{ML}}$ following Equation (26-44).

Because maximum-likelihood estimates are asymptotically efficient (Lesson 11), once we have determined $\hat{\overline{\mathbf{K}}}_{\mathrm{ML}}$, the filter in (26-30) and (26-31) will be the steady-state Kalman filter.

The major advantage of this steady-state approximation is that the filter sensitivity equations are greatly simplified. When $\overline{\mathbf{K}}$ and $\overline{\mathcal{N}}$ are treated as matrices of unknown parameters we do not need the predicted and corrected error-covariance matrices to "compute" $\overline{\mathbf{K}}$ and $\overline{\mathcal{N}}$. The sensitivity equations for (26-32), (26-30), and (26-31) are

$$\frac{\partial \tilde{\mathbf{z}}_{\phi}(k+1|k)}{\partial \phi_i} = -\mathbf{H}_{\phi}\frac{\partial \hat{\mathbf{x}}_{\phi}(k+1|k)}{\partial \phi_i} - \frac{\partial \mathbf{H}_{\phi}}{\partial \phi_i}\hat{\mathbf{x}}_{\phi}(k+1|k) \qquad (26\text{-}33)$$

$$\frac{\partial \hat{\mathbf{x}}_{\phi}(k+1|k)}{\partial \phi_i} = \boldsymbol{\Phi}_{\phi}\frac{\partial \hat{\mathbf{x}}_{\phi}(k|k)}{\partial \phi_i} + \frac{\partial \boldsymbol{\Phi}_{\phi}}{\partial \phi_i}\hat{\mathbf{x}}_{\phi}(k|k) + \frac{\partial \boldsymbol{\Psi}_{\phi}}{\partial \phi_i}\mathbf{u}(k) \qquad (26\text{-}34)$$

and

$$\begin{aligned}\frac{\partial \hat{\mathbf{x}}_{\phi}(k+1|k+1)}{\partial \phi_i} = {} & \frac{\partial \hat{\mathbf{x}}_{\phi}(k+1|k)}{\partial \phi_i} + \frac{\partial \overline{\mathbf{K}}_{\phi}}{\partial \phi_i}[\mathbf{z}(k+1) - \mathbf{H}_{\phi}\hat{\mathbf{x}}_{\phi}(k|k+1)] \\ & - \overline{\mathbf{K}}_{\phi}\mathbf{H}_{\phi}\frac{\partial \hat{\mathbf{x}}_{\phi}(k+1|k)}{\partial \phi_i} \\ & - \overline{\mathbf{K}}_{\phi}\frac{\partial \mathbf{H}_{\phi}}{\partial \phi_i}\hat{\mathbf{x}}_{\phi}(k+1|k) \qquad (26\text{-}35)\end{aligned}$$

where $i = 1, 2, \ldots, p$. Note that $\partial \overline{\mathbf{K}}_{\phi}/\partial \phi_i$ is zero for all ϕ_i not in $\overline{\mathbf{K}}_{\phi}$ and is a matrix filled with zeros and a single unity value for ϕ_i in $\overline{\mathbf{K}}_{\phi}$.

There are more elements in $\boldsymbol{\phi}$ than in $\boldsymbol{\theta}$, because $\overline{\mathcal{N}}$ and $\overline{\mathbf{K}}$ have more unknown elements in them than do $\mathbf{Q}$ and $\mathbf{R}$, i.e., $p > d$. Additionally, $\overline{\mathcal{N}}$ does not appear in the filter equations; it only appears in $\overline{L}(\boldsymbol{\phi}|\mathfrak{Z})$. It is, therefore, possible to obtain a closed-form solution for $\hat{\overline{\mathcal{N}}}_{\mathrm{ML}}$.

Theorem 26-2. *A closed-form solution for matrix $\hat{\overline{\mathcal{N}}}_{\mathrm{ML}}$ is*

$$\hat{\overline{\mathcal{N}}}_{\mathrm{ML}} = \frac{1}{N}\sum_{j=1}^{N} \tilde{\mathbf{z}}_{\phi}(j|j-1)\tilde{\mathbf{z}}'_{\phi}(j|j-1) \qquad (26\text{-}36)$$

Proof. To determine $\hat{\bar{\mathcal{N}}}_{ML}$ we must set $\partial \bar{L}(\phi|\mathfrak{X})/\partial \bar{\mathcal{N}}_\phi = \mathbf{0}$ and solve the resulting equation for $\bar{\mathcal{N}}_{ML}$. This is most easily accomplished by applying gradient matrix formulas to (26-29), that are given in Schweppe (1974). Doing this, we obtain

$$\frac{\partial \bar{L}(\phi|\mathfrak{X})}{\partial \bar{\mathcal{N}}} = \sum_{j=1}^{N} [\bar{\mathcal{N}}^{-1}\tilde{\mathbf{z}}_\phi(j|j-1)\tilde{\mathbf{z}}'_\phi(j|j-1)\bar{\mathcal{N}}^{-1}]' - N(\bar{\mathcal{N}}^{-1})' = \mathbf{0} \tag{26-37}$$

whose solution is $\hat{\bar{\mathcal{N}}}_{ML}$ in (26-36). □

Observe that $\hat{\bar{\mathcal{N}}}_{ML}$ is the sample steady-state-covariance matrix of $\tilde{\mathbf{z}}_\phi$; i.e., as

$$\hat{\phi}_{ML} \to \hat{\phi}_T \qquad \hat{\bar{\mathcal{N}}}_{ML} \to \lim_{j\to\infty} \text{cov}\,[\tilde{\mathbf{z}}(j|j-1)]$$

Suppose we are also interested in determining $\hat{\mathbf{Q}}_{ML}$ and $\hat{\mathbf{R}}_{ML}$. How do we obtain these quantities from $\hat{\phi}_{ML}$?

As in Lesson 19, we let $\bar{\mathbf{K}}$, $\bar{\mathbf{P}}$, and $\bar{\mathbf{P}}_1$ denote the steady-state values of $\mathbf{K}(k+1)$, $\mathbf{P}(k+1|k)$, and $\mathbf{P}(k|k)$, respectively, where

$$\bar{\mathbf{K}} = \bar{\mathbf{P}}\mathbf{H}'(\mathbf{H}\bar{\mathbf{P}}\mathbf{H}' + \mathbf{R})^{-1} = \bar{\mathbf{P}}\mathbf{H}'\bar{\mathcal{N}}^{-1} \tag{26-38}$$

$$\bar{\mathbf{P}} = \mathbf{\Phi}\bar{\mathbf{P}}_1\mathbf{\Phi}' + \mathbf{\Gamma}\mathbf{Q}\mathbf{\Gamma}' \tag{26-39}$$

and

$$\bar{\mathbf{P}}_1 = (\mathbf{I} - \bar{\mathbf{K}}\mathbf{H})\bar{\mathbf{P}} \tag{26-40}$$

Additionally, we know that

$$\bar{\mathcal{N}} = \mathbf{H}\bar{\mathbf{P}}\mathbf{H}' + \mathbf{R} \tag{(26-41)}$$

By the invariance property of maximum-likelihood estimates, we know that

$$\hat{\bar{\mathcal{N}}}_{ML} = \widehat{(\mathbf{H}\bar{\mathbf{P}}\mathbf{H}' + \mathbf{R})}_{ML} = \hat{\mathbf{H}}_{ML}\bar{\mathbf{P}}\hat{\mathbf{H}}'_{ML} + \hat{\mathbf{R}}_{ML} \tag{26-42}$$

and

$$\hat{\bar{\mathbf{K}}}_{ML} = \widehat{(\bar{\mathbf{P}}\mathbf{H}'\bar{\mathcal{N}}^{-1})}_{ML} = \bar{\mathbf{P}}\hat{\mathbf{H}}'_{ML}(\hat{\bar{\mathcal{N}}}_{ML})^{-1} \tag{26-43}$$

Solving (26-43) for $\bar{\mathbf{P}}\hat{\mathbf{H}}'_{ML}$ and substituting the resulting expression into (26-42), we obtain the following solution for $\hat{\mathbf{R}}_{ML}$,

$$\hat{\mathbf{R}}_{ML} = (\mathbf{I} - \hat{\mathbf{H}}_{ML}\hat{\bar{\mathbf{K}}}_{ML})\hat{\bar{\mathcal{N}}}_{ML} \tag{26-44}$$

No closed-form solution exists for $\hat{\mathbf{Q}}_{ML}$. Substituting $\hat{\mathbf{\Phi}}_{ML}$, $\hat{\mathbf{H}}_{ML}$, $\hat{\bar{\mathbf{K}}}_{ML}$, and $\hat{\bar{\mathcal{N}}}_{ML}$ into (26-38)–(26-40), and combining (26-39) and (26-40), we obtain the following two equations

$$\bar{\mathbf{P}} = \hat{\mathbf{\Phi}}_{ML}(\mathbf{I} - \hat{\bar{\mathbf{K}}}_{ML}\hat{\mathbf{H}}_{ML})\bar{\mathbf{P}}\hat{\mathbf{\Phi}}'_{ML} + \widehat{(\mathbf{\Gamma}\mathbf{Q}\mathbf{\Gamma}')}_{ML} \tag{26-45}$$

and

$$\bar{\mathbf{P}}\hat{\mathbf{H}}'_{ML} = \hat{\bar{\mathbf{K}}}_{ML}\hat{\bar{\mathcal{N}}}_{ML} \tag{26-46}$$

These equations must be solved simultaneously for $\bar{\mathbf{P}}$ and $\widehat{(\mathbf{\Gamma Q \Gamma'})}_{ML}$ using iterative numerical techniques. For details, see Mehra (1970a).

Note finally, that the best for which we can hope by this approach is not $\hat{\mathbf{\Gamma}}_{ML}$ and $\hat{\mathbf{Q}}_{ML}$, but only $\widehat{(\mathbf{\Gamma Q \Gamma'})}_{ML}$. This is due to the fact that, when $\mathbf{\Gamma}$ and $\mathbf{Q}$ are both unknown, there will be an ambiguity in their determination, i.e., the term $\mathbf{\Gamma w}(k)$ which appears in our basic state-variable model [for which $\mathbf{E}\{\mathbf{w}(k)\mathbf{w}'(k)\} = \mathbf{Q}$] cannot be distinguished from the term $\mathbf{w}_1(k)$, for which

$$\mathbf{E}\{\mathbf{w}_1(k)\mathbf{w}_1'(k)\} = \mathbf{Q}_1 = \mathbf{\Gamma Q \Gamma'} \tag{26-47}$$

This observation is also applicable to the original problem formulation wherein we obtained $\hat{\boldsymbol{\theta}}_{ML}$ directly; i.e., when both $\mathbf{\Gamma}$ and $\mathbf{Q}$ are unknown, we should really choose

$$\boldsymbol{\theta} = \text{col (elements of } \mathbf{\Phi}, \mathbf{\Psi}, \mathbf{H}, \mathbf{\Gamma Q \Gamma'}, \text{and } \mathbf{R}) \tag{26-48}$$

In summary, when our basic state-variable model is time-invariant and stationary, we can first obtain $\hat{\boldsymbol{\phi}}_{ML}$ by maximizing $\bar{L}(\boldsymbol{\phi}|\mathfrak{Z})$ given in (26-29), subject to the constraints of the simple filter in (26-30), (26-31), and (26-32). A mathematical programming method must be used to obtain those elements of $\hat{\boldsymbol{\phi}}_{ML}$ associated with $\hat{\mathbf{\Phi}}_{ML}$, $\hat{\mathbf{\Psi}}_{ML}$, $\hat{\mathbf{H}}_{ML}$, and $\hat{\bar{\mathbf{K}}}_{ML}$. The closed-form solution, given in (26-36), is used for $\hat{\bar{\mathcal{N}}}_{ML}$. Finally, if we want to reconstruct $\hat{\mathbf{R}}_{ML}$ and $\widehat{(\mathbf{\Gamma Q \Gamma'})}_{ML}$, we use (26-44) for the former and must solve (26-45) and (26-46) for the latter.

Example 26-1 (Mehra, 1971)

The following fourth-order system, which represents the short period dynamics and the first bending mode of a missile, was simulated:

$$\begin{pmatrix} x_1(k+1) \\ x_2(k+1) \\ x_3(k+1) \\ x_4(k+1) \end{pmatrix} = \begin{pmatrix} 0 & 0 & 0 & 0 \\ 0 & 1 & 0 & 0 \\ 0 & 0 & 1 & 0 \\ -\alpha_1 & -\alpha_2 & -\alpha_3 & -\alpha_4 \end{pmatrix} \begin{pmatrix} x_1(k) \\ x_2(k) \\ x_3(k) \\ x_4(k) \end{pmatrix} + \begin{pmatrix} 0 \\ 1 \\ 0 \\ 1 \end{pmatrix} w(k) \tag{26-49}$$

$$z(k+1) = x_1(k+1) + v(k+1) \tag{26-50}$$

For this model, it was assumed that $\mathbf{x}(0) = \mathbf{0}$, $q = 1.0$, $r = 0.25$, $\alpha_1 = -0.656$, $\alpha_2 = 0.784$, $\alpha_3 = -0.18$, and $\alpha_4 = 1.0$.

Using measurements generated from the simulation, maximum-likelihood estimates were obtained for $\boldsymbol{\phi}$, where

$$\boldsymbol{\phi} = \text{col}\,(\alpha_1, \alpha_2, \alpha_3, \alpha_4, \bar{\mathcal{N}}, \bar{k}_1, \bar{k}_2, \bar{k}_3, \bar{k}_4) \tag{26-51}$$

In (26-51), $\bar{\mathcal{N}}$ is a scalar because, in this example, $z(k)$ is a scalar. Additionally, it was assumed that $\mathbf{x}(0)$ was known exactly. According to Mehra (1971, pg. 30) "The starting values for the maximum likelihood scheme were obtained using a correlation technique given in Mehra (1970b). The results of successive iterations are shown in Table 26-1. The variances of the estimates obtained from the matrix of second partial derivatives (i.e., the Hessian matrix of $\bar{L}$) are also given. For comparison purposes, results obtained by using 1000 data points and 100 data points are given." □

TABLE 26-1 Parameter Estimates for Missile Example

Iteration	$L \times 10^{-3}$	$\overline{\mathcal{N}}$	α_1	α_2	α_3	α_4	$\bar{k}_1$	$\bar{k}_2$	$\bar{k}_3$	$\bar{k}_4$
0										
Results from correlation technique, Mehra (1970b).										
1	−1.0706	2.3800	−0.5965	0.8029	−0.1360	0.8696	0.6830	0.2837	0.4191	0.8207
M.L. estimates using 1000 points										
2	−1.0660	2.3811	−0.5938	0.8029	−0.1338	0.8759	0.6803	0.2840	0.4200	0.8312
3	−1.0085	2.4026	−0.6054	0.7452	−0.1494	0.9380	0.6304	0.2888	0.4392	1.0311
4	−0.9798	2.4409	−0.6036	0.8161	−0.1405	0.8540	0.6801	0.3210	0.6108	1.1831
5	−0.9785	2.4412	−0.5999	0.8196	−0.1370	0.8580	0.6803	0.3214	0.6107	1.1835
6	−0.9771	2.4637	−0.6014	0.8086	−0.1503	0.8841	0.7068	0.3479	0.6059	1.2200
7	−0.9769	2.4603	−0.6023	0.8130	−0.1470	0.8773	0.7045	0.3429	0.6106	1.2104
8	−0.9744	2.5240	−0.6313	0.8105	−0.1631	0.9279	0.7990	0.3756	0.6484	1.2589
9	−0.9743	2.5241	−0.6306	0.8108	−0.1622	0.9296	0.7989	0.3749	0.6480	1.2588
10	−0.9734	2.5270	−0.6374	0.7961	−0.1630	0.9505	0.7974	0.3568	0.6378	1.2577
11	−0.9728	2.5313	−0.6482	0.7987	−0.1620	0.9577	0.8103	0.3443	0.6403	1.2351
12	−0.9720	2.5444	−0.6602	0.7995	−0.1783	0.9866	0.8487	0.3303	0.6083	1.2053
13	−0.9714	2.5600	−0.6634	0.7919	−0.2036	1.0280	0.8924	0.3143	0.6014	1.2054
14	−0.9711	2.5657	−0.6624	0.7808	−0.2148	1.0491	0.9073	0.3251	0.6122	1.2200
M.L. estimates using 100 points										
30	−0.9659	2.620	−0.6094	0.7663	−0.1987	1.0156	1.24	0.136	0.454	1.103
Actual values										
	−0.94	2.557	−0.6560	0.7840	−0.1800	1.0000	0.8937	0.2957	0.6239	1.2510
Estimates of standard deviation using 1000 points										
		0.0317	0.0277	0.0247	0.0275	0.0261	0.0302	0.0323	0.0302	0.029
Estimates of standard deviation using 100 points										
		0.149	0.104	0.131	0.084	0.184	0.303	0.092	0.082	0.09

Source: Mehra (1971, pg. 30), © 1971, AIAA.

PROBLEMS

26-1. Obtain the sensitivity equations for $\partial \mathbf{K}_\theta(k+1)/\partial\theta_i$, $\partial \mathbf{P}_\theta(k+1|k)/\partial\theta_i$ and $\partial \mathbf{P}_\theta(k+1|k+1)/\partial\theta_i$. Explain why the sensitivity system for $\partial \hat{\mathbf{x}}_\theta(k+1|k)/\partial\theta_i$ and $\partial \hat{\mathbf{x}}_\theta(k+1|k+1)/\partial\theta_i$ is *linear*.

26-2. Compute formulas for $\mathbf{g}_i$ and $\mathbf{H}_i$. Then simplify $\mathbf{H}_i$ to a *pseudo-Hessian*.

26-3. In the first-order system $x(k+1) = ax(k) + w(k)$, and $z(k+1) = x(k+1) + v(k+1)$, $k = 1, 2, \ldots, N$, a is an unknown parameter that is to be estimated. Sequences $w(k)$ and $v(k)$ are, as usual, mutually uncorrelated and white, and, $w(k) \sim N(w(k); 0, 1)$ and $v(k) \sim N(v(k); 0, 1/2)$. Explain, using equations and a flowchart, how parameter a can be estimated using a MLE.

26-4. Repeat the preceding problem where all conditions are the same except that now $w(k)$ and $v(k)$ are correlated, and $\mathbf{E}\{w(k)v(k)\} = 1/4$.

26-5. We are interested in estimating the parameters a and r in the following first-order system:

$$x(k+1) + ax(k) = w(k)$$
$$z(k) = x(k) + v(k) \qquad k = 1, 2, \ldots, N$$

Signals $w(k)$ and $v(k)$ are mutually uncorrelated, white, and Gaussian, and, $\mathbf{E}\{w^2(k)\} = 1$ and $\mathbf{E}\{n^2(k)\} = r$.

(a) Let $\boldsymbol{\theta} = \text{col}\,(a,r)$. What is the equation for the log-likelihood function?

(b) Prepare a macro flow chart that depicts the sequence of calculations required to maximize $L(\boldsymbol{\theta}|\mathfrak{Z})$. Assume an optimization algorithm is used which requires gradient information about $L(\boldsymbol{\theta}|\mathfrak{Z})$.

(c) Write out the Kalman filter sensitivity equations for parameters a and r.

26-6. Develop the sensitivity equations for the case considered in Lesson 11, i.e., for the case where the only uncertainty present in the state-variable model is measurement noise. Begin with $L(\boldsymbol{\theta}|\mathfrak{Z})$ in (11-42).

26-7. Refer to Problem 24-7. Explain, using equations and a flowchart, how to obtain MLE's of the unknown parameters for:

(a) equation for the unsteady operation of a synchronous motor, in which C and p are unknown;

(c) Duffing's equation, in which C, α, and β are unknown;

(c) Van der Pol's equation, in which ϵ is unknown; and

(d) Hill's equation, in which a and b are unknown.

Lesson 27

Kalman-Bucy Filtering

INTRODUCTION

The Kalman-Bucy filter is the continuous-time counterpart to the Kalman filter. It is a continuous-time minimum-variance filter that provides state estimates for continuous-time dynamical systems that are described by linear, (possibly) time-varying, and (possibly) nonstationary ordinary differential equations.

The Kalman-Bucy filter (KBF) can be derived in a number of different ways, including the following three:

1. Use a formal limiting procedure to obtain the KBF from the KF (e.g., Meditch, 1969).
2. Begin by assuming the optimal estimator is a linear transformation of *all* measurements. Use a calculus of variations argument or the orthogonality principle to obtain the Wiener-Hopf integral equation. Embedded within this equation is the filter kernal. Take the derivative of the Wiener-Hopf equation to obtain a differential equation which is the KBF (Meditch, 1969).
3. Begin by assuming a linear differential equation structure for the KBF, one that contains an unknown time-varying gain matrix that weights the difference between the measurement made at time t and the estimate of that measurement. Choose the gain matrix that minimizes the mean-squared error (Athans and Tse, 1967).

We shall briefly describe the first and third approaches, but first we must define our continuous-time model and formally state the problem we wish to solve.

SYSTEM DESCRIPTION

Our continuous-time system is described by the following state-variable model,

$$\dot{\mathbf{x}}(t) = \mathbf{F}(t)\mathbf{x}(t) + \mathbf{G}(t)\mathbf{w}(t) \tag{27-1}$$

$$\mathbf{z}(t) = \mathbf{H}(t)\mathbf{x}(t) + \mathbf{v}(t) \tag{27-2}$$

where $\mathbf{x}(t)$ is $n \times 1$, $\mathbf{w}(t)$ is $p \times 1$, $\mathbf{z}(t)$ is $m \times 1$, and $\mathbf{v}(t)$ is $m \times 1$. For simplicity, we have omitted a known forcing function term in state equation (27-1). Matrices $\mathbf{F}(t)$, $\mathbf{G}(t)$, and $\mathbf{H}(t)$ have dimensions which conform to the dimensions of the vector quantities in this state-variable model. Disturbance $\mathbf{w}(t)$ and measurement noise $\mathbf{v}(t)$ are zero-mean white noise processes, which are assumed to be uncorrelated, i.e., $\mathbf{E}\{\mathbf{w}(t)\} = \mathbf{0}$, $\mathbf{E}\{\mathbf{v}(t)\} = \mathbf{0}$,

$$\mathbf{E}\{\mathbf{w}(t)\mathbf{w}'(\tau)\} = \mathbf{Q}(t)\delta(t - \tau) \tag{27-3}$$

$$\mathbf{E}\{\mathbf{v}(t)\mathbf{v}'(\tau)\} = \mathbf{R}(t)\delta(t - \tau) \tag{27-4}$$

and

$$\mathbf{E}\{\mathbf{w}(t)\mathbf{v}'(\tau)\} = \mathbf{0} \tag{27-5}$$

Equations (27-3), (27-4), and (27-5) apply for $t \geq t_0$. Additionally, $\mathbf{R}(t)$ is continuous and positive definite, whereas $\mathbf{Q}(t)$ is continuous and positive semidefinite. Finally, we asume that the initial state vector $\mathbf{x}(t_0)$ may be random, and if it is, it is uncorrelated with both $\mathbf{w}(t)$ and $\mathbf{v}(t)$. The statistics of a random $\mathbf{x}(t_0)$ are

$$\mathbf{E}\{\mathbf{x}(t_0)\} = \mathbf{m}_\mathbf{x}(t_0) \tag{27-6}$$

and

$$\text{cov}\,\{\mathbf{x}(t_0)\} = \mathbf{P}_\mathbf{x}(t_0) \tag{27-7}$$

Measurements $\mathbf{z}(t)$ are assumed to be made for $t_0 \leq t \leq \tau$.

If $\mathbf{x}(t_0)$, $\mathbf{w}(t)$, and $\mathbf{v}(t)$ are jointly Gaussian for all $t \in [t_0,\tau]$, then the KBF will be the optimal estimator of state vector $\mathbf{x}(t)$. We will not make any distributional assumptions about $\mathbf{x}(t_0)$, $\mathbf{w}(t)$, and $\mathbf{v}(t)$ in this lesson, being content to establish the *linear optimal estimator* of $\mathbf{x}(t)$.

NOTATION AND PROBLEM STATEMENT

Our notation for a continuous-time estimate of $\mathbf{x}(t)$ and its associated estimation error parallels our notation for the comparable discrete-time quantities, i.e., $\hat{\mathbf{x}}(t|t)$ denotes the optimal estimate of $\mathbf{x}(t)$ which uses all the measurements $\mathbf{z}(t)$, where $t \geq t_0$, and

$$\tilde{\mathbf{x}}(t|t) = \mathbf{x}(t) - \hat{\mathbf{x}}(t|t) \tag{27-8}$$

The mean-squared state estimation error is

$$J[\tilde{\mathbf{x}}(t|t)] = \mathbf{E}\{\tilde{\mathbf{x}}'(t|t)\tilde{\mathbf{x}}(t|t)\} \quad (27\text{-}9)$$

We shall determine $\hat{\mathbf{x}}(t|t)$ that minimizes $J[\tilde{\mathbf{x}}(t|t)]$, subject to the constraints of our state-variable model and data set.

THE KALMAN-BUCY FILTER

The solution to the problem stated in the preceding section is the Kalman-Bucy Filter, the structure of which is summarized in the following:

Theorem 27-1. *The KBF is described by the vector differential equation*

$$\dot{\hat{\mathbf{x}}}(t|t) = \mathbf{F}(t)\hat{\mathbf{x}}(t|t) + \mathbf{K}(t)[\mathbf{z}(t) - \mathbf{H}(t)\hat{\mathbf{x}}(t|t)] \quad (27\text{-}10)$$

where $t \geq t_0$, $\hat{\mathbf{x}}(t_0|t_0) = \mathbf{m}_{\mathbf{x}}(t_0)$,

$$\mathbf{K}(t) = \mathbf{P}(t|t)\mathbf{H}'(t)\mathbf{R}^{-1}(t) \quad (27\text{-}11)$$

and

$$\dot{\mathbf{P}}(t|t) = \mathbf{F}(t)\mathbf{P}(t|t) + \mathbf{P}(t|t)\mathbf{F}'(t) - \mathbf{P}(t|t)\mathbf{H}'(t)\mathbf{R}^{-1}(t)\mathbf{H}(t)\mathbf{P}(t|t) + \mathbf{G}(t)\mathbf{Q}(t)\mathbf{G}'(t) \quad (27\text{-}12)$$

Equation (27-12), which is a matrix Riccati differential equation, is initialized by $\mathbf{P}(t_0|t_0) = \mathbf{P}_{\mathbf{x}}(t_0)$. □

Matrix $\mathbf{K}(t)$ is the *Kalman-Bucy gain matrix*, and $\mathbf{P}(t|t)$ is the state-estimation-error covariance matrix, i.e.,

$$\mathbf{P}(t|t) = \mathbf{E}\{\tilde{\mathbf{x}}(t|t)\tilde{\mathbf{x}}'(t|t)\} \quad (27\text{-}13)$$

Equation (27-10) can be rewritten as

$$\dot{\hat{\mathbf{x}}}(t|t) = [\mathbf{F}(t) - \mathbf{K}(t)\mathbf{H}(t)]\hat{\mathbf{x}}(t|t) + \mathbf{K}(t)\mathbf{z}(t) \quad (27\text{-}14)$$

which makes it very clear that the KBF is a time-varying filter that processes the measurements linearly to produce $\hat{\mathbf{x}}(t|t)$.

The solution to (27-14) is

$$\hat{\mathbf{x}}(t|t) = \mathbf{\Phi}(t,t_0)\hat{\mathbf{x}}(t_0|t_0) + \int_{t_0}^{t} \mathbf{\Phi}(t,\tau)\mathbf{K}(\tau)\mathbf{z}(\tau)d\tau \quad (27\text{-}15)$$

where the state transition matrix $\mathbf{\Phi}(t,\tau)$ is the solution to the matrix differential equation

$$\left.\begin{aligned} \dot{\mathbf{\Phi}}(t,\tau) &= [\mathbf{F}(t) - \mathbf{K}(t)\mathbf{H}(t)]\mathbf{\Phi}(t,\tau) \\ \mathbf{\Phi}(t,t) &= \mathbf{I} \end{aligned}\right\} \quad (27\text{-}16)$$

If $\hat{\mathbf{x}}(t_0|t_0) = \mathbf{0}$, then

$$\hat{\mathbf{x}}(t|t) = \int_{t_0}^{t} \mathbf{\Phi}(t,\tau)\mathbf{K}(\tau)\mathbf{z}(\tau)d\tau = \int_{t_0}^{t} \mathbf{A}(t,\tau)\mathbf{z}(\tau)d\tau \tag{27-17}$$

where the filter kernel $\mathbf{A}(t,\tau)$ is

$$\mathbf{A}(t,\tau) = \mathbf{\Phi}(t,\tau)\mathbf{K}(\tau) \tag{27-18}$$

The second approach to deriving the KBF, mentioned in the introduction to this chapter, begins by assuming that $\hat{\mathbf{x}}(t|t)$ can be expressed as in (27-17) where $\mathbf{A}(t,\tau)$ is unknown. The mean-squared estimation error is minimized to obtain the following Wiener-Hopf integral equation:

$$\mathbf{E}\{\mathbf{x}(t)\mathbf{z}'(\sigma)\} - \int_{t_0}^{t} \mathbf{A}(t,\tau)\mathbf{E}\{\mathbf{z}(\tau)\mathbf{z}'(\sigma)\}d\tau = \mathbf{0} \tag{27-19}$$

where $t_0 \leq \sigma \leq t$. When this equation is converted into a differential equation, one obtains the KBF described in Theorem 27-1. For the details of this derivation of Theorem 27-1 see Meditch, 1969, Chapter 8.

DERIVATION OF KBF USING A FORMAL LIMITING PROCEDURE

Kalman filter Equation (17-11), expressed as

$$\hat{\mathbf{x}}(k+1|k+1) = \mathbf{\Phi}(k+1,k)\hat{\mathbf{x}}(k|k) + \mathbf{K}(k+1)[\mathbf{z}(k+1) - \mathbf{H}(k+1)\mathbf{\Phi}(k+1,k)\hat{\mathbf{x}}(k|k)] \tag{27-20}$$

can also be written as

$$\hat{\mathbf{x}}(t+\Delta t|t+\Delta t) = \mathbf{\Phi}(t+\Delta t,t)\hat{\mathbf{x}}(t|t) + \mathbf{K}(t+\Delta t)[\mathbf{z}(t+\Delta t) - \mathbf{H}(t+\Delta t)\mathbf{\Phi}(t+\Delta t,t)\hat{\mathbf{x}}(t|t)] \tag{27-21}$$

where we have let $t_k = t$ and $t_{k+1} = t + \Delta t$. In Example 24-3 we showed that

$$\mathbf{\Phi}(t+\Delta t,t) \simeq \mathbf{I} + \mathbf{F}(t)\Delta t + 0(\Delta t^2) \tag{27-22}$$

and

$$\mathbf{Q}_d(t) \simeq \mathbf{G}(t)\mathbf{Q}(t)\mathbf{G}'(t)\Delta t + 0(\Delta t^2) \tag{27-23}$$

Observe that $\mathbf{Q}_d(t)$ can also be written as

$$\mathbf{Q}_d(t) \simeq [\mathbf{G}(t)\Delta t]\left[\frac{\mathbf{Q}(t)}{\Delta t}\right][\mathbf{G}(t)\Delta t]' + 0(\Delta t^2) \tag{27-24}$$

and, if we express $\mathbf{w}_d(k)$ as $\mathbf{\Gamma}(k+1,k)\mathbf{w}(k)$, so that $\mathbf{Q}_d(k) = \mathbf{\Gamma}(k+1,k)\mathbf{Q}(k)\mathbf{\Gamma}'(k+1,k)$, then

$$\mathbf{\Gamma}(t+\Delta t,t) \simeq \mathbf{G}(t)\Delta t + 0(\Delta t^2) \tag{27-25}$$

and

$$\mathbf{Q}(k = t) \Rightarrow \frac{\mathbf{Q}(t)}{\Delta t} \tag{27-26}$$

Equation (27-26) means that we replace $\mathbf{Q}(k = t)$ in the KF by $\mathbf{Q}(t)/\Delta t$. Note that we have encountered a bit of a notational problem here, because we have used $\mathbf{w}(k)$ [and its associated covariance, $\mathbf{Q}(k)$] to denote the disturbance in our discrete-time model, and $\mathbf{w}(t)$ [and its associated intensity, $\mathbf{Q}(t)$] to denote the disturbance in our continuous-time model.

Without going into technical details, we shall also replace $\mathbf{R}(t + \Delta t)$ in the KF by $\mathbf{R}(t + \Delta t)/\Delta t$, i.e.,

$$\mathbf{R}(k + 1 = t + \Delta t) \Rightarrow \mathbf{R}(t + \Delta t)/\Delta t \tag{27-27}$$

See Meditch (1969, pp. 139–142) for an explanation.

Substituting (27-22) into (27-21), and omitting all higher-order terms in Δt, we find that

$$\hat{\mathbf{x}}(t + \Delta t|t + \Delta t) = [\mathbf{I} + \mathbf{F}(t)\Delta t]\hat{\mathbf{x}}(t|t) + \mathbf{K}(t + \Delta t)\{\mathbf{z}(t + \Delta t) - \mathbf{H}(t + \Delta t)[\mathbf{I} + \mathbf{F}(t)\Delta t]\hat{\mathbf{x}}(t|t)\} \tag{27-28}$$

from which it follows that

$$\lim_{\Delta t \to 0} \frac{\hat{\mathbf{x}}(t + \Delta t|t + \Delta t) - \hat{\mathbf{x}}(t|t)}{\Delta t} = \mathbf{F}(t)\hat{\mathbf{x}}(t|t) + \lim_{\Delta t \to 0} \mathbf{K}(t + \Delta t)\{\mathbf{z}(t + \Delta t) - \mathbf{H}(t + \Delta t)[\mathbf{I} + \mathbf{F}(t)\Delta t]\hat{\mathbf{x}}(t|t)\}/\Delta t$$

or

$$\dot{\hat{\mathbf{x}}}(t|t) = \mathbf{F}(t)\hat{\mathbf{x}}(t|t) + \lim_{\Delta t \to 0} \mathbf{K}(t + \Delta t)\{\mathbf{z}(t + \Delta t) - \mathbf{H}(t + \Delta t)[\mathbf{I} + \mathbf{F}(t)\Delta t]\hat{\mathbf{x}}(t|t)\}/\Delta t \tag{27-29}$$

Under suitable regularity conditions, which we shall assume are satisfied here, we can replace the limit of a product of functions by the product of limits, i.e.,

$$\lim_{\Delta t \to 0} \mathbf{K}(t + \Delta t)\{\mathbf{z}(t + \Delta t) - \mathbf{H}(t + \Delta t)[\mathbf{I} + \mathbf{F}(t)\Delta t]\hat{\mathbf{x}}(t|t)\}/\Delta t$$
$$= \lim_{\Delta t \to 0} \frac{\mathbf{K}(t + \Delta t)}{\Delta t} \lim_{\Delta t \to 0} \{\mathbf{z}(t + \Delta t) - \mathbf{H}(t + \Delta t)[\mathbf{I} + \mathbf{F}(t)\Delta t]\hat{\mathbf{x}}(t|t)\} \tag{27-30}$$

The second limit on the right-hand side of (27-30) is easy to evaluate, i.e.,

$$\lim_{\Delta t \to 0} \{\mathbf{z}(t + \Delta t) - \mathbf{H}(t + \Delta t)[\mathbf{I} + \mathbf{F}(t)\Delta t]\hat{\mathbf{x}}(t|t)\} = \mathbf{z}(t) - \mathbf{H}(t)\hat{\mathbf{x}}(t|t) \tag{27-31}$$

In order to evaluate the first limit on the right-hand side of (27-30), we first substitute $\mathbf{R}(t + \Delta t)/\Delta t$ for $\mathbf{R}(k + 1 = t + \Delta\ )$ in (17-12), to obtain

$$\mathbf{K}(t + \Delta t) = \mathbf{P}(t + \Delta t|t)\mathbf{H}'(t + \Delta t) [\mathbf{H}(t + \Delta t)\mathbf{P}(t + \Delta t|t)\mathbf{H}'(t + \Delta t)\Delta t + \mathbf{R}(t + \Delta t)]^{-1}\Delta t \tag{27-32}$$

Then, we substitute $\mathbf{Q}(t)/\Delta t$ for $\mathbf{Q}(k = t)$ in (17-13), to obtain

$$\mathbf{P}(t + \Delta t|t) = \mathbf{\Phi}(t + \Delta t,t)\mathbf{P}(t|t)\mathbf{\Phi}'(t + \Delta t,t) + \mathbf{\Gamma}(t + \Delta t,t)\frac{\mathbf{Q}(t)}{\Delta t}\mathbf{\Gamma}'(t + \Delta t,t) \tag{27-33}$$

We leave it to the reader to show that

$$\lim_{\Delta t \to 0} \mathbf{P}(t + \Delta t|t) = \mathbf{P}(t|t) \tag{27-34}$$

hence,

$$\lim_{\Delta t \to 0} \frac{\mathbf{K}(t + \Delta t)}{\Delta t} = \mathbf{P}(t|t)\mathbf{H}'(t)\mathbf{R}^{-1}(t) \triangleq \mathbf{K}(t) \tag{27-35}$$

Combining (27-29), (27-30), (27-31), and (27-35), we obtain the KBF in (27-10) and the KB gain matrix in (27-11).

In order to derive the matrix differential equation for $\mathbf{P}(t|t)$, we begin with (17-14), substitute (27-33) along with the expansions of $\mathbf{\Phi}(t + \Delta t,t)$ and $\mathbf{\Gamma}(t + \Delta t,t)$ into that equation, and use the fact that $\mathbf{K}(t + \Delta t)$ has no zero-order terms in Δt, to show that

$$\begin{aligned}\mathbf{P}(t + \Delta t|t + \Delta t) &= \mathbf{P}(t + \Delta t|t) - \mathbf{K}(t + \Delta t)\mathbf{H}(t + \Delta t)\mathbf{P}(t + \Delta t|t)\\ &= \mathbf{P}(t|t) + [\mathbf{F}(t)\mathbf{P}(t|t) + \mathbf{P}(t|t)\mathbf{F}'(t)\\ &\quad + \mathbf{G}(t)\mathbf{Q}(t)\mathbf{G}'(t)]\Delta t\\ &\quad - \mathbf{K}(t + \Delta t)\mathbf{H}(t + \Delta t)\mathbf{P}(t + \Delta t|t)\end{aligned} \tag{27-36}$$

Consequently,

$$\lim_{\Delta t \to 0} \frac{\mathbf{P}(t + \Delta t|t + \Delta t) - \mathbf{P}(t|t)}{\Delta t} = \dot{\mathbf{P}}(t|t) = \mathbf{F}(t)\mathbf{P}(t|t) + \mathbf{P}(t|t)\mathbf{F}'(t) + \mathbf{G}(t)\mathbf{Q}(t)\mathbf{G}'(t) - \lim_{\Delta t \to 0} \frac{\mathbf{K}(t + \Delta t)\mathbf{H}(t + \Delta t)\mathbf{P}(t + \Delta t|t)}{\Delta t} \tag{27-37}$$

or finally, using (27-35)

$$\dot{\mathbf{P}}(t|t) = \mathbf{F}(t)\mathbf{P}(t|t) + \mathbf{P}(t|t)\mathbf{F}'(t) + \mathbf{G}(t)\mathbf{Q}(t)\mathbf{G}'(t) - \mathbf{P}(t|t)\mathbf{H}'(t)\mathbf{R}^{-1}(t)\mathbf{H}(t)\mathbf{P}(t|t) \tag{27-38}$$

This completes the derivation of the KBF using a formal limiting procedure. It is also possible to obtain continuous-time smoothers by means of this procedure (e.g., see Meditch, 1969, Chapter 7).

DERIVATION OF KBF WHEN STRUCTURE OF THE FILTER IS PRESPECIFIED

In this derivation of the KBF we begin by assuming that the filter has the following structure,

$$\dot{\hat{\mathbf{x}}}(t|t) = \mathbf{F}(t)\hat{\mathbf{x}}(t|t) + \mathbf{K}(t)[\mathbf{z}(t) - \mathbf{H}\hat{\mathbf{x}}(t|t)] \tag{27-39}$$

Our objective is to find the matrix function $\mathbf{K}(\tau)$, $t_0 \leq t \leq \tau$, that minimizes the following mean-squared error,

$$J[\mathbf{K}(\tau)] = \mathbf{E}\{\mathbf{e}'(\tau)\mathbf{e}(\tau)\} \qquad \tau \geq t_0 \tag{27-40}$$

where

$$\mathbf{e}(\tau) = \mathbf{x}(\tau) - \hat{\mathbf{x}}(\tau|\tau) \tag{27-41}$$

This optimization problem is a fixed-time, free-end-point (i.e., τ is fixed but $\mathbf{e}(\tau)$ is not fixed) problem in the calculus of variations (e.g., Kwakernaak and Sivan, 1972; Athans and Falb, 1965; and Bryson and Ho, 1969).

It is straightforward to show that $\mathbf{E}\{\mathbf{e}(\tau)\} = \mathbf{0}$, so that $\mathbf{E}\{[\mathbf{e}(\tau) - \mathbf{E}\{\mathbf{e}(\tau)\}][\mathbf{e}(\tau) - \mathbf{E}\{\mathbf{e}(\tau)\}]'\} = \mathbf{E}\{\mathbf{e}(\tau)\mathbf{e}'(\tau)\}$. Letting

$$\mathbf{P}(t|t) = \mathbf{E}\{\mathbf{e}(t)\mathbf{e}'(t)\} \tag{27-42}$$

we know that $J[\mathbf{K}(\tau)]$ can be reexpressed as

$$J[\mathbf{K}(\tau)] = \text{tr}\,\mathbf{P}(\tau|\tau) \tag{27-43}$$

We leave it to the reader to derive the following state equation for $\mathbf{e}(t)$, and its associated covariance equation,

$$\dot{\mathbf{e}}(t) = [\mathbf{F}(t) - \mathbf{K}(t)\mathbf{H}(t)]\mathbf{e}(t) + \mathbf{G}(t)\mathbf{w}(t) - \mathbf{K}(t)\mathbf{v}(t) \tag{27-44}$$

and

$$\dot{\mathbf{P}}(t|t) = [\mathbf{F}(t) - \mathbf{K}(t)\mathbf{H}(t)]\mathbf{P}(t|t) + \mathbf{P}(t|t)[\mathbf{F}(t) - \mathbf{K}(t)\mathbf{H}(t)]' + \mathbf{G}(t)\mathbf{Q}(t)\mathbf{G}'(t) + \mathbf{K}(t)\mathbf{R}(t)\mathbf{K}'(t) \tag{27-45}$$

where $\mathbf{e}(t_0) = \mathbf{0}$ and $\mathbf{P}(t_0|t_0) = \mathbf{P}_{\mathbf{x}}(t_0)$.

Our optimization problem for determining $\mathbf{K}(t)$ is: *given the matrix differential equation (27-45), satisfied by the error-covariance matrix* $\mathbf{P}(t|t)$, *a terminal time* τ, *and the cost functional* $J[\mathbf{K}(\tau)]$ *in (27-43), determine the matrix* $\mathbf{K}(t)$, $t_0 \leq t \leq \tau$ *that minimizes* $J[\mathbf{K}(t)]$.

The elements $p_{ij}(t|t)$ of $\mathbf{P}(t|t)$ may be viewed as the state variables of a dynamical system, and the elements $k_{ij}(t)$ of $\mathbf{K}(t)$ may be viewed as the control variables in an optimal control problem. The cost functional is then a terminal time penalty function on the state variables $p_{ij}(t|t)$. Euler-Lagrange equations associated with a free end-point problem can be used to determine the optimal gain matrix $\mathbf{K}(t)$.

To do this we define a set of costate variables $\sigma_{ij}(t)$ that correspond to the $p_{ij}(t|t)$, $i,j = 1, 2, \ldots, n$. Let $\mathbf{\Sigma}(t)$ be an $n \times n$ costate matrix that is associated with $\mathbf{P}(t|t)$, i.e., $\mathbf{\Sigma}(t) = (\sigma_{ij}(t))_{ij}$. Next, we introduce the Hamiltonian function $\mathcal{H}(\mathbf{K},\mathbf{P},\mathbf{\Sigma})$ where for notational convenience we have omitted the dependence of $\mathbf{K}$, $\mathbf{P}$, and $\mathbf{\Sigma}$ on t, and,

$$\mathcal{H}(\mathbf{K},\mathbf{P},\mathbf{\Sigma}) = \sum_{i=1}^{n}\sum_{j=1}^{n} \sigma_{ij}(t)\dot{p}_{ij}(t|t) = \text{tr}\,[\dot{\mathbf{P}}(t|t)\mathbf{\Sigma}'(t)] \tag{27-46}$$

Substituting (27-45) into (27-46), we see that

$$\begin{aligned}\mathcal{H}(\mathbf{K},\mathbf{P},\boldsymbol{\Sigma}) = &\operatorname{tr}[\mathbf{F}(t)\mathbf{P}(t|t)\boldsymbol{\Sigma}'(t)] \\ &- \operatorname{tr}[\mathbf{K}(t)\mathbf{H}(t)\mathbf{P}(t|t)\boldsymbol{\Sigma}'(t)] \\ &+ \operatorname{tr}[\mathbf{P}(t|t)\mathbf{F}'(t)\boldsymbol{\Sigma}'(t)] \\ &- \operatorname{tr}[\mathbf{P}(t|t)\mathbf{H}'(t)\mathbf{K}'(t)\boldsymbol{\Sigma}'(t)] \\ &+ \operatorname{tr}[\mathbf{G}(t)\mathbf{Q}(t)\mathbf{G}'(t)\boldsymbol{\Sigma}'(t)] \\ &+ \operatorname{tr}[\mathbf{K}(t)\mathbf{R}(t)\mathbf{K}'(t)\boldsymbol{\Sigma}'(t)]\end{aligned} \tag{27-47}$$

The Euler-Lagrange equations for our optimization problem are:

$$\left.\frac{\partial\mathcal{H}(\mathbf{K},\mathbf{P},\boldsymbol{\Sigma})}{\partial\mathbf{K}}\right|_* = \mathbf{0} \tag{27-48}$$

$$\dot{\boldsymbol{\Sigma}}^*(t) = -\left.\frac{\partial\mathcal{H}(\mathbf{K},\mathbf{P},\boldsymbol{\Sigma})}{\partial\mathbf{P}}\right|_* \tag{27-49}$$

$$\dot{\mathbf{P}}^*(t|t) = \left.\frac{\partial\mathcal{H}(\mathbf{K},\mathbf{P},\boldsymbol{\Sigma})}{\partial\boldsymbol{\Sigma}}\right|_* \tag{27-50}$$

and

$$\boldsymbol{\Sigma}^*(\tau) = \frac{\partial}{\partial\mathbf{P}}\operatorname{tr}\mathbf{P}(\tau|\tau) \tag{27-51}$$

In these equations starred quantities denote optimal quantities, and $|_*$ denotes the replacement of **K**, **P**, and $\boldsymbol{\Sigma}$ by $\mathbf{K}^*$, $\mathbf{P}^*$, and $\boldsymbol{\Sigma}^*$ *after* the appropriate derivative has been calculated. Note, also, that the derivatives of $\mathcal{H}(\mathbf{K},\mathbf{P},\boldsymbol{\Sigma})$ are derivatives of a scalar quantity with respect to a matrix (e.g., **K**, **P**, or $\boldsymbol{\Sigma}$). The calculus of gradient matrices (e.g., Schweppe, 1974; or Athans and Schweppe, 1965) can be used to evaluate these derivatives. The results are:

$$-\boldsymbol{\Sigma}^*\mathbf{P}^{*\prime}\mathbf{H}' - \boldsymbol{\Sigma}^{*\prime}\mathbf{P}^*\mathbf{H}' + \boldsymbol{\Sigma}^*\mathbf{K}^*\mathbf{R} + \boldsymbol{\Sigma}^{*\prime}\mathbf{K}^*\mathbf{R} = \mathbf{0} \tag{27-52}$$

$$\dot{\boldsymbol{\Sigma}}^* = -\boldsymbol{\Sigma}^*(\mathbf{F} - \mathbf{K}^*\mathbf{H}) - (\mathbf{F} - \mathbf{K}^*\mathbf{H})'\boldsymbol{\Sigma}^* \tag{27-53}$$

$$\dot{\mathbf{P}}^* = (\mathbf{F} - \mathbf{K}^*\mathbf{H})\mathbf{P}^* + \mathbf{P}^*(\mathbf{F} - \mathbf{K}^*\mathbf{H})' + \mathbf{GQG}' + \mathbf{K}^*\mathbf{R}\mathbf{K}^{*\prime} \tag{27-54}$$

and

$$\boldsymbol{\Sigma}^*(\tau) = \mathbf{I} \tag{27-55}$$

Our immediate objective is to obtain an expression for $\mathbf{K}^*(t)$.

Fact. *Matrix* $\boldsymbol{\Sigma}^*(\mathrm{t})$ *is symmetric and positive definite.* □

We leave the proof of this fact as an exercise for the reader. Using this fact, and the fact that covariance matrix $\mathbf{P}^*(t)$ is symmetric, we are able to express (27-52) as

$$2\boldsymbol{\Sigma}^*(\mathbf{K}^*\mathbf{R} - \mathbf{P}^*\mathbf{H}') = \mathbf{0} \tag{27-56}$$

Because $\mathbf{\Sigma}^* > 0$, $(\mathbf{\Sigma}^*)^{-1}$ exists so that (27-56) has for its only solution

$$\mathbf{K}^*(t) = \mathbf{P}^*(t|t)\mathbf{H}'(t)\mathbf{R}^{-1}(t) \tag{27-57}$$

which is the Kalman-Bucy gain matrix stated in Theorem 27-1. In order to obtain the covariance equation associated with $\mathbf{K}^*(t)$, substitute (27-57) into (27-54). The result is (27-12).

This completes the derivation of the KBF when the structure of the filter is prespecified.

STEADY-STATE KBF

If our continuous-time system is time-invariant and stationary, then, when certain system-theoretic conditions are satisfied (see, e.g., Kwakernaak and Sivan, 1972), $\dot{\mathbf{P}}(t|t) \to \mathbf{0}$ in which case $\mathbf{P}(t|t)$ has a steady-state value, denoted $\overline{\mathbf{P}}$. In this case, $\mathbf{K}(t) \to \overline{\mathbf{K}}$, where

$$\overline{\mathbf{K}} = \overline{\mathbf{P}}\mathbf{H}'\mathbf{R}^{-1} \tag{27-58}$$

$\overline{\mathbf{P}}$ is the solution of the algebraic Riccati equation

$$\mathbf{F}\overline{\mathbf{P}} + \overline{\mathbf{P}}\mathbf{F}' - \overline{\mathbf{P}}\mathbf{H}'\mathbf{R}^{-1}\mathbf{H}\overline{\mathbf{P}} + \mathbf{G}\mathbf{Q}\mathbf{G}' = \mathbf{0} \tag{27-59}$$

and the steady-state KBF is asymptotically stable, i.e., the eigenvalues of $\mathbf{F} - \overline{\mathbf{K}}\mathbf{H}$ all lie in the left-half of the complex s-plane.

Example 27-1

Here we examine the steady-state KBF for the simplest second-order system, the double integrator,

$$\ddot{x}(t) = w(t) \tag{27-60}$$

and

$$z(t) = x(t) + v(t) \tag{27-61}$$

in which $w(t)$ and $v(t)$ are mutually uncorrelated white noise processes, with intensities q and r, respectively. With $x_1(t) = x(t)$ and $x_2(t) = \dot{x}(t)$, this system is expressed in state-variable format as

$$\begin{pmatrix} \dot{x}_1 \\ \dot{x}_2 \end{pmatrix} = \begin{pmatrix} 0 & 1 \\ 0 & 0 \end{pmatrix}\begin{pmatrix} x_1 \\ x_2 \end{pmatrix} + \begin{pmatrix} 0 \\ 1 \end{pmatrix} w \tag{27-62}$$

$$z = (1 \quad 0)\begin{pmatrix} x_1 \\ x_2 \end{pmatrix} + v \tag{27-63}$$

The algebraic Riccati equation for this system is

$$\begin{aligned}
&\begin{pmatrix} 0 & 1 \\ 0 & 0 \end{pmatrix}\begin{pmatrix} \overline{p}_{11} & \overline{p}_{12} \\ \overline{p}_{12} & \overline{p}_{22} \end{pmatrix} + \begin{pmatrix} \overline{p}_{11} & \overline{p}_{12} \\ \overline{p}_{12} & \overline{p}_{22} \end{pmatrix}\begin{pmatrix} 0 & 0 \\ 1 & 0 \end{pmatrix} \\
&\qquad - \begin{pmatrix} \overline{p}_{11} & \overline{p}_{12} \\ \overline{p}_{12} & \overline{p}_{22} \end{pmatrix}\begin{pmatrix} 1 \\ 0 \end{pmatrix}\frac{1}{r}(1 \quad 0)\begin{pmatrix} \overline{p}_{11} & \overline{p}_{12} \\ \overline{p}_{12} & \overline{p}_{22} \end{pmatrix} \\
&\qquad + \begin{pmatrix} 0 \\ 1 \end{pmatrix} q (0 \quad 1) = \begin{pmatrix} 0 & 0 \\ 0 & 0 \end{pmatrix}
\end{aligned} \tag{27-64}$$

which leads to the following three algebraic equations:

$$2\bar{p}_{12} - \frac{1}{r}\bar{p}_{11}^2 = 0 \tag{27-65a}$$

$$\bar{p}_{22} - \frac{1}{r}\bar{p}_{11}\bar{p}_{12} = 0 \tag{27-65b}$$

and

$$-\frac{1}{r}\bar{p}_{12}^2 + q = 0 \tag{27-65c}$$

It is straightforward to show that the unique solution of these nonlinear algebraic equations, for which $\bar{\mathbf{P}} > 0$, is

$$\bar{p}_{12} = (qr)^{1/2} \tag{27-66a}$$

$$\bar{p}_{11} = \sqrt{2}\, q^{1/4} r^{3/4} \tag{27-66b}$$

$$\bar{p}_{22} = \sqrt{2}\, q^{3/4} r^{1/4} \tag{27-66c}$$

The steady-state KB gain matrix is computed from (27-58), as

$$\bar{\mathbf{K}} = \bar{\mathbf{P}}\mathbf{H}'\frac{1}{r} = \frac{1}{r}\begin{pmatrix}\bar{p}_{11}\\ \bar{p}_{12}\end{pmatrix} = \begin{pmatrix}\sqrt{2}\,(q/r)^{1/4}\\ (q/r)^{1/2}\end{pmatrix} \tag{27-67}$$

Observe that, just as in the discrete-time case, the single-channel KBF depends only on the ratio q/r.

Although we only needed $\bar{p}_{11}$ and $\bar{p}_{12}$ to compute $\bar{\mathbf{K}}$, $\bar{p}_{22}$ is an important quantity, because

$$\bar{p}_{22} = \lim_{t\to\infty} \mathbf{E}\{[\dot{x}(t) - \hat{\dot{x}}(t|t)]^2\} \tag{27-68}$$

Additionally,

$$\bar{p}_{11} = \lim_{t\to\infty} \mathbf{E}\{[x(t) - \hat{x}(t|t)]^2\} \tag{27-69}$$

Using (27-66b) and (27-66c), we find that

$$\bar{p}_{22} = (q/r)^{1/2}\,\bar{p}_{11} \tag{27-70}$$

If $q/r > 1$ (i.e., $\overline{\text{SNR}}$ possibly greater than unity), we will always have larger errors in estimation of $\dot{x}(t)$ than in estimation of $x(t)$. This is not too surprising because our measurement depends only on $x(t)$, and both $w(t)$ and $v(t)$ affect the calculation of $\hat{\dot{x}}(t|t)$.

The steady-state KBF is characterized by the eigenvalues of matrix $\mathbf{F} - \bar{\mathbf{K}}\mathbf{H}$, where

$$\mathbf{F} - \bar{\mathbf{K}}\mathbf{H} = \begin{pmatrix}-\sqrt{2}\,(q/r)^{1/4} & 1\\ -(q/r)^{1/2} & 0\end{pmatrix} \tag{27-71}$$

These eigenvalues are solutions of the equation

$$s^2 + \sqrt{2}\,(q/r)^{1/4}\, s + (q/r)^{1/2} = 0 \tag{27-72}$$

When this equation is expressed in the normalized form

$$s^2 + 2\zeta\omega_n s + \omega_n^2 = 0$$

we find that

$$\omega_n = (q/r)^{1/4} \tag{27-73}$$

and

$$\zeta = 0.707 \tag{27-74}$$

thus, the steady-state KBF for the simple double integrator system is damped at 0.707. The filter's poles lie on the 45° line depicted in Figure 27-1. They can be moved along this line by adjusting the ratio q/r; hence, once again, we may view q/r as a filter tuning parameter. □

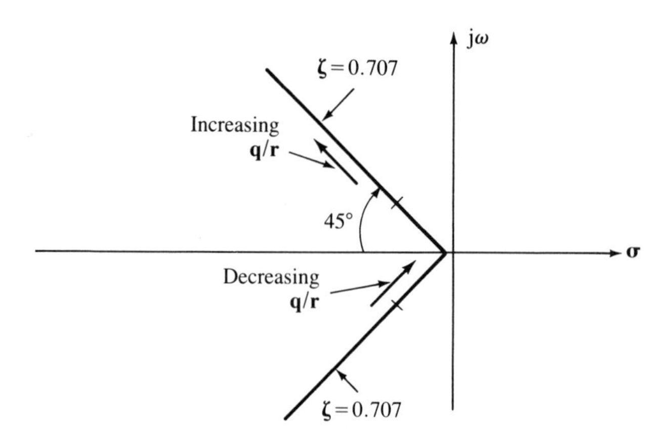

Figure 27-1 Eigenvalues of steady-state KBF lie along ±45 degree lines. Increasing q/r moves them farther away from the origin, whereas decreasing q/r moves them closer to the origin.

AN IMPORTANT APPLICATION FOR THE KBF

Consider the system

$$\left.\begin{aligned}\dot{\mathbf{x}}(t) &= \mathbf{F}(t)\mathbf{x}(t) + \mathbf{B}(t)\mathbf{u}(t) + \mathbf{G}(t)\mathbf{w}(t)\\ \mathbf{x}(t_0) &= \mathbf{x}_0\end{aligned}\right\} \tag{27-75}$$

for $t \geq t_0$ where $\mathbf{x}_0$ is a random initial condition vector with mean $\mathbf{m}_\mathbf{x}(t_0)$ and covariance matrix $\mathbf{P}_\mathbf{x}(t_0)$. Measurements are given by

$$\mathbf{z}(t) = \mathbf{H}(t)\mathbf{x}(t) + \mathbf{v}(t) \tag{27-76}$$

for $t \geq t_0$. The joint random process col $[\mathbf{w}(t), \mathbf{v}(t)]$ is a white noise process with intensity

$$\begin{pmatrix}\mathbf{Q}(t) & 0\\ 0 & \mathbf{R}(t)\end{pmatrix} \qquad t \geq t_0$$

The *controlled variable* can be expressed as

$$\boldsymbol{\xi}(t) = \mathbf{D}(t)\mathbf{x}(t) \qquad t \geq t_0 \tag{27-77}$$

The *stochastic linear optimal output feedback regulator problem* is the problem of finding the functional

$$\mathbf{u}(t) = \mathbf{f}[\mathbf{z}(\tau) \qquad t_0 \leq \tau \leq t] \tag{27-78}$$

for $t_0 \leq t \leq t_1$ such that the objective function

$$J[\mathbf{u}] = \mathbf{E}\left\{\frac{1}{2}\mathbf{x}'(t_1)\mathbf{W}_1\mathbf{x}(t_1) + \frac{1}{2}\int_{t_0}^{t}[\boldsymbol{\xi}'(\tau)\mathbf{W}_3\boldsymbol{\xi}(\tau) + \mathbf{u}'(\tau)\mathbf{W}_2\mathbf{u}(\tau)]d\tau\right\} \tag{27-79}$$

is minimized. Here $\mathbf{W}_1$, $\mathbf{W}_2$, and $\mathbf{W}_3$ are symmetric weighting matrices, and, $\mathbf{W}_1 \geq 0$, $\mathbf{W}_2 > 0$, and $\mathbf{W}_3 > 0$ for $t_0 \leq t \leq t_1$.

In the control theory literature, this problem is also known as the linear-quadratic-Gaussian regulator problem (i.e., the LQG problem; see, Athans, 1971, for example). We state the structure of the solution to this problem, without proof, next.

The optimal control, $\mathbf{u}^*(t)$, which minimizes $J[\mathbf{u}]$ in (27-79) is

$$\mathbf{u}^*(t) = -\mathbf{F}^0(t)\hat{\mathbf{x}}(t|t) \tag{27-80}$$

where $\mathbf{F}^0(t)$ is an optimal gain matrix, computed as

$$\mathbf{F}^0(t) = \mathbf{W}_2^{-1}\mathbf{B}(t)\mathbf{P}_c(t) \tag{27-81}$$

where $\mathbf{P}_c(t)$ is the solution of the control Riccati equation

$$\left.\begin{aligned} -\dot{\mathbf{P}}_c(t) &= \mathbf{F}'(t)\mathbf{P}_c(t) + \mathbf{P}_c(t)\mathbf{F}(t) - \mathbf{P}_c(t)\mathbf{B}(t)\mathbf{W}_2^{-1}\mathbf{B}'(t)\mathbf{P}_c(t) \\ &\quad + \mathbf{D}'(t)\mathbf{W}_3\mathbf{D}(t) \\ \mathbf{P}_c(t_1) &\text{ given} \end{aligned}\right\} \tag{27-82}$$

and $\hat{\mathbf{x}}(t|t)$ is the output of a KBF, properly modified to account for the control term in the state equation, i.e.,

$$\dot{\hat{\mathbf{x}}}(t|t) = \mathbf{F}(t)\hat{\mathbf{x}}(t|t) + \mathbf{B}(t)\mathbf{u}^*(t) + \mathbf{K}(t)[\mathbf{z}(t) - \mathbf{H}(t)\hat{\mathbf{x}}(t|t)] \tag{27-83}$$

We see that the KBF plays an essential role in the solution of the LQG problem.

PROBLEMS

27-1. Explain the replacement of covariance matrix $\mathbf{R}(k + 1 = t + \Delta t)$ by $\mathbf{R}(t + \Delta t)/\Delta t$ in (27-27).

27-2. Show that $\lim_{\Delta t \to 0} \mathbf{P}(t + \Delta t|t) = \mathbf{P}(t|t)$.

27-3. Derive the state equation for error $\mathbf{e}(t)$, given in (27-44), and its associated covariance equation (27-45).

27-4. Prove that matrix $\boldsymbol{\Sigma}^*(t)$ is symmetric and positive definite.

Lesson A

Sufficient Statistics and Statistical Estimation of Parameters

INTRODUCTION

In this lesson,* we discuss the usefulness of the notion of sufficient statistics in statistical estimation of parameters. Specifically, we discuss the role played by sufficient statistics and exponential families in maximum-likelihood and uniformly minimum-variance unbiased (UMVU) parameter estimation.

CONCEPT OF SUFFICIENT STATISTICS

The notion of a sufficient statistic can be explained intuitively (Ferguson, 1967), as follows. We observe $\mathfrak{Z}(N)$ ($\mathfrak{Z}$ for short), where $\mathfrak{Z} = \text{col}\,[\mathbf{z}(1), \mathbf{z}(2), \ldots, \mathbf{z}(N)]$, in which $\mathbf{z}(1), \ldots, \mathbf{z}(N)$ are independent and identically distributed random vectors, each having a density function $p(\mathbf{z}(i)|\boldsymbol{\theta})$, where $\boldsymbol{\theta}$ is unknown. Often the information in $\mathfrak{Z}$ can be represented equivalently in a statistic, $T(\mathfrak{Z})$, whose dimension is independent of N, such that $T(\mathfrak{Z})$ contains all of the information about $\boldsymbol{\theta}$ that is originally in $\mathfrak{Z}$. Such a statistic is known as a *sufficient statistic*.

Example A-1

Consider a sampled sequence of N manufactured cars. For each car we record whether it is defective or not. The observed sample can be represented as $\mathfrak{Z} = \text{col}\,[z(1), \ldots,$

* This lesson was written by Dr. Rama Chellappa, Department of Electrical Engineering-Systems, Unversity of Southern California, Los Angeles, CA 90089.

$z(N)]$, where $z(i) = 0$ if the ith car is not defective and $z(i) = 1$ if the ith car is defective. The total number of observed defective cars is

$$T(\mathfrak{X}) = \sum_{i=1}^{N} z(i)$$

This is a statistic that maps many different values of $z(1), \ldots, z(N)$ into the same value of $T(\mathfrak{X})$. It is intuitively clear that, if one is interested in estimating the proportion θ of defective cars, nothing is lost by simply recording and using $T(\mathfrak{X})$ in place of $z(1), \ldots, z(N)$. The particular sequence of ones and zeros is irrelevant. Thus, as far as estimating the proportion of defective cars, $T(\mathfrak{X})$ contains all the information contained in $\mathfrak{X}$. □

An advantage associated with the concept of a sufficient statistic is dimensionality reduction. In Example A-1, the dimensionality reduction is from N to 1.

Definition A-1. *A statistic* $T(\mathfrak{X})$ *is sufficient for vector parameter* $\boldsymbol{\theta}$, *if and only if the distribution of* $\mathfrak{X}$, *conditioned on* $T(\mathfrak{X}) = t$, *does not involve* $\boldsymbol{\theta}$. □

Example A-2

This example illustrates the application of Definition A-1 to identify a sufficient statistic for the model in Example A-1. Let θ be the probability that a car is defective. Then $z(1), z(2), \ldots, z(N)$ is a record of N Bernoulli trials with probability θ; thus, $Pr(t) = \theta^t(1-\theta)^{N-t}$, $0 < \theta < 1$, where $t = \sum_{i=1}^{N} z(i)$, and $z(i) = 1$ or 0. The conditional distribution of $z(1), \ldots, z(N)$, given $\sum_{i=1}^{N} z(i) = t$, is

$$\mathbf{P}[\mathfrak{X}|T = t] = \frac{\mathbf{P}[\mathfrak{X}, T = t]}{\mathbf{P}(T = t)} = \frac{\theta^t(1-\theta)^{N-t}}{\binom{N}{t}\theta^t(1-\theta)^{N-t}} = \frac{1}{\binom{N}{t}}$$

which is independent of θ; hence, $T(\mathfrak{X}) = \sum_{i=1}^{N} z(i)$ is sufficient. Any one-to-one function of $T(\mathfrak{X})$ is also sufficient. □

This example illustrates that deriving a sufficient statistic using Definition A-1 can be quite difficult. An equivalent definition of sufficiency, which is easy to apply, is given in the following:

Theorem A-1 (Factorization Theorem). *A necessary and sufficient condition for* $T(\mathfrak{X})$ *to be sufficient for* $\boldsymbol{\theta}$ *is that there exists a factorization*

$$p(\mathfrak{X}|\boldsymbol{\theta}) = g(T(\mathfrak{X}), \boldsymbol{\theta})h(\mathfrak{X}) \tag{A-1}$$

where the first factor in (A-1) may depend on $\boldsymbol{\theta}$, *but depends on* $\mathfrak{X}$ *only through* $T(\mathfrak{X})$, *whereas the second factor is independent of* $\boldsymbol{\theta}$. □

The proof of this theorem is given in Ferguson (1967) for the continuous case and Duda and Hart (1973) for the discrete case.

Example A-3 (Continuation of Example A-2)

In Example A-2, the probability distribution of samples $z(1), \ldots, z(N)$ is

$$\mathbf{P}[\mathfrak{X}|\theta] = \theta^t(1-\theta)^{N-t} \tag{A-2}$$

where the total number of defective cars is $t = \sum_{i=1}^{N} z(i)$, and $z(i)$ is either 0 or 1. Equation (A-2) can be written equivalently as

$$\mathbf{P}[\mathfrak{X}|\theta] = \exp\left[t \ln \frac{\theta}{1-\theta} + N \ln(1-\theta)\right] \tag{A-3}$$

Comparing (A-3) with (A-1), we conclude that

$$h(\mathfrak{X}) = 1$$

$$g(T(\mathfrak{X}),\theta) = \exp\left[t \ln \frac{\theta}{1-\theta} + N \ln(1-\theta)\right]$$

and

$$T(\mathfrak{X}) = t = \sum_{i=1}^{N} z(i)$$

Using the Factorization Theorem, it was easy to determine $T(\mathfrak{X})$. □

Example A-4

Let $\mathfrak{X} = \text{col}\,[z(1), \ldots, z(N)]$ be a random sample drawn from a univariate Gaussian distribution, with unknown mean μ, and, known variance $\sigma^2 > 0$. Then

$$p(\mathfrak{X}) = \exp\left[\frac{\mu}{\sigma^2}\sum_{i=1}^{N} z(i) - \frac{N\mu^2}{2\sigma^2}\right] h(\mathfrak{X})$$

where

$$h(\mathfrak{X}) = \exp\left[-\frac{N}{2}\ln 2\pi\sigma^2 - \frac{1}{2\sigma^2}\sum_{i=1}^{N} z^2(i)\right]$$

Based on the Factorization Theorem we identify $T(\mathfrak{X}) = \sum_{i=1}^{N} z(i)$ as a sufficient statistic for μ. □

Because the concept of sufficient statistics involves reduction of data, it is worthwhile to know how far such a reduction can be done for a given problem. The dimension of the smallest set of statistics that is still sufficient for the parameters is called a *minimal sufficient statistic*. See Barankin (1959) and Datz (1959) for techniques useful in identifying a minimal sufficient statistic.

EXPONENTIAL FAMILIES OF DISTRIBUTIONS

It is of interest to study families of distributions, $p(z(i)|\boldsymbol{\theta})$ for which, irrespective of the sample size N, there exists a sufficient statistic of fixed dimension. The exponential families of distributions have this property. For example, the family of normal distributions $N(\mu,\sigma^2)$, with σ^2 known and μ unknown, is an exponential family which, as we have seen in Example A-4, has a one-dimensional sufficient statistic for μ, that is equal to $\sum_{i=1}^{N} z(i)$. As Bickel and Doksum (1977) state,

Definition A-2 (Bickel and Doksum, 1977). *If there exist real-valued functions* $\mathbf{a}(\boldsymbol{\theta})$, *and* $b(\boldsymbol{\theta})$ *on parameter space* $\boldsymbol{\Theta}$, *and real-valued functions* $\mathbf{T}(\mathbf{z})$ [$\mathbf{z}$ *is short for* $\mathbf{z}(i)$] *and* $h(\mathbf{z})$ *on* $\mathbf{R}^N$, *such that the density function* $p(\mathbf{z}|\boldsymbol{\theta})$ *can be written as*

$$p(\mathbf{z}|\boldsymbol{\theta}) = \exp\left[\mathbf{a}'(\boldsymbol{\theta})\mathbf{T}(\mathbf{z}) + b(\boldsymbol{\theta}) + h(\mathbf{z})\right] \tag{A-4}$$

then $p(\mathbf{z}|\boldsymbol{\theta})$, $\boldsymbol{\theta} \in \boldsymbol{\Theta}$, *is said to be a one-parameter exponential family of distributions.* □

The Gaussian, Binomial, Beta, Rayleigh, and Gamma distributions are examples of such one-parameter exponential families.

In a one parameter exponential family, $\mathbf{T}(\mathbf{z})$ is sufficient for $\boldsymbol{\theta}$. The family of distributions obtained by sampling from one-parameter exponential families is also a one-parameter exponential family. For example, suppose that $\mathbf{z}(1), \ldots, \mathbf{z}(N)$ are independent and identically distributed with common density $p(\mathbf{z}|\boldsymbol{\theta})$; then,

$$p(\mathfrak{Z}|\boldsymbol{\theta}) = \exp\left[\mathbf{a}'(\boldsymbol{\theta}) \sum_{i=1}^{N} \mathbf{T}(\mathbf{z}(i)) + Nb(\boldsymbol{\theta}) + \sum_{i=1}^{N} h(\mathbf{z}(i))\right] \tag{A-5}$$

The sufficient statistic $\mathbf{T}(\mathfrak{Z})$ for this situation is

$$\mathbf{T}(\mathfrak{Z}) = \sum_{i=1}^{N} \mathbf{T}(\mathbf{z}(i))$$

Example A-5

Let $\mathbf{z}(1), \ldots, \mathbf{z}(N)$ be a random sample from a multivariate Gaussian distribution with unknown $d \times 1$ mean vector $\boldsymbol{\mu}$ and known covariance matrix $\mathbf{P}_{\boldsymbol{\mu}}$. Then [$\mathbf{z}$ is short for $\mathbf{z}(i)$]

$$p(\mathbf{z}|\boldsymbol{\mu}) = \exp\left[\mathbf{a}'(\boldsymbol{\mu})\mathbf{T}(\mathbf{z}) + b(\boldsymbol{\mu}) + h(\mathbf{z})\right]$$

where

$$\mathbf{a}'(\boldsymbol{\mu}) = \boldsymbol{\mu}'\mathbf{P}_{\boldsymbol{\mu}}^{-1}$$

$$b(\boldsymbol{\mu}) = -\frac{1}{2}\boldsymbol{\mu}'\mathbf{P}_{\boldsymbol{\mu}}^{-1}\boldsymbol{\mu}$$

$$\mathbf{T}(\mathbf{z}) = \mathbf{z}$$

and

$$h(\mathbf{z}) = \exp\left[-\frac{1}{2}\mathbf{z}'\mathbf{P}_{\boldsymbol{\mu}}^{-1}\mathbf{z} - \frac{d}{2}\ln 2\pi - \frac{1}{2}\ln\det\mathbf{P}_{\boldsymbol{\mu}}\right]$$

Additionally,

$$p(\mathfrak{X}|\boldsymbol{\mu}) = \exp\,[\boldsymbol{\mu}'\mathbf{P}_{\mu}^{-1}\,\mathbf{T}(\mathfrak{X}) + Nb(\boldsymbol{\mu}) + h(\mathfrak{X})]$$

where

$$\mathbf{T}(\mathfrak{X}) = \sum_{i=1}^{N} \mathbf{z}(i)$$

and

$$h(\mathfrak{X}) = \exp\left[-\frac{1}{2}\sum_{i=1}^{N} \mathbf{z}'(i)\,\mathbf{P}_{\mu}^{-1}\,\mathbf{z}(i) - \frac{Nd}{2}\ln 2\pi - \frac{N}{2}\ln\det \mathbf{P}_{\mu}\right] \quad \square$$

The notion of a one-parameter exponential family of distributions as stated in Bickel and Doksum (1977) can easily be extended to m parameters and vector observations in a straightforward manner.

Definition A-3. *If there exist real matrices* $\mathbf{A}_1, \ldots, \mathbf{A}_m$, *a real function* b *of* $\boldsymbol{\theta}$, *where* $\boldsymbol{\theta} \in \boldsymbol{\Theta}$, *real matrices* $\mathbf{T}_i(\mathbf{z})$ *and a real function* $h(\mathbf{z})$, *such that the density function* $p(\mathbf{z}|\boldsymbol{\theta})$ *can be written as*

$$p(\mathbf{z}|\boldsymbol{\theta}) = \exp\left\{\mathrm{tr}\left[\sum_{i=1}^{m} \mathbf{A}_i(\boldsymbol{\theta})\mathbf{T}_i(\mathbf{z})\right]\right\}\exp\{b(\boldsymbol{\theta}) + h(\mathbf{z})\} \tag{A-6}$$

then $p(\mathbf{z}|\boldsymbol{\theta})$, $\boldsymbol{\theta} \in \boldsymbol{\Theta}$ *is said to be an* m-*parameter exponential family of distributions.* $\square$

Example A-6

The family of d-variate normal distributions $N(\boldsymbol{\mu},\mathbf{P}_{\mu})$, where both $\boldsymbol{\mu}$ and $\mathbf{P}_{\mu}$ are unknown, is an example of a 2-parameter exponential family in which $\boldsymbol{\theta}$ contains $\boldsymbol{\mu}$ and the elements of $\mathbf{P}_{\mu}$. In this case

$$\mathbf{A}_1(\boldsymbol{\theta}) = \mathbf{a}_1(\boldsymbol{\theta}) = \mathbf{P}_{\mu}^{-1}\,\boldsymbol{\mu}$$

$$\mathbf{T}_1(\mathbf{z}) = \mathbf{z}'$$

$$\mathbf{A}_2(\boldsymbol{\theta}) = -\frac{1}{2}\mathbf{P}_{\mu}^{-1}$$

$$\mathbf{T}_2(\mathbf{z}) = \mathbf{z}\mathbf{z}'$$

$$b(\boldsymbol{\theta}) = -\frac{1}{2}\boldsymbol{\mu}'\mathbf{P}_{\mu}^{-1}\,\boldsymbol{\mu} - \frac{d}{2}\ln 2\pi - \frac{1}{2}\ln\det \mathbf{P}_{\mu}$$

and

$$h(\mathbf{z}) = 0 \quad \square$$

As is true for a one-parameter exponential family of distributions, if $\mathbf{z}(1), \ldots, \mathbf{z}(N)$ are drawn randomly from an m-parameter exponential family, then $p[\mathbf{z}(1), \ldots, \mathbf{z}(N)|\boldsymbol{\theta}]$ form an m-parameter exponential family with sufficient statistics $\mathbf{T}_1(\mathfrak{X}), \ldots, \mathbf{T}_m(\mathfrak{X})$, where $\mathbf{T}_i(\mathfrak{X}) = \sum_{j=1}^{N} \mathbf{T}_i[\mathbf{z}(j)]$.

EXPONENTIAL FAMILIES AND MAXIMUM-LIKELIHOOD ESTIMATION

Let us consider a vector of unknown parameters $\boldsymbol{\theta}$ that describe a collection of N independent and identically distributed observations $\mathfrak{Z} = \text{col}\,[\mathbf{z}(1), \ldots, \mathbf{z}(N)]$. The maximum-likelihood estimate (MLE) of $\boldsymbol{\theta}$ is obtained by maximizing the likelihood of $\boldsymbol{\theta}$ given the observations $\mathfrak{Z}$. Likelihood is defined in Lesson 11 to be proportional to the value of the probability density of the observations, given the parameters, i.e.,

$$l(\boldsymbol{\theta}|\mathfrak{Z}) \propto p(\mathfrak{Z}|\boldsymbol{\theta})$$

As discussed in Lesson 11, a sufficient condition for $l(\boldsymbol{\theta}|\mathfrak{Z})$ to be maximized is

$$J_0(\hat{\boldsymbol{\theta}}_{\text{ML}}|\mathfrak{Z}) < 0 \tag{A-7}$$

where

$$J_0(\hat{\boldsymbol{\theta}}_{\text{ML}}|\mathfrak{Z}) = \left(\frac{\partial^2 L(\boldsymbol{\theta}|\mathfrak{Z})}{\partial\theta_i \partial\theta_j}\right)_{\boldsymbol{\theta} = \hat{\boldsymbol{\theta}}_{\text{ML}}} \qquad i,j = 1, 2, \ldots, n$$

and $L(\boldsymbol{\theta}|\mathfrak{Z}) = \ln l(\boldsymbol{\theta}|\mathfrak{Z})$. Maximum-likelihood estimates of $\boldsymbol{\theta}$ are obtained by solving the system of n equations

$$\frac{\partial L(\boldsymbol{\theta}|\mathfrak{Z})}{\partial\theta_i} = 0 \qquad i = 1, 2, \ldots, n \tag{A-8}$$

for $\hat{\boldsymbol{\theta}}_{\text{ML}}$ and checking whether the solution to (A-8) satisfies (A-7).

When this technique is applied to members of exponential families, $\hat{\boldsymbol{\theta}}_{\text{ML}}$ can be obtained by solving a set of algebraic equations. The following theorem paraphrased from Bickel and Doksum (1977) formalizes this technique for vector observations.

Theorem A-2 (Bickel and Doksum, 1977). *Let* $\text{p}(\mathbf{z}|\boldsymbol{\theta}) = \exp[\mathbf{a}'(\boldsymbol{\theta})\mathbf{T}(\mathbf{z}) + \text{b}(\boldsymbol{\theta}) + \text{h}(\mathbf{z})]$ *and let* $\mathcal{A}$ *denote the interior of the range of* $\mathbf{a}(\boldsymbol{\theta})$. *If the equation*

$$\mathbf{E}_{\boldsymbol{\theta}}\{\mathbf{T}(\mathbf{z})\} = \mathbf{T}(\mathbf{z}) \tag{A-9}$$

has a solution $\hat{\boldsymbol{\theta}}(\mathbf{z})$ *for which* $\mathbf{a}[\hat{\boldsymbol{\theta}}(\mathbf{z})] \in \mathcal{A}$, *then* $\hat{\boldsymbol{\theta}}(\mathbf{z})$ *is the unique MLE of* $\boldsymbol{\theta}$. □

The proof of this theorem can be found in Bickel and Doksum (1977).

Example A-7 (Continuation of Example A-5)

In this case

$$\mathbf{T}(\mathfrak{Z}) = \sum_{i=1}^{N} \mathbf{z}(i)$$

and

$$\mathbf{E}_{\boldsymbol{\mu}}\{\mathbf{T}(\mathfrak{Z})\} = N\boldsymbol{\mu}$$

hence (A-9) becomes

$$\sum_{i=1}^{N} \mathbf{z}(i) = N\boldsymbol{\mu}$$

whose solution, $\hat{\boldsymbol{\mu}}$, is

$$\hat{\boldsymbol{\mu}} = \frac{1}{N}\sum_{i=1}^{N} \mathbf{z}(i)$$

which is the well-known MLE of $\boldsymbol{\mu}$. □

Theorem A-2 can be extended to the m-parameter exponential family case by using Definition A-3. We illustrate the applicability of this extension using the example given below.

Example A-8 (see, also, Example A-6)

Let $\mathfrak{Z} = \text{col}\,[\mathbf{z}(1), \ldots, \mathbf{z}(N)]$ be randomly drawn from $p(\mathbf{z}|\boldsymbol{\theta}) = N(\boldsymbol{\mu}, \mathbf{P}_{\boldsymbol{\mu}})$, where both $\boldsymbol{\mu}$ and $\mathbf{P}_{\boldsymbol{\mu}}$ are unknown, so that $\boldsymbol{\theta}$ contains $\boldsymbol{\mu}$ and the elements of $\mathbf{P}_{\boldsymbol{\mu}}$. Vector $\boldsymbol{\mu}$ is $d \times 1$ and matrix $\mathbf{P}_{\boldsymbol{\mu}}$ is $d \times d$, symmetric and positive definite. We express $p(\mathfrak{Z}|\boldsymbol{\theta})$ as

$$\begin{aligned}
p(\mathfrak{Z}|\boldsymbol{\theta}) &= (2\pi)^{-Nd/2}(\det \mathbf{P}_{\boldsymbol{\mu}})^{-N/2} \\
&\quad \exp\left[-\frac{1}{2}\sum_{i=1}^{N}(\mathbf{z}(i) - \boldsymbol{\mu})'\mathbf{P}_{\boldsymbol{\mu}}^{-1}(\mathbf{z}(i) - \boldsymbol{\mu})\right] \\
&= (2\pi)^{-Nd/2}(\det \mathbf{P}_{\boldsymbol{\mu}})^{-N/2}\exp\left\{-\frac{1}{2}\text{tr}\left[\mathbf{P}_{\boldsymbol{\mu}}^{-1}\right.\right. \\
&\quad \left.\left.\left(\sum_{i=1}^{N}(\mathbf{z}(i) - \boldsymbol{\mu})(\mathbf{z}(i) - \boldsymbol{\mu})'\right)\right]\right\} \\
&= (2\pi)^{-Nd/2}(\det \mathbf{P}_{\boldsymbol{\mu}})^{-N/2}\exp\left\{-\frac{1}{2}\text{tr}\left[\mathbf{P}_{\boldsymbol{\mu}}^{-1}\sum_{i=1}^{N}\mathbf{z}(i)\mathbf{z}'(i)\right.\right. \\
&\quad \left.\left. - 2\boldsymbol{\mu}\sum_{i=1}^{N}\mathbf{z}'(i) + N\boldsymbol{\mu}\boldsymbol{\mu}'\right]\right\}
\end{aligned}$$

Using Theorem A-1 or Definition A-3 it can be seen that $\sum_{i=1}^{N}\mathbf{z}(i)$ and $\sum_{i=1}^{N}\mathbf{z}(i)\mathbf{z}'(i)$ are sufficient for $(\boldsymbol{\mu}, \mathbf{P}_{\boldsymbol{\mu}})$. Letting

$$\mathbf{T}_1(\mathfrak{Z}) = \sum_{i=1}^{N}\mathbf{z}(i)$$

and

$$\mathbf{T}_2(\mathfrak{Z}) = \sum_{i=1}^{N}\mathbf{z}(i)\mathbf{z}'(i)$$

we find that

$$\mathbf{E}_{\boldsymbol{\theta}}\{\mathbf{T}_1(\mathfrak{Z})\} = N\boldsymbol{\mu}$$

and

$$\mathbf{E}_\theta\{\mathbf{T}_2(\mathfrak{Z})\} = N(\mathbf{P}_\mu + \boldsymbol{\mu}\boldsymbol{\mu}')$$

Applying (A-9) to both $\mathbf{T}_1(\mathfrak{Z})$ and $\mathbf{T}_2(\mathfrak{Z})$, we obtain

$$N\boldsymbol{\mu} = \sum_{i=1}^{N} \mathbf{z}(i)$$

and

$$N(\mathbf{P}_\mu + \boldsymbol{\mu}\boldsymbol{\mu}') = \sum_{i=1}^{N} \mathbf{z}(i)\mathbf{z}'(i)$$

whose solutions, $\hat{\boldsymbol{\mu}}$ and $\hat{\mathbf{P}}_\mu$, are

$$\hat{\boldsymbol{\mu}} = \frac{1}{N}\sum_{i=1}^{N} \mathbf{z}(i)$$

and

$$\hat{\mathbf{P}}_\mu = \frac{1}{N}\sum_{i=1}^{N} [\mathbf{z}(i) - \hat{\boldsymbol{\mu}}][\mathbf{z}(i) - \hat{\boldsymbol{\mu}}]'$$

which are the MLE's of $\boldsymbol{\mu}$ and $\mathbf{P}_\mu$. □

Example A-9 (Linear Model)

Consider the linear model

$$\mathfrak{Z}(k) = \mathcal{H}(k)\boldsymbol{\theta} + \mathcal{V}(k)$$

in which $\boldsymbol{\theta}$ is an $n \times 1$ vector of deterministic parameters, $\mathcal{H}(k)$ is deterministic, and $\mathcal{V}(k)$ is a zero-mean white noise sequence, with known covariance matrix $\mathcal{R}(k)$. From (11-25) in Lesson 11, we can express $p(\mathfrak{Z}(k)|\boldsymbol{\theta})$, as

$$p(\mathfrak{Z}(k)|\boldsymbol{\theta}) = \exp[\mathbf{a}'(\boldsymbol{\theta})\mathbf{T}(\mathfrak{Z}(k)) + b(\boldsymbol{\theta}) + h(\mathfrak{Z}(k))]$$

where

$$\begin{aligned}
\mathbf{a}'(\boldsymbol{\theta}) &= \boldsymbol{\theta}' \\
\mathbf{T}(\mathfrak{Z}(k)) &= \mathcal{H}'(k)\mathcal{R}^{-1}(k)\mathfrak{Z}(k) \\
b(\boldsymbol{\theta}) &= -\frac{N}{2}\ln 2\pi - \frac{1}{2}\ln\det\mathcal{R}(k) - \frac{1}{2}\boldsymbol{\theta}'\mathcal{H}'(k)\mathcal{R}^{-1}(k)\mathcal{H}(k)\boldsymbol{\theta}
\end{aligned}$$

and

$$h(\mathfrak{Z}(k)) = -\frac{1}{2}\mathfrak{Z}'(k)\mathcal{R}^{-1}(k)\mathfrak{Z}(k)$$

Observe that

$$\mathbf{E}_\theta\{\mathcal{H}(k)\mathcal{R}^{-1}(k)\mathfrak{Z}(k)\} = \mathcal{H}'(k)\mathcal{R}^{-1}(k)\mathcal{H}(k)\boldsymbol{\theta}$$

hence, applying (A-9), we obtain

$$\mathcal{H}(k)\mathcal{R}^{-1}(k)\mathcal{H}(k)\hat{\boldsymbol{\theta}} = \mathcal{H}'(k)\mathcal{R}^{-1}(k)\mathfrak{Z}(k)$$

whose solution, $\hat{\boldsymbol{\theta}}(k)$, is

$$\hat{\boldsymbol{\theta}}(k) = [\mathcal{H}'(k)\mathcal{R}^{-1}(k)\mathcal{H}(k)]^{-1}\mathcal{H}'(k)\mathcal{R}^{-1}(k)\mathcal{Z}(k)$$

which is the well-known expression for the MLE of $\boldsymbol{\theta}$ (see Theorem 11-3). The case when $\mathbf{R}(k) = \sigma^2\mathbf{I}$, where σ^2 is unknown can be handled in a manner very similar to that in Example A-8. □

SUFFICIENT STATISTICS AND UNIFORMLY MINIMUM-VARIANCE UNBIASED ESTIMATION

In this section we discuss how sufficient statistics can be used to obtain uniformly minimum-variance unbiased (UMVU) estimates. Recall, from Lesson 6, that an estimate $\hat{\theta}$ of parameter θ is said to be unbiased if

$$\mathbf{E}\{\hat{\theta}\} = \theta \tag{A-10}$$

Among such unbiased estimates, we can often find one estimate, denoted θ^*, which improves all other estimates in the sense that

$$\text{var}\,(\theta^*) \leq \text{var}\,(\hat{\theta}) \tag{A-11}$$

When (A-11) is true for all (admissible) values of θ, θ^* is known as the UMVU estimate of θ. The UMVU estimator is obtained by choosing the estimator which has the minimum variance among the class of unbiased estimators. If the estimator is constrained further to be a *linear* function of the observations, then it becomes the BLUE which was discussed in Lesson 9.

Suppose we have an estimate, $\hat{\theta}(\mathcal{Z})$, of parameter θ that is based on observations $\mathcal{Z} = \text{col}\,[z(1), \ldots, z(N)]$. Assume further that $p(\mathcal{Z}|\theta)$ has a finite-dimensional sufficient statistic, $T(\mathcal{Z})$, for θ. Using $T(\mathcal{Z})$, we can construct an estimate $\theta^*(\mathcal{Z})$ which is at least as good as, or even better, than $\hat{\theta}$ by the celebrated Rao-Blackwell Theorem (Bickel and Doksum, 1977). We do this by computing the conditional expectation of $\hat{\theta}(\mathcal{Z})$, i.e.,

$$\theta^*(\mathcal{Z}) = \mathbf{E}\{\hat{\theta}(\mathcal{Z})|T(\mathcal{Z})\} \tag{A-12}$$

Estimate $\theta^*(\mathcal{Z})$ is "better than $\hat{\theta}$" in the sense that $\mathbf{E}\{[\theta^*(\mathcal{Z}) - \theta]^2\} < \mathbf{E}\{[\hat{\theta}(\mathcal{Z}) - \theta]^2\}$. Because $T(\mathcal{Z})$ is sufficient, the conditional expectation $\mathbf{E}\{\hat{\theta}(\mathcal{Z})|T(\mathcal{Z})\}$ will not depend on θ; hence, $\theta^*(\mathcal{Z})$ is a function of $\mathcal{Z}$ only. Application of this conditioning technique can only improve an estimate such as $\hat{\theta}(\mathcal{Z})$; it does not guarantee that $\theta^*(\mathcal{Z})$ will be the UMVU estimate. To obtain the UMVU estimate using this conditioning technique, we need the additional concept of *completeness*.

Definition A-4 [Lehmann (1959; 1980); Bickel and Doksum (1977)]. *A sufficient statistic* $\mathrm{T}(\mathcal{Z})$ *is said to be complete, if the only real-valued function,* g, *defined on the range of* $\mathrm{T}(\mathcal{Z})$, *which satisfies* $\mathbf{E}_\theta\{\mathrm{g}(\mathrm{T})\} = 0$ *for all* θ, *is the function* $\mathrm{g}(\mathrm{T}) = 0$. □

Completeness is a property of the family of distributions of $T(\mathfrak{X})$ generated as θ varies over its range. The concept of a complete sufficient statistic, as stated by Lehmann (1983), can be viewed as an extension of the notion of sufficient statistics in reducing the amount of useful information required for the estimation of θ. Although a sufficient statistic achieves data reduction, it may contain some additional information not required for the estimation of θ. For instance, it may be that $\mathbf{E}_\theta[g(T(\mathfrak{X}))]$ is a constant independent of θ for some nonconstant function g. If so, we would like to have $\mathbf{E}_\theta[g(T(\mathfrak{X}))] = c$, (constant independent of θ) imply that $g(T(\mathfrak{X})) = c$. By subtracting c from $\mathbf{E}_\theta\,[g(T(\mathfrak{X}))]$, one arrives at Definition A-4. Proving completeness using Definition A-4 can be cumbersome. In the special case when $p(z(k)|\theta)$ is a one-parameter exponential family, i.e., when

$$p(z(k)|\theta) = \exp\,[a(\theta)T(z(k)) + b(\theta) + h(z(k))] \tag{A-13}$$

the completeness of $T(z(k))$ can be verified by checking if the range of $a(\theta)$ has an open interval (Lehmann, 1959).

Example A-10

Let $\mathfrak{X} = \text{col}\,[z(1), \ldots, z(N)]$ be a random sample drawn from a univariate Gaussian distribution whose mean μ is unknown, and whose variance $\sigma^2 > 0$ is known. From Example A-5, we know that the distribution of $\mathfrak{X}$ forms a one-parameter exponential family, with $T(\mathfrak{X}) = \sum_{i=1}^N z(i)$ and $a(\mu) = \mu/\sigma^2$. Because $a(\mu)$ ranges over an open interval as μ varies from $-\infty$ to $+\infty$, $T(\mathfrak{X}) = \sum_{i=1}^N z(i)$ is complete and sufficient.

The same conclusion can be obtained using Definition A-4 as follows. We must show that the Gaussian family of probability distributions (with μ unknown and σ^2 fixed) is complete. Note that the sufficient statistic $T(\mathfrak{X}) = \sum_{i=1}^N z(i)$ (see Example A-5) is Gaussian with mean $N\mu$ and variance $N^2\sigma^2$. Suppose g is a function such that $\mathbf{E}_\mu\{g(T)\} = 0$ for all $-\infty < \mu < \infty$; then,

$$\int_{-\infty}^{\infty} \frac{g(T)}{N\sigma\sqrt{2\pi}} \exp\left[-\frac{1}{2\sigma^2N^2}(T - N\mu)^2\right] dT$$

$$= \int_{-\infty}^{\infty} \sqrt{2\pi}\, g(v\sigma N + N\mu) \exp\left(-\frac{v^2}{2}\right) dv = 0 \tag{A-14}$$

implies $g(\cdot) = 0$ for all values of the argument of g. □

Other interesting examples that prove completeness for families of distributions are found in Lehmann (1959).

Once a complete and sufficient statistic $T(\mathfrak{X})$ is known for a given parameter estimation problem, the Lehmann-Scheffe Theorem, given next, can be used to obtain a unique UMVU estimate. This theorem is paraphrased from Bickel and Doksum (1977).

Theorem A-3 [Lehmann-Scheffe Theorem (e.g., Bickel and Doksum, 1977)]. *If a complete and sufficient statistic*, $\mathrm{T}(\mathfrak{X})$, *exists for* θ, *and* $\hat{\theta}$ *is an unbiased estimator of* θ, *then* $\theta^*(\mathfrak{X}) = \mathbf{E}\{\hat{\theta}|\mathrm{T}(\mathfrak{X})\}$ *is an UMVU estimator of* θ. *If*

Variance $[\theta^*(\mathfrak{X})] < \infty$ *for all* θ, *then* $\theta^*(\mathfrak{X})$ *is the unique UMVU estimate of* θ. □

A proof of this theorem can be found in Bickel and Doksum (1977).

This theorem can be applied in two ways to determine an UMVU estimator [Bickel and Doksum (1977), and Lehmann (1959)].

Method 1. Find a statistic of the form $h(T(\mathfrak{X}))$ such that

$$\mathbf{E}\{h(T(\mathfrak{X}))\} = \theta \tag{A-15}$$

where $T(\mathfrak{X})$ is a complete and sufficient statistic for θ. Then, $h(T(\mathfrak{X}))$ is an UMVU estimator of θ. This follows from the fact that

$$\mathbf{E}\{h(T(\mathfrak{X}))|T(\mathfrak{X})\} = h(T(\mathfrak{X}))$$

Method 2. Find an unbiased estimator, $\hat{\theta}$, of θ; then, $\mathbf{E}\{\hat{\theta}|T(\mathfrak{X})\}$ is an UMVU estimator of θ for a complete and sufficient statistic $T(\mathfrak{X})$.

Example A-11 (Continuation of Example A-10)

We know that $T(\mathfrak{X}) = \sum_{i=1}^{N} z(i)$ is a complete and sufficient statistic for μ. Furthermore, $1/N \sum_{i=1}^{N} z(i)$ is an unbiased estimator of μ; hence, we obtain the well-known result from Method 1, that the sample mean, $1/N \sum_{i=1}^{N} z(i)$, is an UMVU estimate of μ. Because this estimator is linear, it is also the BLUE of μ. □

Example A-12 (Linear Model)

As in Example A-9, consider the linear model

$$\mathfrak{X}(k) = \mathcal{H}(k)\boldsymbol{\theta} + \mathcal{V}(k) \tag{A-16}$$

where $\boldsymbol{\theta}$ is a deterministic but unknown $n \times 1$ vector of parameters, $\mathcal{H}(k)$ is deterministic, and $\mathbf{E}\{\mathcal{V}(k)\} = \mathbf{0}$. Additionally, assume that $\mathcal{V}(k)$ is Gaussian with known covariance matrix $\mathcal{R}(k)$. Then, the statistic $\mathbf{T}(\mathfrak{X}(k)) = \mathcal{H}'(k)\mathcal{R}^{-1}\mathfrak{X}(k)$ is sufficient (see Example A-9). That it is also complete can be seen by using Theorem A-4. To obtain UMVU estimate $\boldsymbol{\theta}$, we need to identify a function $h[\mathbf{T}(\mathfrak{X}(k))]$ such that $\mathbf{E}\{h[\mathbf{T}(\mathfrak{X}(k))]\} = \boldsymbol{\theta}$. The structure of $h[\mathbf{T}(\mathfrak{X}(k))]$ is obtained by observing that

$$\mathbf{E}\{\mathbf{T}(\mathfrak{X}(k))\} = \mathbf{E}\{\mathcal{H}'(k)\mathcal{R}^{-1}(k)\mathfrak{X}(k)\} = \mathcal{H}'(k)\mathcal{R}^{-1}(k)\mathcal{H}(k)\boldsymbol{\theta}$$

hence,

$$[\mathcal{H}'(k)\mathcal{R}^{-1}(k)\mathcal{H}(k)]^{-1}\mathbf{E}\{\mathbf{T}(\mathfrak{X}(k))\} = \boldsymbol{\theta}$$

Consequently, the UMVU estimator of $\boldsymbol{\theta}$ is

$$[\mathcal{H}'(k)\mathcal{R}^{-1}(k)\mathcal{H}(k)]^{-1}\mathcal{H}'(k)\mathcal{R}^{-1}(k)\mathfrak{X}(k)$$

which agrees with Equation (9-26). □

We now generalize the discussions given above to the case of an m-parameter exponential family, and scalar observations. This theorem is paraphrased from Bickel and Doksum (1977).

Theorem A-4 [Bickel and Doksum (1977) and Lehmann (1959)]. *Let* $p(z|\boldsymbol{\theta})$ *be an* m-*parameter exponential family given by*

$$p(z|\boldsymbol{\theta}) = \exp\left[\sum_{i=1}^{m} a_i(\boldsymbol{\theta})T_i(z) + b(\theta) + h(z)\right]$$

where $a_1, \ldots, a_m$ *and* b *are real-valued functions of* θ, and $T_1, \ldots, T_m$ *and* h *are real-valued functions of* z. *Suppose that the range of* $\mathbf{a} = \text{col}\,[a_1(\boldsymbol{\theta}), \ldots, a_m(\boldsymbol{\theta})]$ *has an open* m-*rectangle* [*if* $(x_1, y_1), \ldots, (x_m, y_m)$ *are* m *open intervals, the set* $\{(s_1, \ldots, s_m)\colon x_i < s_i < y_i,\ 1 \le i \le m\}$ *is called the open* m-*rectangle*], *then* $\mathbf{T}(z) = \text{col}\,[T_1(z), \ldots, T_m(z)]$ *is complete as well as sufficient.* □

Example A-13 (This example is taken from Bickel and Doksum, 1977, pp. 123–124)

As in Example A-6, let $\mathfrak{Z} = \text{col}\,[z(1), \ldots, z(N)]$ be a sample from a $N(\mu,\sigma^2)$ population where both μ and σ^2 are unknown. As a special case of Example A-6, we observe that the distribution of $\mathfrak{Z}$ forms a two-parameter exponential family where $\boldsymbol{\theta} = \text{col}\,(\mu,\sigma^2)$. Because $\text{col}\,[a_1(\boldsymbol{\theta}), a_2(\boldsymbol{\theta})] = \text{col}\,(\mu/\sigma^2, -\tfrac{1}{2}\sigma^2)$ ranges over the lower halfplane, as $\boldsymbol{\theta}$ ranges over $\text{col}\,[(-\infty,\infty), (0,\infty)]$, the conditions of Theorem A-4 are satisfied. As a result, $\mathbf{T}(\mathfrak{Z}) = \text{col}\,[\sum_{i=1}^{N} z(i), \sum_{i=1}^{N} z^2(i)]$ is complete and sufficient. □

Theorem A-3 also generalizes in a straightforward manner to:

Theorem A-5. *If a complete and sufficient statistic* $\mathbf{T}(\mathfrak{Z}) = \text{col}\,(T_1(\mathfrak{Z}), \ldots, T_m(\mathfrak{Z}))$ *exists for* $\boldsymbol{\theta}$, *and* $\hat{\boldsymbol{\theta}}$ *is an unbiased estimator of* $\boldsymbol{\theta}$, *then* $\boldsymbol{\theta}^*(\mathfrak{Z}) = \mathbf{E}\{\hat{\boldsymbol{\theta}}|\mathbf{T}(\mathfrak{Z})\}$ *is an UMVU estimator of* $\boldsymbol{\theta}$. *If the elements of the covariance matrix of* $\boldsymbol{\theta}^*(\mathfrak{Z})$ *are* $< \infty$ *for all* $\boldsymbol{\theta}$, *then* $\boldsymbol{\theta}^*(\mathfrak{Z})$ *is the unique UMVU estimate of* $\boldsymbol{\theta}$. □

The proof of this theorem is a straightforward extension of the proof of Theorem A-3, which can be found in Bickel and Doksum (1977).

Example A-14 (Continuation of Example A-13)

In Example A-13 we saw that $\text{col}\,[T_1(\mathfrak{Z}), T_2(\mathfrak{Z})] = \text{col}\,[\sum_{i=1}^{N} z(i), \sum_{i=1}^{N} z^2(i)]$ is sufficient and complete for both μ and σ^2. Furthermore, since

$$\bar{z} = \frac{1}{N}\sum_{i=1}^{N} z(i)$$

and

$$\bar{\sigma}^2 = \frac{1}{N-1}\sum_{i=1}^{N} [z(i) - \bar{z}]^2$$

are unbiased estimators of μ and σ^2, respectively, we use the extension of Method 1 to the vector parameter case to conclude that $\bar{z}$ and $\bar{\sigma}^2$ are UMVU estimators of μ and σ^2. □

It is not always possible to identify a function $h(T(\mathfrak{Z}))$ that is an unbiased estimator of $\boldsymbol{\theta}$. Examples that use the conditioning Method 2 to obtain

UMVU estimators are found, for example, in Bickel and Doksum (1977) and Lehman (1980).

PROBLEMS

A-1. Suppose $z(1), \ldots, z(N)$ are independent random variables, each uniform on $[0,\theta]$, where $\theta > 0$ is unknown. Find a sufficient statistic for θ.

A-2. Suppose we have two independent observations from the Cauchy distribution,

$$p(z) = \frac{1}{\pi}\frac{1}{1 + (z - \theta)^2} \qquad -\infty < z < \infty$$

Show that no sufficient statistic exists for θ.

A-3. Let $z(1), z(2), \ldots, z(N)$ be generated by the first-order auto-regressive process,

$$z(i) = \theta z(i - 1) + \sqrt{\beta}\, w(i)$$

where $\{w(i), i = 1, \ldots, N\}$ is an independent and identically distributed Gaussian noise sequence with zero mean and unit variance. Find a sufficient statistic for θ and β.

A-4. Suppose that $T(\mathfrak{Z})$ is sufficent for $\boldsymbol{\theta}$, and that $\hat{\boldsymbol{\theta}}(\mathfrak{Z})$ is a maximum-likelihood estimate of $\boldsymbol{\theta}$. Show that $\hat{\boldsymbol{\theta}}(\mathfrak{Z})$ depends on $\mathfrak{Z}$ only through $T(\mathfrak{Z})$.

A-5. Using Theorem A-2, derive the maximum-likelihood estimator of θ when observations $z(1), \ldots, z(N)$ denote a sample from

$$p(z(i)|\theta) = \theta e^{-\theta z(i)} \qquad z(i) \geq 0,\ \theta > 0$$

A-6. Show that the family of Bernoulli distributions, with unknown probability of success $p\ (0 \leq p \leq 1)$, is complete.

A-7. Show that the family of uniform distributions on $(0,\theta)$, where $\theta > 0$ is unknown, is complete.

A-8. Let $z(1), \ldots, z(N)$ be independent and identically distributed samples, where $p(z(i)|\theta)$ is a Bernoulli distribution with unknown probability of success $p\ (0 \leq p \leq 1)$. Find a complete sufficient statistic, T; the UMVU estimate $\phi(T)$ of p; and, the variance of $\phi(T)$.

A-9. [Taken from Bickel and Doksum (1977)]. Let $z(1), z(2), \ldots, z(N)$ be an independent and identically distributed sample from $N(\mu,1)$. Find the UMVU estimator of $p_\mu[z(1) \geq 0]$.

A-10. [Taken from Bickel and Doksum (1977)]. Suppose that T_1 and T_2 are two UMVU estimates of θ with finite variances. Show that $T_1 = T_2$.

A-11. In Example A-12 prove that $T(\mathfrak{Z}(k))$ is complete.

Appendix A

Glossary of Major Results

Equations (3-10) and (3-11)	Batch formulas for $\hat{\boldsymbol{\theta}}_{WLS}(k)$ and $\hat{\boldsymbol{\theta}}_{LS}(k)$.
Theorem 4-1	Information form of recursive LSE.
Lemma 4-1	Matrix inversion lemma.
Theorem 4-2	Covariance form of recursive LSE.
Theorem 5-1	Multistage LSE.
Theorem 6-1	Necessary and sufficient conditions for a linear batch estimator to be unbiased.
Theorem 6-2	Sufficient condition for a linear recursive estimator to be unbiased.
Theorem 6-3	Cramer-Rao inequality for a scalar parameter.
Corollary 6-1	Achieving the Cramer-Rao lower bound.
Theorem 6-4	Cramer-Rao inequality for a vector of parameters.
Corollary 6-2	Inequality for error-variance of ith parameter.
Theorem 7-1	Mean-squared convergence implies convergence in probability.
Theorem 7-2	Conditions under which $\hat{\theta}(k)$ is a consistent estimator of θ.
Theorem 8-1	Sufficient conditions for $\hat{\boldsymbol{\theta}}_{WLS}(k)$ to be an unbiased estimator of $\boldsymbol{\theta}$.
Theorem 8-2	A formula for cov $[\tilde{\boldsymbol{\theta}}_{WLS}(k)]$.

Corollary 8-1	A formula for cov $[\tilde{\theta}_{LS}(k)]$ under special conditions on the measurement noise.
Theorem 8-3	An unbiased estimator of σ_v^2.
Theorem 8-4	Sufficient conditions for $\hat{\boldsymbol{\theta}}_{LS}(k)$ to be a consistent estimator of $\boldsymbol{\theta}$.
Theorem 8-5	Sufficient conditions for $\hat{\sigma}_v^2(k)$ to be a consistent estimator of σ_v^2.
Equation (9-22)	Batch formula for $\hat{\boldsymbol{\theta}}_{BLU}(k)$.
Theorem 9-1	The relationship between $\hat{\boldsymbol{\theta}}_{BLU}(k)$ and $\hat{\boldsymbol{\theta}}_{WLS}(k)$.
Corollary 9-1	When all the results obtained in Lessons 3, 4 and 5 for $\hat{\boldsymbol{\theta}}_{WLS}(k)$ can be applied to $\hat{\boldsymbol{\theta}}_{BLU}(k)$.
Theorem 9-2	When $\hat{\boldsymbol{\theta}}_{BLU}(k)$ equals $\hat{\boldsymbol{\theta}}_{LS}(k)$ (Gauss-Markov Theorem).
Theorem 9-3	A formula for cov $[\tilde{\boldsymbol{\theta}}_{BLU}(k)]$.
Corollary 9-2	The equivalence between $\mathbf{P}(k)$ and cov $[\tilde{\boldsymbol{\theta}}_{BLU}(k)]$.
Theorem 9-4	Most efficient estimator property of $\hat{\boldsymbol{\theta}}_{BLU}(k)$.
Corollary 9-3	When $\hat{\boldsymbol{\theta}}_{LS}(k)$ is a most efficient estimator of $\boldsymbol{\theta}$.
Theorem 9-5	Invariance of $\hat{\boldsymbol{\theta}}_{BLU}(k)$ to scale changes.
Theorem 9-6	Information form of recursive BLUE.
Theorem 9-7	Covariance form of recursive BLUE.
Definition 10-1	Likelihood defined.
Theorem 10-1	Likelihood ratio of combined data from statistically independent sets of data.
Theorem 11-1	Large-sample properties of maximum-likelihood estimates.
Theorem 11-2	Invariance property of MLE's.
Theorem 11-3	Condition under which $\hat{\boldsymbol{\theta}}_{ML}(k) = \hat{\boldsymbol{\theta}}_{BLU}(k)$.
Corollary 11-1	Conditions under which $\hat{\boldsymbol{\theta}}_{ML}(k) = \hat{\boldsymbol{\theta}}_{BLU}(k) = \hat{\boldsymbol{\theta}}_{LS}(k)$, and, resulting estimator properties.
Theorem 12-1	A formula for $p(\mathbf{x}\|\mathbf{y})$ when $\mathbf{x}$ and $\mathbf{y}$ are jointly Gaussian.
Theorem 12-2	Properties of $\mathbf{E}\{\mathbf{x}\|\mathbf{y}\}$ when $\mathbf{x}$ and $\mathbf{y}$ are jointly Gaussian.
Theorem 12-3	Expansion formula for $\mathbf{E}\{\mathbf{x}\|\mathbf{y},\mathbf{z}\}$ when $\mathbf{x}$, $\mathbf{y}$, and $\mathbf{z}$ are jointly Gaussian, and $\mathbf{y}$ and $\mathbf{z}$ are statistically independent.
Theorem 12-4	Expansion formula for $\mathbf{E}\{\mathbf{x}\|\mathbf{y},\mathbf{z}\}$ when $\mathbf{x}$, $\mathbf{y}$ and $\mathbf{z}$ are jointly Gaussian and $\mathbf{y}$ and $\mathbf{z}$ are not necessarily statistically independent.

Theorem 13-1	A formula for $\hat{\boldsymbol{\theta}}_{MS}(k)$ (The Fundamental Theorem of Estimation Theory).
Corollary 13-1	A formula for $\hat{\boldsymbol{\theta}}_{MS}(k)$ when $\boldsymbol{\theta}$ and $\mathcal{Z}(k)$ are jointly Gaussian.
Corollary 13-2	A linear mean-squared estimator of $\boldsymbol{\theta}$ in the non-Gaussian case.
Corollary 13-3	Orthogonality principle.
Theorem 13-2	When $\hat{\boldsymbol{\theta}}_{MAP}(k) = \hat{\boldsymbol{\theta}}_{MS}(k)$.
Theorem 14-1	Conditions under which $\hat{\boldsymbol{\theta}}_{MS}(k) = \hat{\boldsymbol{\theta}}_{BLU}(k)$.
Theorem 14-2	Condition under which $\hat{\boldsymbol{\theta}}_{MS}(k) = \hat{\boldsymbol{\theta}}^{a}_{BLU}(k)$.
Theorem 14-3	Condition under which $\hat{\boldsymbol{\theta}}_{MAP}(k) = \hat{\boldsymbol{\theta}}^{a}_{BLU}(k)$.
Theorem 15-1	Expansion of a joint probability density function for a first-order Markov process.
Theorem 15-2	Calculation of conditional expectation for a first-order Markov process.
Theorem 15-3	Interpretation of Gaussian white noise as a special first-order Markov process.
Equations (15-17) & (15-18)	The basic state-variable model.
Theorem 15-4	Conditions under which $\mathbf{x}(k)$ is a Gauss-Markov sequence.
Theorem 15-5	Recursive equations for computing $\mathbf{m}_{\mathbf{x}}(k)$ and $\mathbf{P}_{\mathbf{x}}(k)$.
Theorem 15-6	Formulas for computing $\mathbf{m}_{\mathbf{z}}(k)$ and $\mathbf{P}_{\mathbf{z}}(k)$.
Equations (16-4) & (16-11)	Single-stage predictor formulas for $\hat{\mathbf{x}}(k\|k-1)$ and $\mathbf{P}(k\|k-1)$.
Theorem 16-1	Formula for and properties of general state predictor, $\hat{\mathbf{x}}(k\|j)$, $k > j$.
Theorem 16-2	Representations and properties of the innovations process.
Theorem 17-1	Kalman filter formulas and properties of resulting estimates and estimation error.
Theorem 19-1	Steady-state Kalman filter.
Theorem 19-2	Equivalence of steady-state Kalman filter and infinite length digital Wiener filter.
Theorem 20-1	Single-state smoother formula for $\hat{\mathbf{x}}(k\|k+1)$.
Corollary 20-1	Relationship between single-stage smoothing gain matrix and Kalman gain matrix.
Corollary 20-2	Another way to express $\hat{\mathbf{x}}(k\|k+1)$.

Theorem 20-2	Double-stage smoother formula for $\hat{\mathbf{x}}(k\|k+2)$.
Corollary 20-3	Relationship between double-stage smoothing gain matrix and Kalman gain matrix.
Corollary 20-4	Two other ways to express $\hat{\mathbf{x}}(k\|k+2)$.
Theorem 21-1	Formulas for a useful fixed-interval smoother of $\mathbf{x}(k)$, $\hat{\mathbf{x}}(k\|N)$, and its error-covariance matrix, $\mathbf{P}(k\|N)$.
Theorem 21-2	Formulas for a most useful two-pass fixed-interval smoother of $\mathbf{x}(k)$ and its associated error-covariance matrix.
Theorem 21-3	Formulas for a most useful fixed-point smoothed estimator of $\mathbf{x}(k)$, $\hat{\mathbf{x}}(k\|k+l)$ where $l = 1, 2, \ldots$, and its associated error-covariance matrix, $\mathbf{P}(k\|k+l)$.
Theorem 22-1	Conditions under which a single-channel state-variable model is equivalent to a convolutional sum model.
Theorem 22-2	Recursive minimum-variance deconvolution formulas.
Theorem 22-3	Steady-state MVD filter, and zero phase nature of $\hat{\mu}_s(k\|N)$.
Theorem 22-4	Equivalence between steady-state MVD filter and Berkhout's infinite impulse response digital Wiener deconvolution filter.
Theorem 22-5	Maximum-likelihood deconvolution results.
Theorem 22-6	Structure of minimum-variance waveshaper.
Theorem 22-7	Recursive fixed-interval waveshaping results.
Theorem 23-1	How to handle biases that may be present in a state-variable model.
Theorem 23-2	Predictor-corrector Kalman filter for the correlated noise case.
Corollary 23-1	Recursive predictor formulas for the correlated noise case.
Corollary 23-2	Recursive filter formulas for the correlated noise case.
Equations (24-1) & (24-2)	Nonlinear state-variable model.
Equations (24-23) & (24-30)	Perturbation state-variable model.
Theorem 24-1	Solution to a time-varying continuous-time state equation.

Equations (24-39) & (24-44)	Discretized state-variable model.
Theorem 25-1	A consequence of relinearizing about $\hat{\mathbf{x}}(k\|k)$.
Equations (25-22) & (25-27)	Extended Kalman filter prediction and correction equations.
Theorem 26-1	Formula for the log-likelihood function of the basic state-variable model.
Theorem 26-2	Closed-form formula for the maximum-likelihood estimate of the steady-state value of the innovation's covariance matrix.
Theorem 27-1	Kalman-Bucy filter equations.
Definition A-1	Sufficient statistic defined.
Theorem A-1	Factorization theorem.
Theorem A-2	A method for computing the unique maximum-likelihood estimator of $\boldsymbol{\theta}$ that is associated with exponential families of distributions.
Theorem A-3	Lehmann-Scheffe Theorem. Provides a uniformly minimum-variance unbiased estimator of θ.
Theorem A-4	Method for determining whether or not $T(\mathbf{z})$ is complete as well as sufficient when $p(\mathbf{z}\|\boldsymbol{\theta})$ is an m-parameter exponential family.
Theorem A-5	Provides a uniformly minimum-variance unbiased estimator of vector $\boldsymbol{\theta}$.

References

AGUILERA, R., J. A. DEBREMAECKER, and S. HERNANDEZ. 1970. "Design of recursive filters." *Geophysics*, Vol. 35, pp. 247–253.

ANDERSON, B. D. O., and J. B. MOORE. 1979. *Optimal Filtering*. Englewood Cliffs, NJ: Prentice-Hall.

AOKI, M. 1967. *Optimization of Stochastic Systems—Topics in Discrete-Time Systems*. NY: Academic Press.

ÅSTRÖM, K. J. 1968. "Lectures on the identification problem—the least squares method." Rept. No. 6806, Lund Institute of Technology, Division of Automatic Control.

ATHANS, M. 1971. "The role and use of the stochastic linear-quadratic-Gaussian problem in control system design." *IEEE Trans. on Automatic Control*, Vol. AC-16, pp. 529–552.

ATHANS, M., and P. L. FALB. 1965. *Optimal Control: An Introduction to the Theory and Its Applications*. NY: McGraw-Hill.

ATHANS, M., and F. SCHWEPPE. 1965. "Gradient matrices and matrix calculations." MIT Lincoln Labs., Lexington, MA, Tech. Note 1965-53.

ATHANS, M., and E. TSE. 1967. "A direct derivation of the optimal linear filter using the maximum principle." *IEEE Trans. on Automatic Control*, Vol. AC-12, pp. 690–698.

ATHANS, M., R. P. WISHNER, and A. BERTOLINI. 1968. "Suboptimal state estimation for continuous-time nonlinear systems from discrete noisy measurements." *IEEE Trans. on Automatic Control*, Vol. AC-13, pp. 504–514.

BARANKIN, E. W., and M. KATZ, JR. 1959. "Sufficient Statistics of Minimal Dimension." *Sankhya*, Vol. 21, pp. 217–246.

BARANKIN, E. W. 1961. "Application to Exponential Families of the Solution to the Minimal Dimensionality Problem for Sufficient Statistics." *Bull. Inst. Internat. Stat.*, Vol. 38, pp. 141–150.

BARD, Y. 1970. "Comparison of gradient methods for the solution of nonlinear parameter estimation problems." *SIAM J. Numerical Analysis*, Vol. 7, pp. 157–186.

BERKHOUT, A. G. 1977. "Least-squares inverse filtering and wavelet deconvolution." *Geophysics*, Vol. 42, pp. 1369–1383.

BICKEL, P. J., and K. A. DOKSUM. 1977. *Mathematical Statistics: Basic Ideas and Selected Topics*. San Francisco: Holden-Day, Inc.

BIERMAN, G. J. 1973a. "A comparison of discrete linear filtering algorithms." *IEEE Trans. on Aerospace and Electronic Systems*, Vol. AES-9, pp. 28–37.

BIERMAN, G. J. 1973b. "Fixed-interval smoothing with discrete measurements." *Int. J. Control*, Vol. 18, pp. 65–75.

BIERMAN, G. J. 1977. *Factorization Methods for Discrete Sequential Estimation*. NY: Academic Press.

BRYSON, A. E., JR., and M. FRAZIER. 1963. "Smoothing for linear and nonlinear dynamic systems." TDR 63-119, pp. 353–364, Aero. Sys. Div., Wright-Patterson Air Force Base, Ohio.

BRYSON, A. E., JR., and D. E. JOHANSEN. 1965. "Linear filtering for time-varying systems using measurements containing colored noise." *IEEE Trans. on Automatic Control*, Vol. AC-10, pp. 4–10.

BRYSON, A. E., JR., and Y. C. HO. 1969. *Applied Optimal Control*. Waltham, MA: Blaisdell.

CHEN, C. T. 1970. *Introduction to Linear System Theory*. NY: Holt.

CHI, C. Y. 1983. "Single-channel and multichannel deconvolution." Ph.D. dissertation, Univ. of Southern Califiornia, Los Angeles, CA.

CHI, C. Y., and J. M. MENDEL. 1984. "Performance of minimum-variance deconvolution filter." *IEEE Trans. on Acoustics, Speech and Signal Processing*, Vol. ASSP-32, pp. 1145–1153.

CRAMER, H. 1946. *Mathematical Methods of Statistics*. Princeton, NJ: Princeton Univ. Press.

DAI, G-Z., and J. M. MENDEL. 1986. "General Problems of Minimum-Variance Recursive Waveshaping." *IEEE Trans. on Acoustics, Speech and Signal Processing*, Vol. ASSP-34.

DONGARRA, J. J., J. R. BUNCH, C. B. MOLER, and G. W. STEWART. 1979. *LINPACK User's Guide*, Philadelphia: SIAM.

DUDA, R. D., and P. E. HART. 1973. *Pattern Classification and Scene Analysis*. NY: Wiley Interscience.

EDWARDS, A. W. F. 1972. *Likelihood*. London: Cambridge Univ. Press.

FAURRE, P. L. 1976. "Stochastic Realization Algorithms." in *System Identification: Advances and Case Studies* (eds., R. K. Mehra and D. G. Lainiotis), pp. 1–25. NY: Academic Press.

FERGUSON, T. S. 1967. *Mathematical Statistics: A Decision Theoretic Approach*. NY: Academic Press.

FRASER, D. 1967. "Discussion of optimal fixed-point continuous linear smoothing (by J. S. Meditch)." *Proc. 1967 Joint Automatic Control Conf.*, p. 249, Univ. of PA, Philadelphia.

GOLDBERGER, A. S. 1964. *Econometric Theory*. NY: John Wiley.

GRAYBILL, F. A. 1961. *An Introduction to Linear Statistical Models*. Vol. 1, NY: McGraw-Hill.

GUPTA, N. K., and R. K. MEHRA. 1974. "Computational aspects of maximum likelihood estimation and reduction of sensitivity function calculations." *IEEE Trans. on Automatic Control*, Vol. AC-19, pp. 774–783.

GURA, I. A., and A. B. BIERMAN. 1971. "On computational efficiency of linear filtering algorithms." *Automatica*, Vol. 7, pp. 299–314.

HAMMING, R. W. 1983. *Digital Filters*, 2nd Edition. Englewood Cliffs, NJ: Prentice-Hall.

HO, Y. C. 1963. "On the stochastic approximation method and optimal filtering." *J. of Math. Anal. and Appl.*, Vol. 6, pp. 152–154.

JAZWINSKI, A. H. 1970. *Stochastic Processes and Filtering Theory*. NY: Academic Press.

KAILATH, T. 1968. "An innovations approach to least-squares estimation—Part 1: Linear filtering in additive white noise." *IEEE Trans. on Automatic Control*, Vol. AC-13, pp. 646–655.

KAILATH, T. K. 1980. *Linear Systems*. Englewood Cliffs, NJ: Prentice-Hall.

KALMAN, R. E. 1960. "A new approach to linear filtering and prediction problems." *Trans. ASME J. Basic Eng. Series D*, Vol. 82, pp. 35–46.

KALMAN, R. E., and R. BUCY. 1961. "New results in linear filtering and prediction theory." *Trans. ASME, J. Basic Eng., Series D*, Vol. 83, pp. 95–108.

KASHYAP, R. L., and A. R. RAO. 1976. *Dynamic Stochastic Models from Empirical Data*. NY: Academic Press.

KELLY, C. N., and B. D. O. ANDERSON. 1971. "On the stability of fixed-lag smoothing algorithms." *J. Franklin Inst.*, Vol. 291, pp. 271–281.

KMENTA, J. 1971. *Elements of Econometrics*. NY: MacMillan.

KOPP, R. E., and R. J. ORFORD. 1963. "Linear regression applied to system identification for adaptive control systems." *AIAA J.*, Vol. 1, pp. 2300.

KUNG, S. Y. 1978. "A new identification and model reduction algorithm via singular value decomposition." Paper presented at the 12th Annual Asilomar Conference on Circuits, Systems, and Computers, Pacific Grove, CA.

KWAKERNAAK, H., and R. SIVAN. 1972. *Linear Optimal Control Systems*. NY: Wiley-Interscience.

LAUB, A. J. 1979. "A Schur method for solving algebraic Riccati equations." *IEEE Trans. on Automatic Control*, Vol. AC-24, pp. 913–921.

LEHMANN, E. L. 1959. *Testing Statistical Hypotheses*. NY: John Wiley.

LEHMANN, E. L. 1980. *Theory of Point Estimation*. NY: John Wiley.

LJUNG, L. 1976. "Consistency of the Least-Squares Identification Method." *IEEE Trans. on Automatic Control*. Vol. AC-21, pp. 779–781.

LJUNG, L. 1979. "Asymptotic behavior of the extended Kalman filter as a parameter estimator for linear systems." *IEEE Trans. on Automatic Control*, Vol. AC-24, pp. 36–50.

MARQUARDT, D. W. 1963. "An algorithm for least-squares estimation of nonlinear parameters." *J. Soc. Indust. Appl. Math.*, Vol. 11, pp. 431–441.

MCLOUGHLIN, D. B. 1980. "Distributed systems—notes." *Proc. 1980 Pre-JACC Tutorial Workshop on Maximum-Likelihood Identification*, San Francisco, CA.

MEDITCH, J. S. 1969. *Stochastic Optimal Linear Estimation and Control.* NY: McGraw-Hill.

MEHRA, R. K. 1970a. "An algorithm to solve matrix equations $\mathbf{PH}^T = \mathbf{G}$ and $\mathbf{P} = \mathbf{\Phi P \Phi}^T + \mathbf{\Gamma\Gamma}^T$." *IEEE Trans. on Automatic Control*, Vol. AC-15.

MEHRA, R. K. 1970b. "On-line identification of linear dynamic systems with applications to Kalman filtering." *Proc. Joint Automatic Control Conference*, Atlanta, GA, pp. 373–382.

MEHRA, R. K. 1971. "Identification of stochastic linear dynamic systems using Kalman filter representation." *AIAA J.*, Vol. 9, pp. 28–31.

MEHRA, R. K., and J. S. TYLER. 1973. "Case studies in aircraft parameter identification." *Proc. 3rd IFAC Symposium on Identification and System Parameter Estimation*, North Holland, Amsterdam.

MENDEL, J. M. 1971. "Computational requirements for a discrete Kalman filter." *IEEE Trans. on Automatic Control*, Vol. AC-16, pp. 748–758.

MENDEL, J. M. 1973. *Discrete Techniques of Parameter Estimation: the Equation Error Formulation.* NY: Marcel Dekker.

MENDEL, J. M. 1975. "Multi-stage least squares parameter estimators." *IEEE Trans. on Automatic Control*, Vol. AC-20, pp. 775–782.

MENDEL, J. M. 1981. "Minimum-variance deconvolution." *IEEE Trans. on Geoscience and Remote Sensing*, Vol. GE-19, pp. 161–171.

MENDEL, J. M. 1983a. "Minimum-variance and maximum-likelihood recursive waveshaping." *IEEE Trans. on Acoustics, Speech, and Signal Processing*, Vol. ASSP-31, pp. 599–604.

MENDEL, J. M. 1983b. *Optimal Seismic Deconvolution: an Estimation Based Approach.* NY: Academic Press.

MENDEL, J. M., and D. L. GIESEKING. 1971. "Bibliography on the linear-quadratic-gaussian problem." *IEEE Trans. on Automatic Control*, Vol. AC-16, pp. 847–869.

MORRISON, N. 1969. *Introduction to Sequential Smoothing and Prediction.* NY: McGraw-Hill.

NAHI, N. E. 1969. *Estimation Theory and Applications.* NY: John Wiley.

OPPENHEIM, A. V., and R. W. SCHAFER. 1975. *Digital Signal Processing.* Englewood Cliffs, NJ: Prentice-Hall.

PAPOULIS, A. 1965. *Probability, Random Variables, and Stochastic Processes.* NY: McGraw-Hill.

PELED, A., and B. LIU. 1976. *Digital Signal Processing: Theory, Design, and Implementation.* NY: John Wiley.

RAUCH, H. E., F. TUNG, and C. T. STRIEBEL. 1965. "Maximum-likelihood estimates of linear dynamical systems." *AIAA J.*, Vol. 3, pp. 1445–1450.

SCHWEPPE, F. C. 1965. "Evaluation of likelihood functions for gaussian signals." *IEEE Trans. on Information Theory*, Vol. IT-11, pp. 61–70.

SCHWEPPE, F. C. 1973. *Uncertain Dynamic Systems*. Englewood Cliffs, NJ: Prentice-Hall.

SHANKS, J. L. 1967. "Recursion filters for digital processing." *Geophysics*, Vol. 32, pp. 32–51.

SORENSON, H. W. 1970. "Least-squares estimation: from Gauss to Kalman." *IEEE Spectrum*, Vol. 7, pp. 63–68.

SORENSON, H. W. 1980. *Parameter Estimation: Principles and Problems*. NY: Marcel Dekker.

SORENSON, H. W., and J. E. SACKS. 1971. "Recursive fading memory filtering." *Information Science*, Vol. 3, pp. 101–119.

STEFANI, R. T. 1967. "Design and simulation of a high performance, digital, adaptive, normal acceleration control system using modern parameter estimation techniques." Rept. No. DAC-60637, Douglas Aircrft Co., Santa Monica, CA.

STEPNER, D. E., and R. K. MEHRA. 1973. "Maximum likelihood identification and optimal input design for identifying aircraft stability and control derivatives." Ch. IV, NASA-CR-2200.

STEWART, G. W. 1973. *Introduction to Matrix Computations*. NY: Academic Press.

TREITEL, S. 1970. "Principles of digital multichannel filtering." *Geophysics*, Vol. XXXV, pp. 785–811.

TREITEL, S., and E. A. ROBINSON. 1966. "The design of high-resolution digital filters." *IEEE Trans. on Geoscience and Electronics*, Vol. GE-4, pp. 25–38.

TUCKER, H. G. 1962. *An Introduction to Probability and Mathematical Statistics*. NY: Academic Press.

TUCKER, H. G. 1967. *A Graduate Gourse in Probability*. NY: Academic Press.

VAN TREES, H. L. 1968. *Detection, Estimation and Modulation Theory*, Vol. 1. NY: John Wiley.

ZACKS, S. 1971. *The Theory of Statistical Inference*. NY: John Wiley.

Index